Mass Spectrometry
Principles and Applications

Second Edition

Edmond de Hoffmann
Université Catholique de Louvain

Vincent Stroobant
Ludwig Institute for Cancer Research

JOHN WILEY & SONS, LTD

Chichester • New York • Weinheim • Brisbane • Singapore • Toronto

Original French language- Spéctometrie de Masse- © Masson, Paris, 1994
Second edition © Dunod, Paris 1999

Reprinted April 2002, February 2003, November 2003, August 2004

Other Wiley Editorial Offices

John Wiley & Sons Inc., 111 River Street, Hoboken, NJ 07030, USA

Jossey-Bass, 989 Market Street, San Francisco, CA 94103-1741, USA

Wiley-VCH Verlag GmbH, Boschstr. 12, D-69469 Weinheim, Germany

John Wiley & Sons Australia Ltd, 33 Park Road, Milton, Queensland 4064, Australia

John Wiley & Sons (Asia) Pte Ltd, 2 Clementi Loop #02-01, Jin Xing Distripark, Singapore
129809

John Wiley & Sons (Canada) Ltd, 22 Worcester Road, Etobicoke, Ontario M9W 1L1

Wiley also publishes its books in a variety of electronic formats. Some content that appears in
print may not be available in electronic books.

Library of Congress Cataloging-in-Publication Data

Hoffmann, Edmond de
[Spectrométrie de masse. English]
Mass spectrometry : principles and applications / Edmond de Hoffmann, Vincent
Stroobant. — 2nd. ed.
p. cm.
Includes bibliographical references and index.
ISBN 0–471–48565–9 (hardcover) — ISBN 0–471–48566–7 (pbk.)
1. Mass spectrometry. I. Stroobant, Vincent. II. Title.

QD96.M3 H6413 2001
543′.0873 — dc21

2001033254

British Library Cataloguing in Publication Data

A catalogue record for this book is available from the British Library

ISBN 0 471 48565 9 (ppc)   ISBN 0 471 48566 7 (pbk)

Typeset in 10.5/12.5pt Times Roman by Laser Words, Chennai, India
Printed and bound in Great Britain by Biddles Ltd, Guildford and King's Lynn
This book is printed on acid-free paper responsibly manufactured from sustainable
forestry, in which at least two trees are planted for each one used for paper production.

Mass Spectrometry

Second Edition

Contents

Preface

The first edition of this book was published in 1996, by three authors. Unfortunately, Jean Charette passed away in November 1998, aged 74. We not only lost a great friend, but also a respected colleague. We dedicate this new edition to his memory.

Following the first studies of J. J. Thomson (1912), mass spectrometry has undergone countless improvements. Since 1958, gas chromatography–mass spectrometry coupling has revolutionized the analysis of volatile compounds. Another revolution occurred in the 80's when the technique was adapted to the study of non-volatile compounds such as peptides, oligosaccharides, phospholipids, bile acids etc. From the discoveries of electrospray and matrix-assisted laser desorption in the last decade, compounds with molecular weights exceeding hundreds of thousands of daltons, such as synthetic polymers, proteins, glycans and polynucleotides, can be analyzed by mass spectrometry.

Combined with the development of tandem mass spectrometry and coupling with other separation techniques, such as HPLC and capillary electrophoresis, the new instruments allow one to obtain significant information from mixtures of natural or synthetic compounds. The enthusiastic — while sometimes incredulous — reactions of organic or life-science researchers when receiving the information obtained, for example, with a sample extracted from a thin layer chromatographic spot, are the best reward for the efforts devoted by the specialist for the development of these new methods in mass spectrometry.

Starting from the very foundations of mass spectrometry, this book presents all the important techniques developed to the present day. It describes many analytical methods based on these techniques and emphasizes their usefulness with numerous examples. The reader will also find the necessary information for the interpretation of data. A series of exercises allows the reader to check his or her understanding of the subject. Numerous references are given for those who wish to go further. Important Internet addresses are also provided. We hope that this new edition will be useful to students, teachers and researchers.

Many colleagues and friends have read the manuscript and their comments have been very helpful. Some of them have made a thorough reading. They deserve a special mention: Magda Claeys, Bruno Domon, Jean-Claude Tabet and François Van Hoof. We also wish to acknowledge the remarkable work done by the scientific editors at Wiley's.

The authors would also like to acknowledge the financial support of the FNRS (Fonds National de la Recherche Scientifique, Brussels).

Many useful comments have been published on the first edition, or sent to the editor or the authors. One of them, sent by Steen Ingemann, was particularly detailed and constructive.

We would like to thank the Catholic University of Louvain and all our colleagues and friends whose help has been invaluable to us.

Edmond de Hoffmann and Vincent Stroobant
Louvain-la-Neuve, June 2001

Introduction

Mass spectrometry's characteristics have raised it to an outstanding position among analytical methods: unequaled sensitivity and detection limits and the diversity of its applications, e.g. atomic physics, reaction physics, reaction kinetics, geochronology, all forms of chemical analysis (especially in biomedicine and ion-molecule reactions) and determination of thermodynamic parameters ($\Delta G^0{}_f$, K_a, etc.).

Mass spectrometry has progressed extremely rapidly during the last decade: production, separation and detection of ions, data acquisition, data reduction, etc. This has led to the development of entirely new instruments and applications.[1-3]

Principles

The first step in the mass spectrometric analysis of compounds is the production of gas-phase ions of the compound, e.g. by electron ionization:

$$M + e^- \longrightarrow M^{\bullet +} + 2e^-$$

This molecular ion normally undergoes fragmentations. Because it is a radical cation with an odd number of electrons, it can fragment to give either a radical and an ion with an even number of electrons, or a molecule and a new radical cation. Let us stress the important difference between these two types of ions and the need to write them correctly:

$$
M^{\bullet +}
\begin{cases}
\nearrow & \underset{\text{even ion}}{EE^+} \; + \; \underset{\text{radical}}{R^{\bullet}} \\[2ex]
\searrow & \underset{\text{odd ion}}{OE^{\bullet +}} \; + \; \underset{\text{molecule}}{N}
\end{cases}
$$

These two types of ions have different chemical properties. Each primary product ion derived from the molecular ion can, in turn, undergo fragmentation, and so on. All these ions are separated in the spectrometer according to their mass-to-charge ratio and are detected in proportion to their abundance. A mass spectrum of the molecule is thus produced, which can be presented as a graph or as a table (Figure 1).

Most of the ions have a charge corresponding to the loss of only one electron. Multiply charged ions also can be obtained: they are detected according to the mass-to-charge ratio, the charge of the electron being taken as a charge unit. The total charge of the ions will be represented by q, the electron charge by e and the number

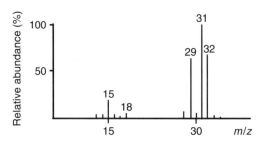

m/z	Relative abundance (%)	m/z	Relative abundance (%)
12	0.33	28	6.3
13	0.72	29	64.0
14	2.4	30	3.8
15	13.0	31	100.0
16	0.21	32	66.0
17	1.0	33	0.73
18	0.9	34	~0.1

Figure 1
Mass spectrum of methanol by electron ionization, (presented as a graph and as a table)

of charges on the ions by z:

$$q = ze \quad \text{and} \quad e = 1.6 \times 10^{-19} \text{ coulomb}$$

where m/z represents the mass-to-charge ratio. When the mass is given in daltons
(Da) or in atomic mass units (u) and the charge in number of electron charges, m/z
is given in thomson (Th). Sometimes, m is given as the mass number and z as the
charge number, both of which are unitless. The atomic mass units u or Da have the
same fundamental definition:

$$1 \text{ u} = 1 \text{ Da} = 1.665402 \times 10^{-27} \text{ kg} \pm 0.59 \text{ ppm}$$

However, they are traditionally used in different contexts: when dealing with mean
isotopic masses, as generally used in stoichiometric calculations, the Da will be
preferred; in mass spectrometry, masses referring to the main isotope of each element
are used and expressed in u.

   Ions provide information concerning the nature and the structure of their precursor
molecule. In the spectrum of a pure compound, the molecular ion, if present, appears
at the highest value of m/z (followed by ions containing heavier isotopes) and gives
the molecular weight of the compound. This molecular ion appears at m/z 32 in
the spectrum of methanol, where the peak at m/z 33 is due to the presence of the
^{13}C isotope, with an intensity that is 1.1% of that of the m/z 32 peak. In the same
spectrum, the peak at m/z 15 indicates the presence of a methyl group. The difference
between 32 and 15, i.e. 17, is characteristic of the loss of a neutral mass of 17 Da by

the molecular ion and is typical of a hydroxyl group. In the same spectrum, the peak at m/z 16 could formally correspond to ions $CH_4^{\bullet+}$, O^+ or even CH_3OH^{2+}, because they all have m/z values equal to 16 at low resolution. However, O^+ is unlikely to occur, and a doubly charged ion for such a small molecule is not stable enough to be observed.

For stoichiometric calculations chemists use the average mass calculated using the atomic weight, which is an average of the isotopes of each element of the molecule. In mass spectrometry, the monoisotopic mass generally is used. This mass is calculated by using the mass of the most abundant isotope for each constituent element. The difference between the average mass and the monoisotopic mass can amount to several daltons, depending on the number of atoms and their isotopic composition. The type of mass determined by mass spectrometry depends largely on the resolution of the analyzer.

Let us consider as an example CH_3Cl. Actually, chlorine atoms are mixtures of two isotopes whose exact masses are 34.968852 and 36.965903. Their relative abundances are 75.77% and 24.23%. The atomic weight of chlorine atoms is the balanced average: $(34.968852 \times 0.7577 + 36.965903 \times 0.2423) = 35.453$ Da. The average mass of CH_3Cl is $12.011 + (3 \times 1.00794) + 35.453 = 50.4878$ Da, whereas its monoisotopic mass is $12.000000 + (3 \times 1.007825) + 34.968852 = 49.992327$ u.

When the mass of CH_3Cl is measured using a mass spectrometer, two isotopic peaks will appear at their respective masses and relative abundances. Thus, two mass-to-charge ratios will be observed with a mass spectrometer. The first peak will be at m/z $(34.968852 + 12.000000 + 3 \times 1.007825) = 49.992327$, rounded to m/z 50. The mass-to-charge value of the second peak will be $(36.96590 + 12.000000 + 3 \times 1.007825) = 51.989365$, rounded to m/z 52. The abundance at this latter m/z value is $(24.23/75.77) = 0.3198$, or 31.98% of that observed at m/z 50. Carbon and hydrogen also are composed of isotopes, but at much lower abundances. They are neglected for this example.

Diagram of a Mass Spectrometer

A mass spectrometer always contains the following elements (Figure 2):

1. A device to introduce the compound that is analyzed, e.g. a gas chromatograph or a direct insertion probe.

2. A source to produce ions from the sample.

3. One or several analyzers to separate the various ions according to their mass-to-charge ratio.

4. A detector to count the ions emerging from the last analyzer and to measure their abundance.

5. A computer to process the data, which produces the mass spectrum in a suitable form and controls the instrument through feedback.

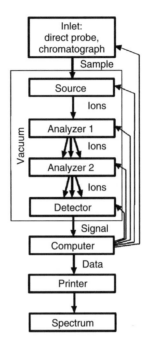

Figure 2
Basic diagram for a mass spectrometer with
two analyzers and feedback control carried
out by a data system

History

A large number of mass spectrometers have been developed according to this fundamental scheme since Thomson's experiments in 1897. Listed here are some of the
highlights of this evolution:[4,5]

1886:   E. Goldstein discovers anode rays (positive gas-phase ions) in gas discharge.[6]

1897:   J.J. Thomson discovers the electron and determines its m/z ratio. *Nobel Prize
        in 1906.*

1898:   W. Wien analyzes the anode rays by magnetic deflection and then establishes
        that these rays carried a positive charge.[7] *Nobel Prize in 1911.*

1901:   R. Kaufmann analyzes cathodic rays using parallel electric and magnetic
        fields.[8]

1909:   R.A. Millikan and H. Fletcher determine the elementary unit of charge.

1912:   J.J. Thomson constructs the first mass spectrometer (then called a parabola
        spectrograph).[9] He obtains mass spectra of O_2, N_2, CO, CO_2 and $COCl_2$. He

observes negative and multiply charged ions. He discovers metastable ions. In 1913, he discovers isotopes 20 and 22 of neon.

1918: A.J. Dempster develops the electron ionization source and the first spectrometer with a sector shaped magnet (180°) and direction focusing.[10]

1919: F.W. Aston develops the first mass spectrometer with velocity focusing.[11] *Nobel Prize in 1922*. He measures mass defects in 1923.[12]

1932: K.T. Bainbridge proves the mass energy equivalence postulated by Einstein.[13]

1934: R. Conrad applies mass spectrometry to organic chemistry.[14]

1934: W.R. Smythe, L.H. Rumbaugh and S.S. West succeed in the first preparative isotope separation.[15]

1940: A.O. Nier isolates uranium-235.[16]

1942: The Consolidated Engineering Corporation builds the first commercial instrument dedicated to organic analysis for the Atlantic Refinery Company.

1945: First recognition of the metastable peaks by J.A. Hipple and E.U. Condon.[17]

1948: A.E. Cameron and D.F. Eggers publish a design and mass spectra for a linear time-of-flight (TOF) mass spectrometer.[18] W. Stephens proposed the concept of this analyzer in 1946.[19]

1949: H. Sommer, H.A. Thomas and J.A. Hipple describe the first application in mass spectrometry of ion cyclotron resonance (ICR).[20]

1952: Theories of quasi-equilibrium (QET[21] and RRKM[22]) explain the monomolecular fragmentation of ions. R.A. Marcus receives the *Nobel Prize in 1992*.

1952: E.G. Johnson and A.O. Nier develop double-focusing instruments.[23]

1953: W. Paul and H.S. Steinwedel describe the quadrupole analyzer and the ion trap or quistor in a patent. W. Paul, H.P. Reinhard and U. Von Zahn, of Bonn University, describe the quadrupole spectrometer in *Zeitschrift für Physik* in 1958. Paul receives the *Nobel Prize in 1989*.[24]

1955: W.L. Wiley and I.H. McLaren of Bendix Corporation made key advances in linear TOF design.[25]

1956: J. Beynon shows the analytical usefulness of high resolution and exact mass determinations in the determination of the ion's elementary composition.[26]

1956: First spectrometers coupled with a gas chromatograph by F.W. McLafferty[27] and R.S. Gohlke.[28]

1957:   Kratos introduces the first commercial mass spectrometer with double focusing.

1958:   Bendix introduces the first commercial linear TOF instrument.

1966:   M.S.B. Munson and F.H. Field discover chemical ionization (CI).[29]

1967:   F.W. McLafferty[30] and K.R. Jennings[31] introduce the collision-induced dissociation (CID) procedure.

1968:   Finnigan introduces the first commercial quadrupole mass spectrometer.

1968:   First mass spectrometers coupled with data processing units.

1969:   H.D. Beckey demonstrates field desorption (FD) mass spectrometry of organic molecules.[32]

1972:   V.I. Karatev, B.A. Mamyrim and D.V. Smikk introduce the reflectron that corrects the kinetic energy distribution of the ions in a TOF mass spectrometer.[33]

1973:   R.G. Cooks, J.H. Beynon, R.M. Caprioli and G.R. Lester publish the book *Metastable Ions*, a landmark in tandem mass spectrometry.[34]

1974:   E.C. Horning, D.I. Carroll, I. Dzidic, K.D. Haegele, M.D. Horning and R.N. Stillwell discover atmospheric pressure chemical ionization (APCI).[35]

1974:   First high-performance liquid chromatraphy/mass spectrometry (HPLC/MS) coupling by P.J. Arpino, M.A. Baldwin and F.W. McLafferty.[36]

1974:   M.D. Comisarov and A.G. Marshall develop Fourier transformer ICR (FTICR) mass spectrometry.[37]

1975:   First commercial gas chromatography/mass spectrometry (GC/MS) instruments with capillary columns.

1976:   R.D. McFarlane and D.F. Torgesson introduce the plasma desorption (PD) source.[38]

1994:   R.G. Cooks and T.L. Kruger propose the kinetic method for thermochemical determination based on measurement of the rates of competitive fragmentations of cluster ions.[39]

1978:   R.A. Yost and C.G. Enke build the first triple quadrupole mass spectrometer, one of the most popular types of tandem mass spectrometry instrument.[40]

1978:   Introduction of lamellar and high-field magnets.

1980:   R.S. Houk, V.A. Fassel, G.D. Flesch, A.L. Gray and E. Taylor demonstrate the potential of inductively coupled plasma (ICP) mass spectrometry.[41]

1981: M. Barber, R.S. Bordoli, R.D. Sedgwick and A.H. Tyler describe the fast atom bombardment (FAB) source.[42]

1982: First complete spectrum of insulin (5750 Da) by FAB[43] and PD.[44]

1982: Finnigan and Sciex introduce the first commercial triple quadrupole mass spectrometers.

1983: C.R. Blakney and M.L. Vestal describe the thermospray (TSP).[45]

1983: G.C. Stafford, P.E. Kelly, J.E. Syka, W.E. Reynolds and J.F.J. Todd describe the development of a gas chromatography (GC) detector based on an ion trap and commercialized by Finnigan under the name Ion Trap™.[46]

1987: M. Karas, D. Bachmann, U. Bahr and F. Hillenkamp discover matrix-assisted laser desorption/ionization (MALDI).[47]

1987: R.D. Smith describes the coupling of capillary electrophoresis (CE) with mass spectrometry.[48]

1988: J. Fenn develops the electrospray for electrospray ionization (ESI).[49] First spectra of proteins above 20 000 Da. He demonstrated the elecrospray's potential as a mass spectrometric technique for small molecules in 1984.[50] The concept of this source was proposed in 1968 by M. Dole.[51]

1991: V. Katta, B.T. Chait, and B. Gamen, Y.T. Li and J.D. Henion demonstrate that specific non-covalent complexes could be detected by mass spectrometry.[52,53]

1991: B. Spengler, D. Kirsch and R. Kaufmann obtain structural information with reflectron TOF mass spectrometry (MALDI post-source decay).[54]

1994: M. Wilm and M. Mann describe the nanoelectrospray source (then called the microelectrospray source).[55]

The progress of experimental methods and the refinements in instruments led to spectacular improvements in resolution, sensitivity, mass range and accuracy. Resolution ($m/\delta m$) developed as follows:

1913: 13 $m/\delta m$, Thomson.[9]

1918: 100 $m/\delta m$, Dempster.[10]

1919: 130 $m/\delta m$, Aston.[11]

1937: 2000 $m/\delta m$, Aston.[56]

1998: 8 000 000 $m/\delta m$, Marshall and co-workers.[57]

A continuous improvement has allowed analysis to reach detection limits at the nano-, pico- and femtomole levels.[58,59] Furthermore, the direct coupling of chromatographic techniques with mass spectrometry has improved these limits to the atto- and zeptomole levels.[60,61]

Regarding the mass range, DNA ions of 10^8 Da were weighed by mass spectrometry.[62] The measurement of the atomic masses has reached an accuracy of better than 10^{-9} u.[63]

In another field, Litherland et al.[64] succeeded in determining a $^{14}C/^{12}C$ ratio of $1 : 10^{15}$ and hence in dating a 40 000-year-old sample with a 1% error.

Ion Free Path

All mass spectrometers must function under high vacuum (low pressure). This is necessary to allow ions to reach the detector without undergoing collisions with other gaseous molecules. Indeed, collisions would produce a deviation of the trajectory and the ion would lose its charge against the walls of the instrument. On the other hand, ion–molecule collisions could produce unwanted reactions and hence increase the complexity of the spectrum. Nevertheless, we will see later that useful techniques use controlled collisions in specific regions of a spectrometer.

According to the kinetic theory of gases, the mean free path L (in m) is given by Equation (1):

$$L = \frac{kT}{\sqrt{2}p\sigma} \tag{1}$$

where k is the Boltzmann constant, T is the temperature (in K), p is the pressure (in Pa) and σ is the collision cross-section (in m^2); $\sigma = \pi d^2$, where d is the sum of the radii of the stationary molecule and the colliding ion (in m). In fact, one can approximate the mean free path of an ion under normal conditions in a mass spectrometer ($k = 1.38 \times 10^{-21}$ J K^{-1}, $T \approx 300$ K, $\sigma \approx 45 \times 10^{-20}$ m^2) using Equation (2) or (3), where L is in centimeters and pressure p is, respectively, in pascal or milliTorr:

$$L = \frac{0.66}{p} \tag{2}$$

$$L = \frac{4.95}{p} \tag{3}$$

Table 1 is a conversion table for pressure units.

In a mass spectrometer, the mean free path should be at least 1 m and hence the maximum pressure should be 66 nbar. In instruments using a high-voltage source, the pressure must be reduced further to prevent the occurrence of discharges. In contrast, some trap-based instruments operate at higher pressure.

However, introducing the sample to a mass spectrometer requires the transfer of the sample at atmospheric pressure into a region of high vacuum without compromising the latter. In the same way, producing efficient ion–molecule collisions

Table 1 Pressure units (the official SI unit is the pascal)

1 pascal (Pa) = 1 newton (N) m^{-2}
1 bar = 10^6 dyn cm^{-2} = 10^5 Pa
1 millibar (mbar) = 10^{-3} bar = 10^2 Pa
1 microbar (μbar) = 10^{-6} bar = 10^{-1} Pa
1 nanobar (nbar) = 10^{-9} bar = 10^{-4} Pa
1 atmosphere (atm) = 1.013 bar = 101 308 Pa
1 Torr = 1 mmHg = 1.333 mbar = 133.3 Pa
1 psi = 1 pound per square inch = 0.07 atm

requires the mean free path to be reduced to around 0.1 mm, implying at least a 60 Pa pressure in the region of the spectrometer. These large differences in pressure are controlled with the help of an efficient pumping system using mechanical pumps in conjunction with turbomolecular, diffusion or cryogenic pumps. The mechanical pumps allow a vacuum of about 10^{-3} Torr to be obtained. Once this vacuum is achieved, the operation of the other pumping systems allows a vacuum as high as 10^{-10} Torr to be reached.

References

1. Siuzdak G., *Proc. Natl. Acad. Sci. USA*, **91**(24), 11290 (1994).
2. Noble D., *Anal. Chem.*, **67**(7), 265A (1995).
3. Burlingame A.L., Boyd R.K. and Gaskell S.J., *Anal. Chem.*, **70**(16), 647R (1998).
4. Borman S., Dagani R., Rawls R.L., *et al.*, *Chem. Eng. News*, **76**, 39 (1998).
5. http://www.chemheritage.org/asms/timeline.htm.
6. Goldstein E., *Berl. Ber.*, **39**, 691 (1886).
7. Wien W., *Verh. Phys. Ges.*, **17**, (1898).
8. Kaufmann R.L., Heinen H.J., Shurmann L.W., *et al.*, in *Microbeam Analysis*, ed. by D.E. Newburg, San Francisco Press, San Francisco, 1979, pp. 63–72.
9. Thomson J.J., *Rays of Positive Electricity and Their Application to Chemical Analysis*, Longmans Green, London, 1913.
10. Dempster A.J., *Phys. Rev.*, **11**, 316 (1918).
11. Aston F.W., *Philos. Mag.*, **38**, 707 (1919).
12. Aston F.W., *Mass Spectra and Isotopes*, 2nd edition, Edward Arnold and Co., London, 1942.
13. Bainbridge K.T., *Phys. Rev.*, **42**, 1 (1932); Bainbridge K.T. and Jordan E.B., *Phys. Rev.*, **50**, 282 (1936).
14. Conrad R., *Trans. Faraday Soc.*, **30**, 215 (1934).
15. Smythe W.R., Rumbaugh L.H. and West S.S., *Phys. Rev.*, **45**, 724 (1934).
16. Nier A.O., *Rev. Sci. Instrum.*, **11**, 252 (1940).
17. Hipple J.A. and Condon E.U., *Phys. Rev.*, **69**, 347 (1946).
18. Cameron A.E. and Eggers D.F., *Rev. Sci. Instrum.*, **19**, 605 (1948).
19. Stephens W., *Phys. Rev.*, **69**, 691 (1946).
20. Sommer H., Thomas H.A. and Hipple J.A., *Phys. Rev.*, **76**, 1877 (1949).
21. Rosenstock H.M., Wallenstein M.B., Warhaftig A.L., *et al.*, *Proc. Natl. Acad. Sci. USA.*, **38**, 667 (1952).

22. Marcus R.A., *J. Chem. Phys.*, **20**, 359 (1952).
23. Johnson E.G. and Nier A.O., *Phys. Rev.*, **91**, 12 (1953).
24. Paul W. and Steinwedel H.S., *Z. Naturforsch.*, **8a**, 448 (1953); Paul W., Reinhard H.P. and von Zahn U., *Z. Phys.*, **152**, 143 (1958).
25. Wiley W.L. and McLaren I.H., *Rev. Sci. Instrum.*, **16**, 1150 (1955).
26. Beynon J., *Mikrochim. Acta*, 437 (1956).
27. McLafferty F.W., *Appl. Spectrosc.*, **11**, 148 (1957).
28. Gohlke R.S., *Anal. Chem.*, **31**, 535 (1959).
29. Munson M.S.B. and Field F.H., *J. Am. Chem. Soc.*, **88**, 2681 (1966).
30. McLafferty F.W. and Bryce T.A., *Chem. Commun.*, 1215 (1967).
31. Jennings K.R., *Int. J. Mass Spectrom. Ion Phys.*, **1**, 227 (1968).
32. Beckey H.D., *Int. J. Mass Spectrom. Ion Phys.*, **2**, 500 (1969).
33. Karataev V.I., Mamyrin B.A. and Smikk D.V., *Sov. Phys.-Tech. Phys.* **16**, 1177 (1972).
34. Cooks R.G., Beynon J.H., Caprioli R.M. *et al.*, *Metastable Ions*, Elsevier, New York, 1973.
35. Horning E.C., Carroll D.I., Dzidic I., *et al.*, *J. Chromatogr. Sci.*, **412**, 725 (1974).
36. Arpino P.J., Baldwin M.A. and McLafferty F.W., *Biomed. Mass Spectrom.*, **1**, 80 (1974).
37. Comisarov M.B. and Marshall A.G., *Chem. Phys. Lett.*, **25**, 282 (1974).
38. McFarlane R.D. and Torgesson D.F., *Science*, **191**, 920 (1976).
39. Cooks R.G. and Kruger T.L., *J. Am. Chem. Soc.*, **99**, 1279 (1977).
40. Yost R.A. and Enke C.G., *J. Am. Chem. Soc.*, **100**, 2274 (1978).
41. Houk R.S., Fassel V.A., Flesch G.D., *et al.*, *Anal. Chem.*, **52**, 2283 (1980).
42. Barber M., Bardoli R.S., Sedgwick R.D., *et al.*, *J. Chem. Soc., Chem. Commun.*, **15**, 325 (1981).
43. McFarlane R.D., *J. Am..Chem. Soc.*, **104**, 2948 (1982).
44. Barber M., Bordoli R., Elliott G., *et al.*, *J. Chem. Soc., Chem. Commun.*, **16**, 936 (1982).
45. Blakney C.R. and Vestal M.L., *Anal. Chem.*, **55**, 750 (1983).
46. Stafford G.C., Kelley P.E., Syka J.E., *et al.*, *Int. J. Mass Spectrom. Ion Processes*, **60**, 85 (1984).
47. Karas M., Bachmann D., Bahr U., *et al.*, *Int. J. Mass Spectrom. Ion Processes*, **78**, 53 (1987).
48. Olivares J.A., Nguyen N.T., Yonker C.R., *et al.*, *Anal. Chem.*, **59**, 1230 (1987).
49. Fenn J.B., Mann M., Meng C.K., *et al.*, *Science*, **246**, 64 (1989).
50. Yamashita M. and Fenn J.B., *J. Chem. Phys.*, **80**, 4451 (1984).
51. Dole M., Mach L.L., Hines R.L., *et al.*, *J. Chem. Phys.*, **49**, 2240 (1968).
52. Katta V. and Chait B.T., *J. Am. Chem. Soc.*, **113**, 8534 (1991).
53. Gamen B., Li Y.T. and Henion J.D., *J. Am. Chem. Soc.*, **113**, 6294 (1991).
54. Spengler B., Kirsch D. and Kaufmann R., *Rapid Comm. Mass Spectrom.*, **5**, 198 (1991).
55. Wilm M. and Mann M., *Proc. 42nd ASMS Conference*, Chicago, IL, 1994, p. 770.
56. Aston F.W., *Proc. R. Soc., London, Ser. A*, **163**, 391 (1937).
57. Shi S.D.H., Hendrickson C.L. and Marshall A.G., *Proc. Natl. Acad. Sci. USA*, **95**, 11532 (1998).
58. Solouki T., Marto J.A., White F.M., *et al.*, *Anal. Chem.*, **67**, 4139 (1995).
59. Morris H.R., Paxton T., Panico M., *et al.*, *J. Protein Chem.*, **16**, 469 (1997).
60. Valaskovic G.A., Kelleher N.L. and McLafferty F.W., *Science*, **273**, 1199 (1996).
61. Wang P.G., Murugaiah V., Yeung B. Vouros P. and Giese R.W., *J. Chromatogr. A*, **721**, 289 (1996).
62. Chen R., Cheng X., Mitchell D., *et al.*, *Anal. Chem.*, **67**, 1159 (1995).
63. Di Fillip F., Natarajan V., Bradley M., *et al.*, *Phys. Scr.*, **T59**, 144 (1995).
64. Litherland A.E., Benkens R.P., Lilius L.R., *et al.*, *Nucl. Instrum. Methods*, **186**, 463 (1981).

1

Ion Sources

In ion sources, the analyzed samples are ionized prior to analysis in the mass spectrometer. A variety of ionization techniques are used for mass spectrometry. The most important considerations are the internal energy transferred during the ionization process and the physicochemical properties of the analyte that can be ionized. Some ionization techniques are very energetic and cause extensive fragmentation. Other techniques are softer and only produce molecular species. Electron ionization and chemical ionization are only suitable for gas-phase ionization and thus their use is limited to compounds sufficiently volatile and thermally stable. However, a large number of compounds are thermally labile or do not have sufficient vapor pressure. Ions of these compounds must be extracted directly from the condensed to the gas phase.

These direct ion sources exist as two types: liquid-phase ion sources and solid-state ion sources. In a liquid-phase ion source the analyte is in solution. This solution is introduced, by nebulization, as droplets into the mass spectrometer through some vacuum pumping stages. Electrospray, thermospray and atmospheric pressure chemical ionization sources correspond to this type. In a solid-state ion source the analyte is an involatile deposit. It is obtained by various preparation methods that frequently involve the introduction of a matrix that can be either a solid or a viscous fluid. This deposit then is irradiated by energetic particles or photons that desorb ions near the surface of the deposit. These ions can be extracted by an electric field and focused towards the analyzer. Matrix-assisted laser desorption, secondary ion mass spectrometry, field desorption and plasma desorption sources all use this strategy to produce ions. Fast atom bombardment uses an involatile liquid matrix.

The ion sources produce ions mainly by ionizing a neutral molecule through electron ejection, electron capture, protonation, deprotonation, adduct formation or the transfer of a charged species from a condensed phase to the gas phase. Ion production often implies gas-phase ion–molecule reactions. A brief description of such reactions is given at the end of the chapter.

1.1 Electron Ionization

An electron ionization (EI) source, formerly called electron impact, was devised by Dempster and improved by Bleakney[1] and Nier.[2] It is widely used in organic mass spectrometry. This ionization technique works well for many gas-phase molecules but induces extensive fragmentation so that the molecular ions are not always observed.

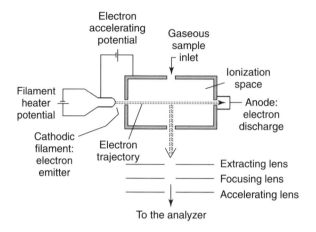

Figure 1.1
Diagram of an electron ionization source

As shown in Figure 1.1, this source consists of a heated filament giving off electrons. The latter are accelerated towards an anode and collide with the gaseous molecules of the analyzed sample injected into the source. Gases and samples with high vapor pressure are introduced directly into the source. Liquids and solids usually are heated to increase the vapor pressure for analysis.

Each electron is associated with a wave whose wavelength λ is given by

$$\lambda = \frac{h}{mv}$$

where m is its mass, v is its velocity and h is Planck's constant. This wavelength is 2.7 Å for a kinetic energy of 20 eV and 1.4 Å for 70 eV. When this wavelength is close to the bond lengths, the wave is disturbed and becomes complex. If one of the frequencies has an energy hv corresponding to a transition in the molecule, an energy transfer that leads to various electronic excitations can occur.[3] When there is enough energy, an electron can be expelled. The electrons do not 'impact' molecules. For this reason, it is recommended that the term 'electron impact' be avoided.

Figure 1.2 displays a typical curve of the number of ions produced by a given electron current, at constant pressure of the sample, when the acceleration potential of the electrons (or their kinetic energy) is varied.[4] At low potentials the energy is lower than the molecule ionization energy. At high potentials, the wavelength becomes too small and molecules become 'transparent' to these electrons. In the case of organic molecules, a wide maximum appears at around 70 eV. At this level, small changes in the electron energy do not significantly affect the pattern of the spectrum.

On average, one ion is produced for every 1000 molecules entering the source under the usual spectrometer conditions, at 70 eV. Furthermore, 10–20 eV are transferred to the molecules during the ionization process. Because approximately 10 eV

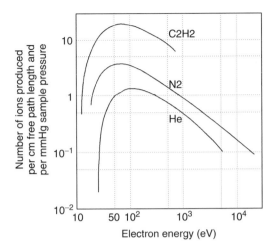

Figure 1.2
Number of ions produced as a function of the electron
energy. A wide maximum appears at around 70 eV

are enough to ionize most organic molecules, the excess energy leads to extensive fragmentation. This fragmentation can be useful because it provides structural
information for structure elucidation of an unknown analyte.

At a given acceleration potential and at constant temperature, the number of ions
I produced per unit time in a volume V is linked to the pressure p and to the electron
current i through the following equation:

$$I = Npi V$$

where N is a constant proportionality coefficient. This equation shows that the sample
pressure is correlated directly with the resulting ionic current. This allows such a
source to be used in quantitative measurements.

Figure 1.3 displays two EI spectra of the same β-lactam compound obtained at
70 and 15 eV. Obviously, at lower energy there is less fragmentation. At first glance
the molecular ion is detected better at low energy. However, the absolute intensity,
in arbitrary units, proportional to the number of detected ions is actually lower:
about 250 units at 70 eV and 150 units at 15 eV. Thus, the increase in relative
intensity due to the lower fragmentation is illusory. Actually, there is a general loss
of intensity due to the decrease in ionization efficiency at lower electron energy. This
will generally be the rule, so the method is not very useful for better detection of
the molecular ion. However, the lowering of the ionization voltage may favor some
fragmentation processes.

A modification implies desorbing the sample from a heated rhenium filament near
the electronic beam. This method is called desorption electron ionization (DEI).

Under conventional EI conditions, the formation of negative ions is inefficient
compared with the formation of positive ions.

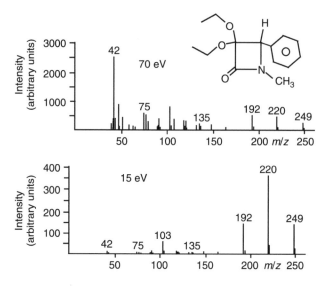

Figure 1.3
Two spectra of the β-lactam illustrated. Although the relative
intensity of the molecular ion peak is greater at lower ionization energy, its absolute intensity, as read from the left-hand
scale, is actually somewhat reduced

1.2 Chemical Ionization

Electron ionization leads to fragmentation of the molecular ion, which sometimes
prevents its detection. Chemical ionization (CI) is a technique that produces ions
with little excess energy. Thus, this technique presents the advantage of yielding a
spectrum with less fragmentation in which the molecular species is easily recognized.
Consequently, chemical ionization is complementary to electron ionization.

Chemical ionization[5] consists of producing ions through a collision of the molecule
to be analyzed with primary ions present in the source. Ion–molecule collisions thus
will be induced in a definite part of the source. In order to do so, the local pressure has
to be sufficient to allow for frequent collisions. We saw that the mean free path could
be calculated from Equation (1) (see Introduction). At a pressure of approximately
60 Pa, the free path is about 0.1 mm. The source then is devised to maintain a local
pressure of that magnitude. A solution consists of introducing into the source a small
box about 1 cm along its side, as shown in Figure 1.4.

Two lateral holes allow for the crossing of electrons and another hole at the bottom
allows the product ions to pass through. Moreover, there is a reagent gas input tube
and an opening for the sample intake. The sample will be introduced by means of a
probe, which will close the opening.

This probe carries the sample within a hollow or contains the end part of a
capillary coming from a chromatograph or carries a filament on which the sample

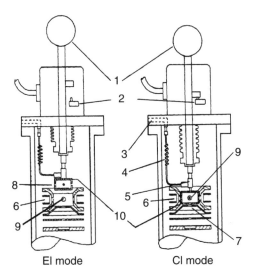

Figure 1.4
Combined EI and CI source. Lowering the box 10 switches from the EI
to CI mode: 1, EI/CI switch (in EI mode the box serves as a pusher); 2,
micro-switch; 3, entrance for the reagent gas; 4, flexible capillary carrying
the reagent gas; 5, diaphragm; 6, filament giving off electrons; 7, path of
the ions towards the analyzer inlet; 8, hole for the ionizing electrons in
CI mode; 9, sample inlet; 10, box with holes, also named 'ion volume'.
(Reproduced from Finnigan MAT 44S documentation, with permission)

was deposited. In the latter case, we will talk about desorption chemical ionization
(DCI). The pumping speed is sufficient to maintain a 60 Pa pressure within the box.
Outside, the usual pressure in a source, about 10^{-3} Pa, will be maintained.

Inside the box, the sample pressure will amount to a small fraction of the reagent
gas pressure. Thus, an electron entering the box will preferentially ionize the reagent
gas molecules through electron ionization. The resulting ion then will mostly collide
with other reagent gas molecules, thus creating an ionization plasma through a series
of reactions. Both positive and negative ions of the substance to be analyzed will be
formed by chemical reactions with ions in this plasma. This causes proton transfer
reactions, hydride abstractions, adduct formations, charge transfers, etc.

This plasma also will contain low-energy electrons, called thermal electrons. These
are either electrons that were used for the first ionization and later slowed, or electrons produced by ionization reactions. These slow electrons may be associated with
molecules, thereby yielding negative ions by electron capture.

Ions produced from a molecule by the abstraction of a proton or a hydride, or the
addition of a proton or other ion are termed 'ions of the molecular species' or, less
often, 'pseudomolecular ions'. They allow the determination of the molecular mass
of the molecules in the sample. The term 'molecular ions' refers to $M^{\bullet+}$ or $M^{\bullet-}$ ions.

1.2.1 Proton transfer

Among the wide variety of possible ionization reactions, the most common is proton transfer. Indeed, when analyte molecules M are introduced in the ionization plasma, the reagent gas ions GH^+ often can transfer a proton to the molecules M and produce protonated molecular ions MH^+. This chemical ionization reaction can be described as an acid–base reaction, the reagent gas ions GH^+ and the analyte molecules being Brönsted acid (proton donor) and Brönsted base (proton acceptor), respectively. The tendency for a reagent ion GH^+ to protonate a particular analyte molecule M may be assessed from their proton affinity values. The proton affinity (PA) is the negative of the enthalpy change for the protonation reaction (see Appendix 6). The observation of protonated molecular ions MH^+ implies that the analyte molecule M has a proton affinity much higher than that of the reagent gas (PA(M) > PA(G)). If the reagent gas has a proton affinity much higher than that of an analyte (PA(G) > PA(M)), proton transfer from GH^+ to M will be energetically too unfavorable.

The selectivity in the type of compound that can be protonated and the internal energy of the resulting protonated molecular ion depend on the relative proton affinities of the reagent gas and the analyte. By thermalizing collisions, this energy depends also on the ion source temperature and pressure. The energetics of the proton transfer can be controlled by using different reagent gases. The most common reagent gases are methane (PA = 5.7 eV), isobutane (PA = 8.5 eV) and ammonia (PA = 9.0 eV). Isobutane and ammonia are not only more selective but protonation of a compound by these reagent gases is considerably less exothermic than protonation by methane. Thus, fragmentation may occur with methane but with isobutane or ammonia the spectrum often presents solely a protonated molecular ion.

The differences between EI and CI spectra are clearly illustrated in Figure 1.5. Indeed, the EI spectrum of butyl methacrylate displays a very low molecular ion at m/z 142. In contrast, its CI spectra exhibit the protonated molecular ion at m/z 143 and very few fragmentations. This example shows also the control of the fragmentation degree in CI by changing the reagent gas. Methane and isobutane CI mass spectra of butyl methacrylate give the protonated molecular ions but the degree of fragmentation is different. With isobutane, the base peak is the protonated molecular ion at m/z 143 whereas with methane the base peak is a fragment ion at m/z 87.

1.2.2 Adduct formation

In chemical ionization plasma, all the ions are liable to associate with polar molecules to form adducts, a kind of gas-phase solvation. The process is favored by the possible formation of hydrogen bonds. For the adduct to be stable, the excess energy must be eliminated, a process that requires a collision with a third partner. The reaction rate equation observed in the formation of these adducts is indeed third order. Ions resulting from the association of a reagent gas molecule G with a protonated molecular ion MH^+ or a fragment ion F^+, or of a protonated molecular ion MH^+ with a neutral molecule, etc., often are found in chemical ionization spectra. Every ion in the plasma may become associated with either a sample molecule or a reagent gas

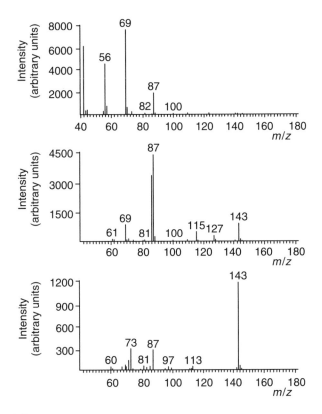

Figure 1.5
The EI (*top*), methane CI (*middle*) and isobutane CI (*bottom*) mass spectra
of butyl methacrylate. The ionization techniques (EI vs. CI) and the reagent
gases (methane vs. isobutane) influence the amount of fragmentation and
the prominence of the protonated molecular ions detected at 143 Th

molecule. Some of these ions are useful in the confirmation of the molecular mass,
such as:

$$MH^+ + M \longrightarrow (2M + H)^+$$

$$F^+ + M \longrightarrow (F + M)^+$$

These associations are often useful to identify a mixture or to determine the molecular
masses of the constituents of the mixture. In fact, a mixture of two species M
and N can give rise to associations such as $(MH + N)^+$, $(F + N)^+$, $(F + M)^+$, etc.
Adducts resulting from neutral species obtained by neutralization of fragments, or
by a neutral loss during a fragmentation, are always at much too low a concentration
to be observed.

It is always useful to examine the peaks appearing beyond the ions of the molecular
species of a substance thought to be pure. If some peaks cannot be explained by
reasonable associations, a mixture must be suspected.

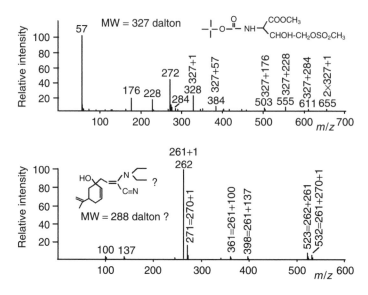

Figure 1.6
Two examples of chemical ionization (isobutane) spectra. The top
spectrum is that of a pure compound. The bottom spectrum is that
of a mixture of two compounds with masses 261 and 270 Th. They
correspond respectively to the loss of hydrogen cyanide and water

Figure 1.6 shows an example of chemical ionization spectra for a pure sample and
for a mixture. When interpreting the results, one must keep in mind always that a
mixture that is observed may result from the presence of several constituents before
vaporization or from their formation after vaporization.

The first spectrum contains the peaks of various adducts of the molecular ion of a
pure compound. The second shows the spectrum of a substance that is initially pure,
as shown by other analysis, but that appears as a mixture in the gas phase because
it loses either hydrogen cyanide or water.

1.2.3 Charge-transfer chemical ionization

Rare gases, nitrogen, carbon monoxide and other gases with high ionization potential
react by charge exchange:

$$Xe + e^- \longrightarrow Xe^{\bullet+} + 2e^-$$

$$Xe^{\bullet+} + M \longrightarrow M^{\bullet+} + Xe$$

A radical cation is obtained, as in electron ionization, but with a smaller energy
content. Less fragmentation is thus observed. In practice, these gases are not used
very often.

1.2.4 Reagent gas

1.2.4.1 Methane as reagent gas

If methane is introduced into the ion volume through the tube, the primary reaction with the electrons will be a classical electron ionization reaction:

$$CH_4 + e^- \longrightarrow CH_4^{•+} + 2e^-$$

This ion will fragment, mainly through the following reactions:

$$CH_4^{•+} \longrightarrow CH_3^+ + H^•$$
$$CH_4^{•+} \longrightarrow CH_2^{•+} + H_2$$

However, mostly it will collide and react with other methane molecules, yielding

$$CH_4^{•+} + CH_4 \longrightarrow CH_5^+ + CH_3^•$$

Other ion–molecule reactions with methane will occur in the plasma, such as

$$CH_3^+ + CH_4 \longrightarrow C_2H_5^+ + H_2$$

A $C_3H_5^+$ ion is formed by the following successive reactions:

$$CH_2^{•+} + CH_4 \longrightarrow C_2H_3^+ + H_2 + H^•$$
$$C_2H_3^+ + CH_4 \longrightarrow C_3H_5^+ + H_2$$

The relative abundance of all these ions will depend on the pressure. Figure 1.7 shows the spectrum of the plasma obtained at 200 μbar (20 Pa). Taking CH_5^+, the most abundant ion, as a reference (100%), $C_2H_5^+$ amounts to 83% and $C_3H_5^+$ to 14%.

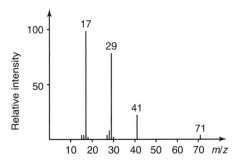

Figure 1.7
Spectrum of methane ionization plasma at 20 Pa. The relative intensities depend on the pressure in the source

Unless it is a saturated hydrocarbon, the sample will react mostly by acquiring a proton in an acid–base type of reaction with one of the plasma ions, e.g.

$$M + CH_5^+ \longrightarrow MH^+ + CH_4$$

A systematic study showed that the main ionizing reactions of molecules containing heteroatoms occurred through acid–base reactions with $C_2H_5^+$ and $C_3H_5^+$. If, however, the sample is a saturated hydrocarbon RH, the ionization reaction will be a hydride abstraction:

$$RH + CH_5^+ \longrightarrow R^+ + CH_4 + H_2$$

Moreover, ion–molecule adduct formation is observed in the case of polar molecules, which is a type of gas-phase solvation, e.g.

$$M + CH_3^+ \longrightarrow (M + CH_3)^+$$

These ions $(MH)^+$, R^+ and $(M + CH_3)^+$ and other adducts of ions with the molecule are termed 'molecular species' or, less often, 'pseudomolecular ions.' They allow the determination of the molecular mass of the molecules in the sample.

1.2.4.2 Isobutane as reagent gas

Isobutane loses an electron upon electron ionization and yields the corresponding radical cation, which will fragment mainly through the loss of a hydrogen radical to yield a *t*-butyl cation, and to a lesser extent through the loss of a methyl radical:

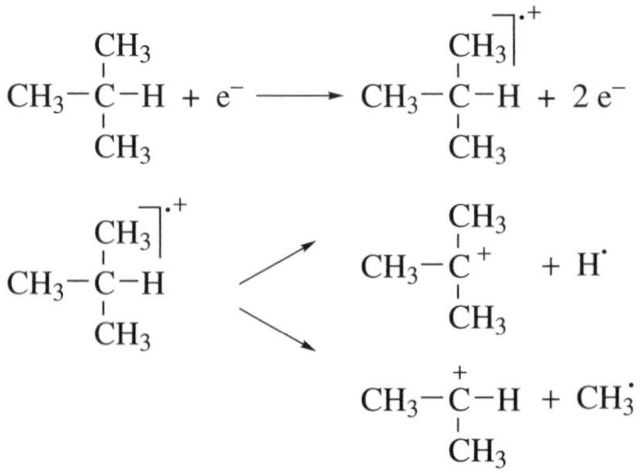

An ion with mass 39 Da also is observed in its spectrum (Figure 1.8), which corresponds to $C_3H_3^+$. Neither its formation mechanism nor its structure is known,

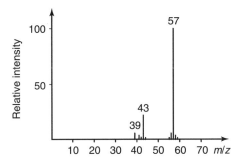

Figure 1.8
Spectrum of the isobutane plasma under chemical ionization conditions at 200 μbar

but it is possible that it is the aromatic cyclopropenium ion:

Here, again, the plasma ions will react mainly through proton transfer to the sample, but polar molecules also will form adducts with the t-butyl ions $(M + 57)^+$ and with $C_3H_3^+$, yielding $(M + 39)^+$ among others.

This isobutane plasma will be very inefficient in ionizing hydrocarbons because the t-butyl cation is relatively stable. This characteristic allows it to be used to detect various substances in mixtures that also contain hydrocarbons.

1.2.4.3 Ammonia as reagent gas

The radical cation generated by electron ionization reacts with an ammonia molecule to yield the ammonium ion and the $NH_2^•$ radical:

$$NH_3^{•+} + NH_3 \longrightarrow NH_4^+ + NH_2^•$$

An ion with mass 35 Da is observed in the plasma (Figure 1.9), which results from the association of an ammonium ion and an ammonia molecule:

$$NH_4^+ + NH_3 \longrightarrow (NH_4 + NH_3)^+$$

This adduct represents 15% of the intensity of the ammonium ion at 200 μbar.

In this gas, the ionization mode will depend on the nature of the sample. The basic molecules, mostly amines, will ionize through a proton transfer:

$$RNH_2 + NH_4^+ \longrightarrow RNH_3^+ + NH_3$$

Polar molecules and those able to form hydrogen bonds while presenting no or little basic character will form adducts. In intermediate cases, two pseudomolecular ions

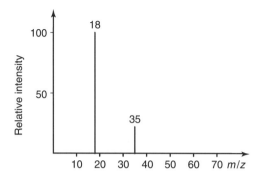

Figure 1.9
Spectrum of an ammonia ionization plasma at
200 µbar

$(M + 1)^+$ and $(M + 18)^+$ will be observed. Compounds that do not correspond to the criteria listed above, e.g. saturated hydrocarbons, will not be ionized efficiently. Alkanes, aromatics, ethers and nitrogen compounds other than amines will be ionized hardly at all. Comparing spectra measured with various reagent gases thus will be very instructive. For example, detection, in the presence of a wealth of saturated hydrocarbons, of a few compounds liable to be ionized is possible, as shown in Figure 1.10.

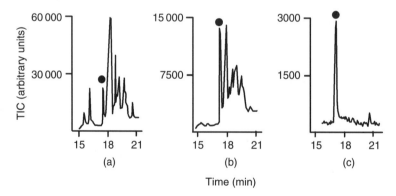

Figure 1.10
Gas chromatography/mass spectrometry total ion current (TIC) traces of butyl methacrylate dissolved in C11–C12 saturated hydrocarbon. (A) Electron ionization. The peak corresponding to butyl methacrylate is marked by a dot. The peaks following are C11 saturated hydrocarbons. (B) Same trace obtained by chemical ionization using methane as the reagent gas. Butyl methacrylate (dot) is still well detected, and the hydrocarbon peaks are attenuated. (C) Chemical ionization using isobutane as the reagent gas. Butyl methacrylate is detected well but the hydrocarbon peaks are virtually non-existent

1.2.5 Negative ion formation

Almost all neutral substances are able to yield positive ions, whereas negative ions require the presence of acidic groups or electronegative elements to produce them. This allows some selectivity for their detection in mixtures. Negative ions can be produced by capture of thermal electrons by the analyte molecule or by ion–molecule reactions between analyte and ions present in the reagent plasma.

All chemical ionization plasmas contain electrons of low energies, issued either directly from the filament but deactivated through collisions, or mostly from primary ionization reactions that produce two low-energy electrons through the ionization reaction. The interaction of electrons with molecules leads to negative ion production by three different mechanisms:[5]

$$AB + e^- \longrightarrow AB^{\bullet -} \text{ (associative resonance capture)}$$

$$AB + e^- \longrightarrow A^{\bullet} + B^- \text{ (dissociative resonance capture)}$$

$$AB + e^- \longrightarrow A^+ + B^- + e^- \text{ (ion pair production)}$$

These electrons can be captured by a molecule. The process can be associative or dissociative. The associative resonance capture that leads to the formation of negative molecular ions needs electrons in the energy range 0–2 eV whereas the dissociative resonance capture is observed with electrons of 0–15 eV and leads to the formation of negative fragment ions.

The associative resonance capture is favored for molecules with several electronegative atoms or with possibilities to stabilize ions by resonance. The energy to remove an electron from the molecular anion by autodetachment is generally very low. Consequently, any excess of energy from the negative molecular ions as it is formed must be removed by collision. Thus, in CI conditions, the reagent gas not only serves for producing thermal electrons but also serves as a source of molecules for collision to stabilize the ions formed.

Ion pair production is observed with a wide range of electron energies above 15 eV. It is principally this process that leads to negative ion production under conventional EI conditions. Ion pair production forms structurally insignificant, very low mass ions with a sensitivity that is 3–4 orders of magnitude lower than that for positive ion production.

The need to detect tetrachlorodioxins with high sensitivity has contributed to the development of negative ion chemical ionization. In some cases, the detection sensitivity with electron capture is better than with positive ions. It can be explained by the high mobility of the electron, which ensures a greater rate of electron attachment than the rate of ion formation involving transfer of a much larger particle. However, electron capture is very dependent on the experimental conditions and thus can be irreproducible.

Note the different behaviors that the electron can adopt toward the molecules. Electrons at thermal equilibrium, i.e. those whose kinetic energy is less than about 1 eV (1 eV = 98 kJ mol^{-1}), can be captured by molecules and yield negative radical anions. Those whose energy lies between unity and a few hundred electronvolts behave as a wave and transfer energy to molecules without any 'collisions'. Finally, molecules will be 'transparent' to electrons of higher energies: here, we enter the field of electron microscopy.

As discussed already, negative ions can be formed also through ion–molecule reactions with one of the plasma ions. These reactions can be an acid–base reaction or an addition reaction through adduct formation. The mixture CH_4–N_2O (75 : 25) is very useful because of the following reactions involving low-energy electrons:

$$N_2O + e^- \longrightarrow N_2O^{\bullet-}$$

$$N_2O^{\bullet-} \longrightarrow N_2 + O^{\bullet-}$$

$$O^{\bullet-} + CH_4 \longrightarrow CH_3^{\bullet} + OH^-$$

This plasma also contains other ions. The advantage derives from the simultaneous presence of thermal electrons allowing the capture of electrons and of a basic ion OH^-, which reacts with acidic compounds in the most classical acid–base reaction. Because of the presence of methane, the same mixture is suitable also for positive ion production.

The energy balance for the formation of negative ions appears as in the following example (PhOH = phenol):[6]

$$
\begin{array}{ll}
H^+ + OH^- \longrightarrow H_2O & \Delta H^\circ = -1634.7 \text{ kJ mol}^{-1} \\
PhOH \longrightarrow PhO^- + H^+ & \Delta H^\circ = +1456 \text{ kJ mol}^{-1} \\
PhOH + OH^- \longrightarrow PhO^- + H_2O & \Delta H^\circ = -178.7 \text{ kJ mol}^{-1}
\end{array}
$$

The exothermicity of the reaction is the result of the formation of H_2O, which is neutral and carries off the excess energy. The product anion will be 'cold'.

However, during a positive ionization, the following will occur:

$$
\begin{array}{ll}
CH_5^+ \longrightarrow CH_4 + H^+ & \Delta H^\circ = +543.5 \text{ kJ mol}^{-1} \\
(C_2H_5)_2S + H^+ \longrightarrow (C_2H_5)_2SH^+ & \Delta H^\circ = -856.7 \text{ kJ mol}^{-1} \\
CH_5^+ + (C_2H_5)_2S \longrightarrow (C_2H_5)_2SH^+ + CH_4 & \Delta H^\circ = -313.2 \text{ kJ mol}^{-1}
\end{array}
$$

In this case, the exothermicity comes mainly from the association of the proton with the molecule to be ionized. The resulting cation will contain an appreciable level of excess energy.

1.2.6 Desorption chemical ionization (DCI)

Baldwin and McLafferty[7] noticed that introducing a sample directly into the chemical ionization plasma on a glass or a metal support allowed the temperature for observation of the mass spectrum to be reduced, sometimes by as much as 150°C. This prevents the pyrolysis of non-volatile samples. A drop of the sample in solution is applied on a rhenium or tungsten wire. The solvent then is evaporated and the

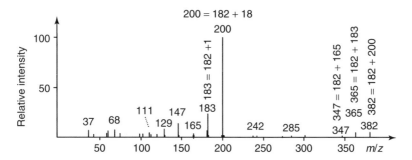

Figure 1.11
The DCI spectrum of mannitol, a non-volatile compound, with H_2O as an reagent gas. Note that water yields radical cation adducts $(M + H_2O)^{\bullet+}$

probe introduced into the mass spectrometer source. The sample is desorbed by rapidly heating the filament by passing a controllable electric current through the wire. The ion formation from these compounds sometimes lasts for only a very short time and the ions of the molecular species are observed for only a few seconds. The spectrum appearance generally varies with the temperature. Figure 1.11 displays the spectrum of mannitol obtained by this DCI technique.

The observed spectrum probably results from the superposition of several phenomena: evaporation of the sample with rapid ionization, direct ionization on the surface of the filament, direct ion desorption and, at higher temperature, pyrolysis followed by ionization. Generally, the molecular species ion is clearly detected in the case of non-volatile compounds. The method can be useful, for example, for tetrasaccharides, small peptides, nucleic acids and other organic salts, which can be detected in either the positive or negative ion mode.

1.3 Fast Atom/Ion Bombardment and Secondary Ion Mass Spectrometry

Secondary ion mass spectrometry (SIMS) analyzes the secondary ions emitted when a surface is irradiated with an energetic primary ion beam.[8,9] Ion sources with primary ion beams of very low current are called static sources because they do not damage the surface of the sample, as opposed to dynamic sources that produce surface erosion. Static SIMS causes less damage to any molecules on the surface than dynamic SIMS and gives spectra that can be similar to those obtained by plasma desorption. This technique is mostly used with solids and is especially useful to study conducting surfaces. High-resolution chemical maps are produced by scanning a tightly focused ionizing beam across the surface.

Fast atom bombardment (FAB)[11] and liquid secondary ion mass spectrometry (LSIMS)[12] are techniques that consist of focusing on the sample a high primary current beam of neutral atoms/molecules or ions, respectively. Essential features of these two ionization techniques are that the sample must be dissolved in a non-volatile liquid matrix. In practice, glycerol is most often used. *m*-Nitrobenzylic alcohol is a good liquid matrix for non-polar compounds. Di- and triethanolamine are efficient in producing negative ions, owing to their basicity. Thioglycerol and a eutectic mixture of dithiothreitol and dithioerythritol (5 : 1 w/w), referred to as 'magic bullet', are alternatives to glycerol.

These techniques use current beams that are, as in dynamic SIMS, high enough to damage the surface. But they produce ions from the surface, as in static SIMS, because convection and diffusion inside the matrix continuously create a fresh layer from the surface for producing new ions. The energetic particles hit the sample solution, inducing a shock wave that ejects ions and molecules from the solution. Ions are accelerated by a potential difference towards the analyzer. These techniques induce no or little ionization. They generally eject into the gas phase ions that were already present in the solution.

Under these conditions, both ion and neutral bombardment are practical techniques. The neutral atom beam at about 5 keV is obtained by ionizing a compound, most often argon but sometimes xenon. Ions are accelerated and focused towards the compound to be analyzed under several kilovolts. They then go through a collision cell where they are neutralized by charge exchange between atoms and ions. Their momentum is sufficient to maintain focusing. The remaining ions then are eliminated from the beam as it passes between electrodes. A diagram of such a source is shown in Figure 1.12. The reaction may be written as follows:

$$Ar_{(rapid)}{}^{\bullet+} + Ar_{(slow)} \longrightarrow Ar_{(slow)}{}^{\bullet+} + Ar_{(rapid)}$$

Using a 'cesium gun', one produces a beam of Cs^+ ions at about 30 keV. It is claimed to give better sensitivity than a neutral atom beam for high molecular

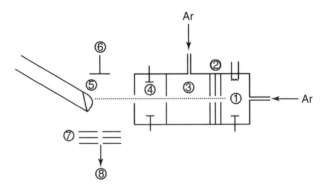

Figure 1.12
Diagram of a FAB gun. (1) Ionization of argon; the resulting ions are accelerated and focused by lenses (2). In (3), the argon ions exchange their charge with neutral atoms, thus becoming rapid neutral atoms. As the beam path passes between the electrodes (4), all ionic species are deflected. Only rapid neutral atoms reach the sample dissolved in a drop of glycerol (5). The ions ejected from the drop are accelerated by the pusher (6) and focused by electrodes (7) towards the analyzer (8)

weights. However, the advantage of using neutral molecules instead of ions lies in the avoidance of an accumulation of charges in the non-conducting samples.

This method is very efficient for producing ions from polar compounds with high molecular weights. Ions up to 10 000 Da and above can be observed, such as peptides and nucleotides. Moreover, it often produces ion beams that can be maintained for long periods of time, sometimes several tens of minutes, which allows several types of

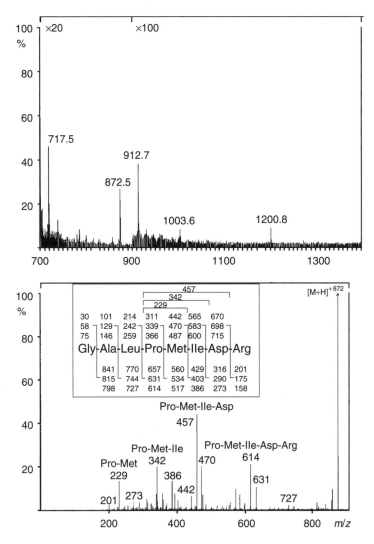

Figure 1.13

(*Top*) The FAB mass spectrum of a mixture of five peptides. The *m/z* of the protonated molecular ion $(M + H)^+$ of each of them is observed. (*Bottom*) Product ion tandem mass spectrum of the $(M + H)^+$ ion with *m/z* 872, giving the sequence of this peptide alone. The values within the frame are the masses of the various possible fragments for the indicated sequence

analysis to be carried out. This advantage is especially appreciated in measurements using multiple analyzers, such as in tandem mass spectrometry (MS/MS). Other desorption techniques, such as DCI or field desorption (FD) (see later), generally give rise to transient signals lasting only a few seconds at most. However, FAB and LSIMS require a matrix such as glycerol, whose ions make the spectrum more complex. Neither DCI nor FD impose this inconvenience.

Figure 1.13 displays an example of FABMS and MS/MS applied to the detection of peptides in a mixture and the sequence determination of one of them.

1.4 Field Desorption

The introduction of field desorption (FD) as a method for the analysis of involatile molecules is principally due to Beckey.[13]

In FD, the sample is deposited, through evaporation of a solution containing also a salt, on a tungsten or rhenium filament covered with carbon needles. A potential difference is set between this filament and an electrode so as to obtain a field that can go up to 10^8 V cm^{-1}. The filament is heated until the sample melts. The ions migrate and accumulate at the tip of the needles, where they end up being desorbed, carrying along molecules of the sample. The technique is demanding and requires an experienced operator. It has now been largely replaced by other desorption techniques. However, it remains an excellent method to ionize high-molecular-mass non-polar compounds.

1.5 Plasma Desorption

Plasma desorption (PD) has been introduced by Mcfarlane and Torgesson.[14] In this ionization technique, the sample deposited on a small aluminized nylon foil is exposed in the source to fission fragments of ^{252}Cf having an energy of several mega-electronvolts.

The shock waves resulting from the bombardment of a few thousand fragments per second induce the desorption of neutrals and ions. This technique has allowed the observation of ions above 10 000 Da.[15] However, nowadays it is of limited use and has been replaced mainly by matrix-assisted laser desorption/ionization (MALDI).

1.6 Laser Desorption and Matrix-assisted Laser Desorption Ionization

Laser desorption (LD) is an efficient method for producing gaseous ions. Generally, laser pulses yielding 10^6–10^{10} W cm^{-2} are focused on a sample surface of about 10^{-3}–10^{-4} cm^2, most often a solid. These laser pulses ablate material from the surface and create a microplasma of ions and neutral molecules, which may react among themselves in the dense vapor phase near the sample surface. The laser pulse realizes both the vaporization and the ionization of the sample.

This technique is used in the study of surfaces and in the analysis of the local composition of samples, such as inclusions in minerals or in cell organelles. It normally allows selective ionization by adjusting the laser wavelength. However, in most conventional infrared LD modes, the laser creates a thermal spike and thus it is not necessary to match the laser wavelength with the sample.

Because the signals are very short, simultaneous detection analyzers or time-of-flight analyzers are required. The probability of obtaining a useful mass spectrum depends critically on the specific physical proprieties of the analyte (e.g. photoabsorption, volatility, etc.). Furthermore, the ions produced are almost always fragmentation products of the original molecule if its mass is above approximately 500 Da. This situation changed dramatically with the development of MALDI.[16,17]

1.6.1 Principle of MALDI

Matrix-assisted laser desorption/ionization (MALDI) is achieved in two steps.[18-20] In the first step, the compound to be analyzed is mixed in a solvent containing small organic molecules in solution, called a matrix, and has a strong absorption at the laser wavelength. This mixture is dried before analysis and any liquid solvents used in preparation of the solution are removed. The result is a 'solid solution' deposit of analyte-doped matrix crystals where the analyte molecules are embedded throughout the matrix so that they are completely isolated from one other.

The second step involves ablation of bulk portions of this solid solution by intense pulses of laser for a short duration. Indeed, the irradiation by the laser induces rapid heating of the crystals by the accumulation of a large amount of energy in the condensed phase through excitation of the matrix molecules. The rapid heating causes localized sublimation of the matrix crystals and expansion of the matrix into the gas phase, entraining intact analyte in the expanding matrix plume. Little internal energy is transferred to the analyte molecules and they may be cooled during the expansion process. Ionization reactions can occur at any time during this process but the origin of ions produced in MALDI is still not fully understood.[21]

Among the chemical and physical ionization pathways suggested for MALDI are gas-phase photoionization, excited-state proton transfer, ion–molecule reactions, desorption of preformed ions, etc. The most widely accepted ion formation mechanism involves gas-phase proton transfer in the expanding plume with photoionized matrix molecules. Figure 1.14 shows a diagram of the MALDI process.

The MALDI process is more sensitive than other laser ionization techniques. Indeed, the number of matrix molecules widely exceed those of the analyte, thus separating its molecules and thereby preventing the formation of sample clusters, which inhibit the appearance of molecular ions. The matrix also serves to minimize sample damage from the laser pulse by absorbing most of the incident energy and increases the efficiency of energy transfer from the laser to the analyte. Thus the sensitivity is highly increased. It is also more universal than other laser ionization techniques. Indeed, it is not necessary to adjust the wavelength to match the absorption frequency of each analyte because it is the matrix that absorbs the laser pulse. Furthermore, because the process is independent of the absorption properties and size

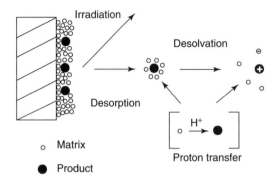

Figure 1.14
Diagram of the principle of MALDI

of the compound to be analyzed, MALDI allows to the desorption and ionization of analytes with very high molecular mass in excess of 100 000 Da. For example, MALDI allows the detection of picomoles of proteins with molecular mass up to 300 000 Da.[22,23]

The technique of MALDI mass spectrometry has become a powerful analytical tool for both synthetic polymers and biopolymers. Typical MALDI spectra include mainly the monocharged molecular species. Some multiply charged ions, some multimers and very few fragments also can be observed. Figure 1.15 shows MALDI spectrum of a monoclonal antibody[24] of about 150 kDa. This figure also presents the MALDI spectrum of a synthetic polymer corresponding to poly(methyl methacrylate) with an average mass of about 7100 Da.

1.6.2 Practical considerations

Different lasers have been used. Ultraviolet (UV) lasers are the most common because of their ease of operation and their low price. Nitrogen lasers ($\lambda = 337$ nm) are considered as the standard, although Nd : YAG lasers ($\lambda = 266$ or 355 nm) also are used. The MALDI technique also can use infrared (IR) lasers such as Er : YAG lasers ($\lambda = 2.94$ μm) or CO_2 lasers ($\lambda = 10.6$ μm). It is not the power density that is the most important parameter to produce significant ion current but the total energy in the laser pulse at a given wavelength.[25] Generally, the power density required corresponds to an energy flux of 20 mJ cm^{-2}. The pulse widths of lasers vary from a few tens of nanoseconds to a few hundred microseconds. It is important to determine the threshold irradiance: the laser pulse power that results in the onset of desorption of the matrix. Molecular species of the analyte are generally observed at slightly higher irradiances but higher laser power leads to more extensive fragmentation and induces a loss of mass resolution. The MALDI spectra obtained with UV and IR lasers are essentially identical. There are only very small differences. Indeed, when the IR laser is used there is less adduct formation and less fragmentation, indicating that IR-MALDI is somewhat cooler.

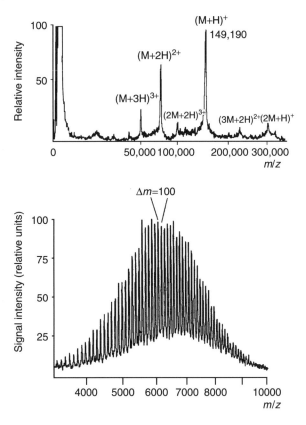

Figure 1.15
The MALDI spectra of a monoclonal antibody (*top*) and poly(methyl methoacrylate) of average mass 7100 Da (*bottom*) (Reproduced (modified) from Ref. 24 and from Finnigan MAT documentation, with permission)

   Matrix selection and optimization of the sample preparation protocol are the most important steps in the analysis. However, they are still empirical procedures. The MALDI matrix selection is based on the laser wavelength used and the class of compound analyzed. Requirements of MALDI matrices are strong absorbance at the laser wavelength, low enough mass to be sublimable, vacuum stability and lack of chemical reactivity. However these general guidelines for matrix selection are not sufficient to predict a good matrix. Indeed, numerous matrix candidates have been inspected and their ability to function as a MALDI matrix has been exemplified. But only very few are good matrices.

   Common UV-MALDI matrices are listed in Table 1.1, along with the class of compounds with which they are used. A relational database of MALDI matrices also is available on Internet.[26] The matrices used with an IR laser, such as urea, caboxylic acids, alcohols and even water, are often closer to the natural solutions than the highly aromatic UV-MALDI matrices.

Table 1.1   Some common UV-MALDI matrices

Matrix	Applications
α-Cyano-4-hydroxycinnamic acid	Peptides, proteins, organic compounds
3,5-Dimethoxy-4-hydroxycinnamic acid (sinapic)	Higher mass biopolymers
2,5-Dihydroxybenzoic acid (gentisic)	Peptides, proteins, carbohydrates
3-Hydroxypicolinic acid	Oligonucleotides
Trihydroxyacetophenone	Oligonucleotides, peptides
5-Chlorosalicylic acid	Water-insoluble polymers

A number of different sample preparation methods have been described in the literature.[27,28] Several of these protocols are accessible on Internet.[29] The original method that is always the most widely used has been called 'dried-droplet'. This method consists of mixing some saturated matrix solution (5–10 µl) with a smaller volume (1–2 µl) of an analyte solution. Then, a droplet (0.5–2 µl) of the resulting mixture is placed on the MALDI probe. The droplet is dried at room temperature and, when the liquid has evaporated completely to form crystals, the sample may be loaded into the mass spectrometer.

The MALDI technique is relatively less sensitive to contamination (salts, buffers, detergents, etc.) compared with other ionization techniques,[30] but prior purification to remove the contaminants leads to improvements in the quality of mass spectra. The analyte must be incorporated into the matrix crystals, and if contaminants can disturb this process then MALDI generally may serve to sequester the analyte from contaminants.

1.6.3 Fragmentations

The MALDI process can lead to fragmentations that occur as a result of the excess energy that is imparted to the analyte during the desorption/ionization process. There are essentially three different types of fragmentation that generate fragment ions in MALDI spectra: fragmentation at the sample surface occurring on a time scale equal to or less than the desorption event is called prompt fragmentation; fragmentation occurring in the source after the desorption event but before the acceleration event is called fast fragmentation; and fragmentation that occurs after the acceleration region of the mass spectrometer is called post-source decay (PSD) fragmentation. Prompt and fast fragmentations lead to product ions that are always apparent in the MALDI spectra, whereas the observation of product ions from PSD fragmentation needs certain instrumental conditions. For example, a MALDI source coupled to a linear time-of-flight analyzer allows prompt and fast detection of fragment ions at their appropriate m/z ratio but the metastable fragment ions cannot be resolved from their precursor ions and thus cause a broadening of the peaks with a concomitant loss of mass resolution and sensitivity.

1.7 Thermospray

The principle of the thermospray (TSP), proposed by Blakney and Vestal[31,32] in 1983, is shown in Figure 1.16.

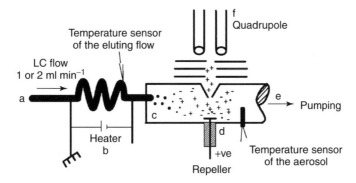

Figure 1.16
Diagram of a thermospray source. The chromatographic effluent comes in
at (a) the transfer line is suddenly heated at (b) and the spray is formed
under vacuum at (c). At (d) the spray goes between a pusher with a positive
potential and a negative cone for positive ions. The ions are thus extracted
from the spray droplets and accelerated towards the spectrometer (f). At
(e) a high-capacity pump maintains the vacuum

A solution containing a salt and the sample to be analyzed is pumped into a steel
capillary, which is heated to a high temperature to allow quick heating of the liquid.
The solution passes through a vacuum chamber as a supersonic beam. A fine droplet
spray occurs, containing ions and solvent and sample molecules. The ions in the
solution are extracted and accelerated towards the analyzer by a repeller and by a
lens-focusing system. They are desorbed from the droplets carrying one or several
solvent molecules or dissolved compounds. It is thus not necessary to vaporize before
ionization: ions go directly from the liquid phase to the vapor phase. To improve
the ion extraction, the droplets at the outlet of the capillary may be charged by a
corona discharge. The droplets stay on their supersonic path to the outlet, where they
are pumped out continuously through an opening located in front of the supersonic
beam. Large vapor volumes from the solvent are thus avoided.

In order to avoid freezing the droplets under vacuum, the liquid must be heated during
injection. This heating is programmed by feedback from a thermocouple that measures
the beam temperature under vacuum. The heating is achieved by having a current going
through the capillary that carries the liquid; this also acts as a heating resistance.

1.8 Electrospray

Electrospray ionization (ESI) and atmospheric pressure chemical ionization (APCI)
are two examples of atmospheric pressure ionization (API) sources. Such sources
ionize the sample at atmospheric pressure and then transfer the ions into the
mass spectrometer. When the sample ionization is performed under atmospheric
pressure,[33,34] an ionization efficiency $10^3 - 10^4$ times as great as in a reduced-pressure
CI source is obtained.

The problem lies in coupling an atmospheric pressure source compartment with an analyzer compartment that must be kept at a very low pressure (10^{-5} Torr). This problem is solved by introducing focusing lenses with very small openings between both compartments or focusing multipole lenses, and by using several differential stages of high-capacity pumps.

Another problem lies in the cooling caused by the sample and the solvent adiabatic expansion that causes the appearance of ion clusters. This is avoided by the introduction of a high-temperature transfer tube (250°C) or by applying a heated, dry nitrogen counter-current or gas curtain.

In the literature, electrospray ionization is abbreviated to either ESI or ES. Because ES is ambiguous we prefer to use ESI. The success of ESI started when Fenn et al.[35] showed that multiply charged ions were obtained from proteins, allowing their molecular weight to be determined with instruments whose mass range is limited to as low as 2000 Th. At the beginning, ESI was considered to be an ionization source dedicated to protein analysis. Later on, its use was extended not only to other polymers and biopolymers but also to the analysis of small polar molecules. It appeared, indeed, that ESI allows very high sensitivity to be reached and is easy to couple to high-performance liquid chromatography (HPLC), micro-HPLC or capillary electrophoresis. Electrospray principles and biological applications were reviewed extensively in 1996,[36] 1997[37] and 2000.[38]

An electrospray[39-44] is produced by applying a strong electric field, under atmospheric pressure, to a liquid passing through a capillary tube with a weak flux

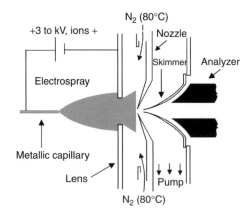

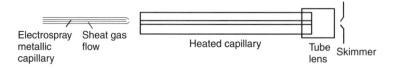

Figure 1.17
Diagram of ESI sources, using skimmers for ion focalization and a curtain of heated nitrogen gas for desolvation (*top*), or with a heated capillary for desolvation (*bottom*)

(normally $1-10$ µl min^{-1}). The electric field is obtained by applying a potential difference of $3-6$ kV between this capillary and the counter-electrode, separated by $0.3-2$ cm, producing electric fields of the order of 10^6 V m^{-1} (Figure 1.17). This field induces a charge accumulation at the liquid surface located at the end of the capillary, which will break to form highly charged droplets. A gas injected coaxially at a low flow rate allows dispersion of the spray to be limited in space. These droplets then pass either through a curtain of heated inert gas, most often nitrogen, or through a heated capillary to remove the last solvent molecules.

The spray starts at an 'onset voltage' that, for a given source, depends on the surface tension of the solvent. In a source that has an onset voltage of 4 kV for water (surface tension 0.073 N m^{-2}), 2.2 kV is estimated for methanol (0.023 N m^{-2}), 2.5 kV for acetonitrile (0.030 N m^{-2}) and 3 kV for dimethylsulfoxide (0.043 N m^{-2}).[45] If one examines with a microscope the nascent drop forming at the tip of the capillary while increasing the voltage, as displayed schematically in Figure 1.18, at low voltages the drop appears spherical and then elongates under the pressure of the accumulated charges at the tip in the stronger electric field; when the surface tension is broken, the shape of the drop changes to a 'Taylor cone' and the spray appears.

Gomez and Tang[46] were able to obtain photographs of droplets formed and dividing in an ESI source. A drawing of a decomposing droplet is displayed in Figure 1.19. From their observations they concluded that breakdown of the droplets can occur before the limit given by the Rayleigh equation is reached, because the

Figure 1.18
Effect of ESI potential on the drop at the tip of the capillary (as observed with binoculars) while increasing the voltage. (*Left*) At low voltage, the drop is almost spherical. (*Center*) At about 1 or 2 kV but below the onset potential the drop elongates under the pressure of the charges accumulating at the tip. (*Right*) At onset voltage, the pressure is higher than the surface tension, the shape of the drop changes at once to a Taylor cone and small droplets are released. The droplets divide, producing the spray

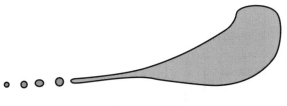

Rayleigh: $q^2 = 8\pi^2 \, \varepsilon_0 \gamma \, D^3$

Figure 1.19
Drawing of a decomposing droplet in an ESI source, according to Ref. 46: q = charge; ε_0 = permittivity of the environment; γ = surface tension and D = diameter of a supposed spherical droplet

droplets are deformed mechanically, thus reducing the repulsion necessary to break down the droplets.

The solvent contained in the droplets evaporates, which causes them to shrink to the point where the repelling coulombic forces come close to their cohesion forces, thereby causing their division. These droplets then undergo a cascade of ruptures, yielding smaller and smaller droplets. Each rupture yields two droplets of similar size, but under the effect of the strong electrical field many smaller, highly charged droplets are produced too. When the electric field on their surface becomes large enough, desorption of ions from the surface occurs.[45] Charges in excess accumulate at the surface of the droplet. In the bulk, analytes as well as electrolytes (whose positive and negative charges are in equal number) are present. Desorption of charged molecules occurs from the surface, which means that sensitivity is higher for compounds whose concentration at the surface is higher, i.e. the more lipophilic. When mixtures of compounds are analyzed, those present at the surface of droplets can mask, even completely, the presence of compounds that are more soluble in the bulk.

The ions obtained from large molecules carry a greater charge if several ionizable sites are present. Typically, a protein will carry one charge per 1000 Da, or less if there are very few basic amino acids. As an example, the electrospray spectrum of phage λ lysozyme is shown in Figure 1.20. Small molecules, say less than 1000 Da, will produce mainly monocharged ions. Electrospray ionization can be used also in the case of molecules without any ionizable site through the formation of sodium, potassium, ammonium, chlorine, acetate or other adducts.

Electrospray has important characteristics: it is able to produce multiply charged ions from large molecules. The formation of ions is a result of the electrochemical process and of the accumulation of charge in the droplets. The electrospray current is limited by the electrochemical process that occurs at the probe tip and is sensitive to concentration rather than to the total amount of sample.

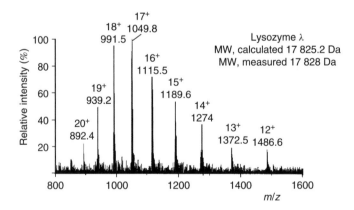

Figure 1.20
Electrospray spectrum of phage λ lysozyme. The *m/z* (in Th) and the number of charges are indicated on each peak. The molecular mass is measured as being $17\,828 \pm 2.0$ Da

1.8.1 Multiply charged ions

Large molecules with several ionizable sites produce multiply charged ions by ESI, as shown for the lysozyme positive ions in Figure 1.20.

Obtaining multiply charged ions is advantageous because it improves the sensitivity at the detector and it allows the analysis of high-molecular-weight molecules using analyzers with a weak nominal mass limit. Indeed, the technical characteristics of mass spectrometers are such that the value being measured is not the mass, but the mass-to-charge ratio m/z.

The ESI mass spectra of biological macromolecules normally correspond to a statistical distribution of consecutive peaks characteristic of multiply charged molecular ions obtained through protonation $(M + zH)^{z+}$ or deprotonation $(M - zH)^{z-}$, with minor, if any, contributions of ions produced by dissociations or fragmentations. However, because the measured apparent mass is actually m/z, to know m one needs to determine the number of charges z.

Consider a positive ion with charge z_1 whose mass-to-charge ratio is measured as being m_1 (Th), issued from a molecular ion with mass M (Da) to which z_1 protons have been added. We then have:

$$z_1 m_1 = M + z_1 m_p$$

where m_p is the mass of the proton. An ion separated from the first one by $(j - 1)$ peaks, in increasing order of mass-to-charge ratio, has a measured ratio of m_2 (Th) and a number of charges $z_1 - j$, so that

$$m_2(z_1 - j) = M + (z_1 - j)m_p$$

These two equations lead to:

$$z_1 = \frac{j(m_2 - m_p)}{(m_2 - m_1)} \quad \text{and} \quad M = z_1(m_1 - m_p)$$

In the case of negative multiply charged ions, analogous equations lead to:

$$z_1 = \frac{j(m_2 + m_p)}{(m_2 - m_1)} \quad \text{and} \quad M = z_1(m_1 + m_p)$$

In the example shown in Figure 1.20 using the peaks at m/z 939.2 and 1372.5 ($j = 6$), we obtain $z_1 = 6(1372.5 - 1.0073)/(1372.5 - 939.2) = 19$ and we can number all the peaks measured according to the number of charges. Mass M can be calculated from their masses. The average value obtained from all of the measured peaks is 17827.9 Da, with a mean error of 2.0 Da. This technique allowed the determination of the molecular masses of proteins above 130 kDa with a detection limit of about 1 pmol using a quadrupole analyzer.

A variety of algorithms have been developed to allow determination of the molecular mass through the transformation of multiply charged peaks present in the ESI

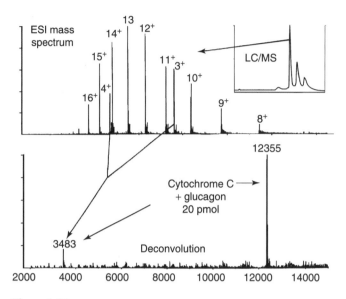

Figure 1.21
Deconvolution of an ESI spectrum of a protein mixture. (Reproduced from Finnigan MAT documentation, with permission)

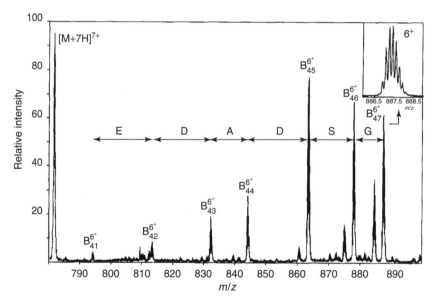

Figure 1.22
Product ion spectrum of the $[M + 7H]^{7+}$ ion from the following peptide:
ALVRQGLAKVAYVYKPNNTHEQHLRKSEAQAKKEKLLNIWSEDNADSGQ.
Notice that fragment ions having lower charge number z may appear at higher m/z values than the precursor, which indeed occurs in the spectrum shown. The inset shows that, owing to the high resolution, the isotopic peaks are observed separated by 1/6 Th, and thus $1/z = 1/6$ or $z = 6$. (Reproduced from Ref. 47 with permission)

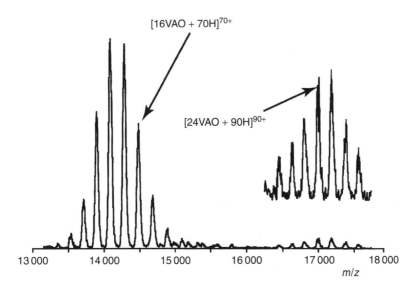

Figure 1.23
Micro-ESI Q–TOF spectrum of assemblies of vanillyl alcohol oxidase
obtained by W.J.H. van Berkel and co-workers.[48] (*Left*) An octamer–dimer
of 1.02 MDa molecular weight. (*Right*) An octamer–trimer of 1.53 MDa.
(Courtesy of Dr A.J.R. Heck)

spectrum into singly charged peaks. Some of them also allow the deconvolution ESI
spectra of mixtures, as is shown in Figure 1.21. However, the complexity of the
spectra obtained for a single compound is such that only simple mixtures can be
analyzed.

At high resolution, the individual peaks with different charge states observed at
low resolution are each split into several peaks corresponding to the isotope distribution. Because neighbor peaks differ by 1 Da, the observed distance between them will
be $1/z$, allowing direct determination of the charge state of the corresponding ion.
This is important for MS/MS spectra of multiply charged ions because the preceding
rules to assign the z value cannot be applied. An example is displayed in Figure 1.22.

The ability of this ionization method for the determination of very high molecular weights is illustrated in Figure 1.23.[48] The spectrum displayed is obtained from
assemblies of vanillyl alcohol oxidase containing, respectively, 16 and 24 proteins.
The spectrum was obtained with a hybrid quadrupole–time-of-flight instrument,
Q–TOF 1 Micromass, equipped with a microelectrospray source. To obtain such
a spectrum, one not only needs a mass spectrometer with sufficient mass range and
resolution but also high skill in protein purification.

1.8.2 Electrochemistry and electric field as origins of multiply charged ions

Charges of ions generated by ESI do not reflect the charge state of compounds in the
analyzed solution but are the result of both charge accumulation in the droplets and

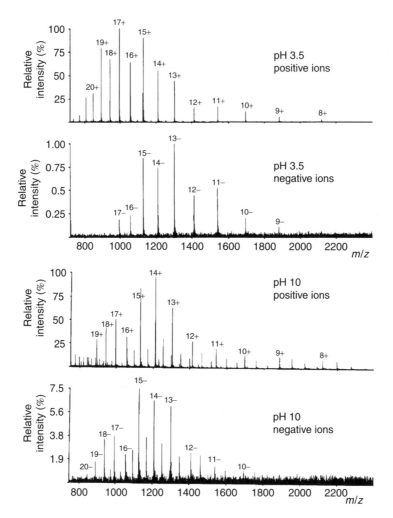

Figure 1.24
Electrospray spectra of myoglobin have been acquired in the positive
and negative ion mode, at pH 3 and pH 10. At pH 3, negative ions
are observed, the most intense ions bearing from 13 to 15 charges. A
calculation based on known pK values shows that in the original solution
only one molecule in about 3500 bears one negative charge. (Reproduced
from Ref. 49 with permission)

charge modification by electrochemical process at the probe tip. This is demonstrated
clearly by a convincing experiment reported by Fenselau and co-workers, and is
illustrated in Figure 1.24.[49] They showed that negative ions of myoglobin can be
observed at pH 3, whereas a calculation based on known pK values predicts that
only one molecule per about 3500 would have one negative charge in the original
solution. The results point to the role of charge accumulation in droplets under the
influence of the electric field on the formation of multiply charged ions. Furthermore,

Table 1.2   Some values of equivalent conductivity

Cation	$\lambda_0{}^+$	Anion	$\lambda_0{}^-$
H_3O^+	350	OH^-	200
$NH_4{}^+$	74	Br^-	78
K^+	74	I^-	77
Na^+	50	Cl^-	76

the 'pumping out' of the negative charges can be performed only if, at the same time, the same number of positive charges is neutralized electrochemically at the probe tip.

Moreover, it is worth noting that the negative ion spectrum of myoglobin at pH 3 shows a better signal-to-noise ratio than the same spectrum at pH 10 (Figure 1.24). This results from the fact that protons have a high electrochemical mobility, and is the first indication of the importance of the reduction process, when negative ions are analyzed, that occurs at the probe tip.

At pH 10, positive ions can be observed too. Additional peaks in the spectra result from a modification of the protein at basic pH (loss of heme group).

Thus, it is worthy of consideration to try to acidify a solution with a view to better detection of negative ions, and vice versa. Indeed, both H_3O^+ and OH^- have high limit equivalent conductivities, as shown in Table 1.2.

1.8.3 Sensitivity to concentration

Another feature of ESI is its sensitivity to concentration, and not to the total quantity of sample injected in the source, as is the case for most other sources. This is shown in Figure 1.25.[50] The intensity of the signals from the two monitored compounds is measured while injecting the total flow from an HPLC column (400 µl min^{-1}), or a part of this flow, after splitting. As can be seen, the sensitivity increases somewhat

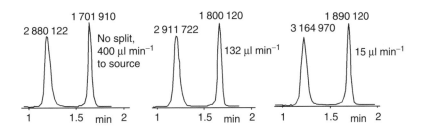

Figure 1.25
High performance liquid chromatography on a 2.1 mm column at 400 µl min^{-1} flow. Two drugs are monitored by selected ion monitoring. (*Left*) A 400 µl min^{-1} flow is injected in the source. (*Centre*) A 132 µl min^{-1} flow is split to the source. (*Right*) A 15 µl min^{-1} flow is split to the source. The integration values are displayed on top of the peaks and show that at reduced flow rates the sensitivity is slightly increased. (Data from Ref. 50)

when the flow entering the source is reduced. This remains true up to flows as low as some tens of nanoliters per minute. When flow rates higher than about 500 µl min^{-1} are used, the sensitivity is reduced. Lower flow rates also allow less analyte and buffer in the source to be injected, reducing contamination. Furthermore, for the same amount of sample, an HPLC column with a lower diameter, and using smaller flow rates, will give an increased sensitivity because the concentration of the sample in the elution solvent is increased.

Based on this concentration dependence, modifications of the technique, called 'micro electrospray' (µESI), or 'nanospray' (nESI), which use much lower flow rates down to some tens of nanoliters per/minute, have been developed using adapted probe tips.[51-53] Detection limits in the range of attomoles (10^{-15} mol) injected have been demonstrated.

1.8.4 Limitation of ion current from the source by the electrochemical process

As may be seen from the electric circuit in Figure 1.26, when positive ions are extracted for analysis electrons have to be provided in the circuit from the capillary. This will occur through oxidation of species in the solution at the capillary tip, mainly ions having sufficient mobility. In other words, the same number of negative charges must be 'pumped' out of the solution as positive charges are extracted to the analyzer. Thus, ESI is truly an electrochemical process, with its dependence on ion concentration and mobility as well as on polarization effects at the probe tip. For the detection of positive ions, electrons have to be provided by the solution and thus oxidation occurs. For negative ions, electrons have to be consumed and thus reduction occurs. These electrochemical reactions occur in the last micrometers of the metallic capillary. A major consequence is that the total number of ions per unit time that can be extracted to the spectrometer is actually limited by the electric

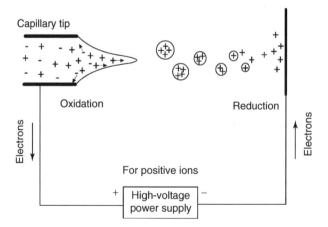

Figure 1.26
Schematic representation of electrochemical process in ESI

current produced by the oxidation or reduction process at the probe tip. This limiting current is not dependent on the flow rate up to very low flow, and explains why ESI is only concentration dependent. In practice, the total ion current is limited to a maximum of about 1 μA.

In ESI, the number of variables is large, including nature of the solvent, flow, nature and size of the capillary, distance to the counter-electrode, etc. Furthermore, the ionization process includes many parameters, such as surface tension, nature of analyte and electrolytes, presence or not of other analytes, electrochemical processes at the probe tip, etc. A paper by Fenn et al.[54] makes a critical comparison of the various theories about ESI. A simplified theory will be presented here for the relation between analyte concentration and abundances of the ions.

Ions, either positive or negative, of an analyte A will be desorbed from the droplets, producing a theoretical ion current $I_A = k_A[A]$, where k_A is a rate constant depending on the nature of A. Let us suppose that another ion B is produced from the buffer, at a rate $I_B = k_B[B]$, and that no other ions are sprayed. The total ion current for these two ions is $I_T = (I_A + I_B)$, but this total ion current is limited by the oxidation, if positive ions are desorbed, or the reduction process that occurs at the probe tip. This limiting current is denoted as I_M, and $I_T = I_M$.

The current for each ion will be proportional to its relative desorption rate, and the pertinent equations are:

$$I_A = I_M \frac{k_A[A]}{k_A[A] + k_B[B]} \qquad I_B = I_M \frac{k_B[B]}{k_A[A] + k_B[B]} \tag{1.1}$$

Let us consider that $[B]$ remains constant but the analyte concentration $[A]$ varies, then two limiting cases are to be considered. First, $k_B[B] \gg k_A[A]$, then

$$I_A \approx I_M \frac{k_A[A]}{k_B[B]} \qquad I_B \approx I_M \frac{k_B[B]}{k_B[B]} \approx I_M \tag{1.2}$$

This means that the intensity detected for A will be proportional to its concentration, but the sensitivity will be inversely proportional to $[B]$.

The other extreme case is, $k_A[A] \gg k_B[B]$ leading to:

$$I_A \approx I_M \frac{k_A[A]}{k_A[A]} \approx I_M \qquad I_B \approx I_M \frac{k_B[B]}{k_A[A]} \tag{1.3}$$

where I_A remains constant and quantitation of $[A]$ is not possible. The intensity of the signal for B will become weaker as $[A]$ increases.

This is shown by an experimental example from Ref. 45 in Figure 1.27. In a solvent containing NH_4^+ and Na^+ ions at constant concentrations, an increasing amount of morphine·HCl chlorhydrate is added. The graph shows the number of amperes at the capillary tip and the intensity monitored at the mass of protonated morphine and the sum of the intensities for the NH_4^+ and Na^+ ions. At low concentrations of morphine·HCl, Equation (1.2) applies and linearity towards the morphine concentration is observed. At high concentrations the intensity for

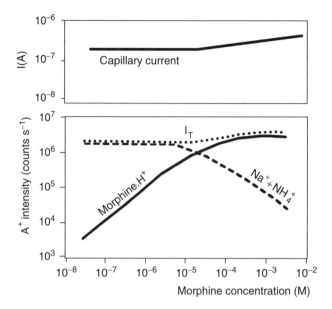

Figure 1.27
Influence of concentration on observed ion abundances when increasing concentrations of morphine.HCl are injected in a solvent containing a constant concentration of sodium and ammonium salts. Linearity is observed at low concentrations, but from about 5×10^{-6} significant curvature is observed (note that the scales are logarithmic). At still higher concentrations the intensity levels out. (Modified from Ref. 45)

morphine is constant and the signal for the other ions diminishes, in agreement with Equation (1.3). At intermediate values Equation (1.1) applies.

1.8.5 Practical considerations

To observe a stable spray, a minimum amount of electrolyte in the solvent is required, but this is so low that normal solvents contain enough electrolytes for this purpose. On the other end, the maximum tolerable total concentration of electrolytes to still have good sensitivity is about 10^{-3} M. Furthermore, volatile electrolytes are preferred to avoid contamination of the source. With most samples, the problem is more to remove the salts rather than to add some. Also, sample dilution often is performed. Very often HPLC methods have to be modified for ESI when they use high concentrations of buffers.

When it is believed that it could be better to add an electrolyte to improve sample detection, one should think about the electrochemical process when selecting it. We have seen, for instance, that adding an acid can improve the detection of negative ions. But the ESI process is not simple and many trials often are needed.

1.9 Atmospheric Pressure Chemical Ionization

Atmospheric pressure chemical ionization (APCI) is an ionization technique that uses gas-phase ion–molecule reactions at atmospheric pressure.[55,56] This is a method analogous to CI (commonly used in gas chromatography/mass spectrometry) where primary ions are produced by corona discharges on a solvent spray. The APCI technique is mainly applied to polar and ionic compounds with moderate molecular weight up to about 1500 Da and generally gives monocharged ions. The principle governing an APCI source is shown in Figure 1.28.

The analyte in solution from a direct inlet probe or a liquid chromatography eluate at a flow rate between 0.2 and 2 ml min^{-1} is introduced directly into a pneumatic nebulizer where it is converted into a thin fog by a high-speed nitrogen beam. Droplets then are displaced by the gas flow through a heated quartz tube called a desolvation/vaporization chamber. The heat transferred to the spray droplets allows vaporization of the mobile phase and of the sample in the gas flow. The temperature of this chamber is controlled, which makes the vaporization conditions independent of the flow and from the nature of the mobile phase. The hot gas (120°C) and the compounds leave this tube. After desolvation, they are carried along a corona discharge electrode where ionization occurs. The ionization processes in APCI are equivalent to the processes that take place in CI but all of these occur under atmospheric pressure. In the positive ion mode, either proton transfer or adduction of reactant gas ion can occur to produce the ions of molecular species, depending on the relative proton affinities of the reactant ions and the gaseous analyte molecules. In the negative mode, the ions of the molecular species are produced either by proton abstraction or adduct formation.

Generally, the evaporated mobile phase acts as the ionizing gas and reactant ions are produced from the effect of a corona discharge on the nebulized solvent. Typically, the corona discharge forms by electron ionization primary ions such as $N_2^{\bullet+}$ or $O_2^{\bullet+}$. Then, these ions collide with vaporized solvent molecules to form secondary reactant gas ions.

The electrons needed for the primary ionization are not produced by a heated filament, because the pressure in that part of the interface is atmospheric pressure and

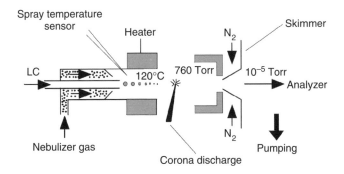

Figure 1.28
Diagram of an APCI source

the filament would burn, but rather using corona discharges or β^- particle emitters. These two electron sources are fairly insensitive to the presence of corrosive or oxidizing gases.

Because the ionization of the substrate occurs at atmospheric pressure and thus with a high collision frequency, it is very efficient. Furthermore, the high frequency of collisions serves to thermalize the reactant species. In the same way, the rapid desolvation and vaporization of the droplets reduce considerably the thermal decomposition of the analyte. The result is ionization yielding predominantly ions of molecular species with few fragmentations.

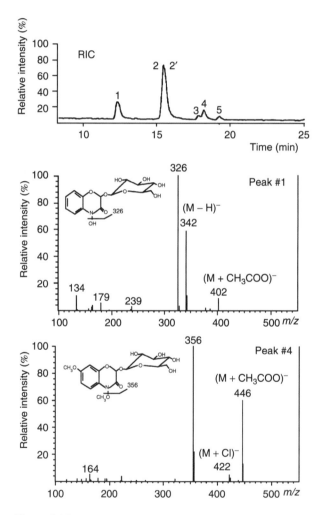

Figure 1.29
The HPLC/APCI analysis of a mixture of glucoconjugated compounds related to DIMBOA. The spectrum from peak 4 does not display the deprotonated molecular species. The molecular mass (387 Da) is deduced from the adducts. The sample is obtained from extracts of maize plants[58]

The ions produced at atmospheric pressure enter the mass spectrometer through a tiny inlet and then are focused towards the analyzer. This inlet must be sufficiently wide to allow the entry of as many ions as possible while keeping a correct vacuum within the instrument to allow analysis. The most common solution to all these constraints consists of using the differential pumping technique on one or several stages, each separated from the others by skimmers.[57] In the intermediate pressure region, an effective declustering of the formed ions occurs.

The APCI technique has become a popular ionization source for applications of coupled high-performance liquid chromatography/mass spectrometry. Figure 1.29 shows an example of an application of HPLC/APCI coupling.[58] It shows the analysis obtained from extracts of maize plants. Six compounds are identified by mass spectrometry. These compounds have been identified as glucoconjugated DIMBOA (2,4-dihydroxy-7-methoxy-1,4-benzoxazin-3-one) and similar molecules that differ by the number of methoxy groups in the benzene ring and/or by N-O-methylation of the hydroxamate function. This example clearly shows the influence of the analyte on the type of observed molecular species. Indeed, the presence of an acidic group in the compound from peak 1 allows mainly the detection of deprotonated molecular ion whereas the compound from peak 4, which does not contain an acid group, leads only to the formation of adduct ions.

1.10 Inorganic Ionization Sources

Mass spectrometry is not only an indispensable tool in organic and biochemical analysis, but it also becomes a powerful technique for inorganic analysis.[59-61] Indeed, over the last 20 years the application of mass spectrometry to inorganic and organometallic compounds has revolutionized the analysis of these compounds. Important advances have been made in the diversification of ionization source, in the commercial availability of the instruments and in the fields of applications.

Mass spectrometry is now widely used for inorganic characterization and micro-surface analysis. Electron ionization is the preferred ionization source for volatile inorganic compounds whereas others that are non-volatile may be analyzed using ionization sources already described, such as SIMS, FD, FAB, LD or ESI.[62-64] The example in Figure 1.30 shows the analysis of orthorhombic sulfur (S_8 ring) and $Cr(CO)_2(dpe)_2$ obtained with an EI source and an ESI source, respectively.

In addition, quantitative and qualitative elemental analysis of inorganic compounds with high accuracy and high sensitivity can be effected by mass spectrometry. For elemental analysis, atomization of the analyzed sample, which corresponds to the transformation of solid matter to atomic vapor, and ionization of the resulting atoms occur in the source. These atoms then are sorted and counted with the help of mass spectrometry. The complete decomposition of the sample in the ionization source into its constituent atoms is necessary because incomplete decomposition results in a complex mass spectrum in which isobaric overlap might cause unsuspected spectral interferences. Furthermore, the distribution of any element in different species leads to a decrease in sensitivity for this element.

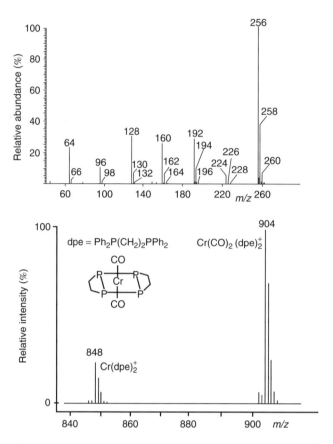

Figure 1.30
Analysis of inorganic compounds by mass spectrometry. (*Top*)
The EI spectrum of orthorhombic sulfur (S_8). (*Bottom*) The ESI
spectrum of $Cr(CO)_2(dpe)_2$ (redrawn using data from Ref. 63)

Four techniques based on mass spectrometry are widely used for multielemental trace analysis of inorganic compounds in a wide range of sample types: the techniques of thermal ionization, spark source, glow discharge and inductively coupled plasma mass spectrometry. In these techniques, atomization and ionization of the analyzed sample are accomplished, respectively, by volatilization from a heated surface, attack by electrical discharge, rare-gas ion sputtering and vaporization in a hot flame produced by inductive coupling.

All of these ionization sources are classical sources that are used also for optical spectroscopy. The only fundamental difference is that these sources are not used for atomization/excitation processes to generate photons but to generate ions.

1.10.1 Thermal ionization source

Thermal ionization is based on the production of atomic or molecular ions at the hot surface of a metal filament.[65,66] In this ionization source, the sample is deposited on

a metal filament (W, Pt or Re) and an electric current is used to heat the metal to
a high temperature. The ions are formed by electron transfer from the atom to the
filament for positive species or from the filament to the atom for negative species.
The analyzed sample can be fixed to the filament by depositing drops of the sample
solution on the filament surface, followed by evaporation of the solvent to complete
dryness or by using electrodeposition methods.

Single, double and triple filaments have been used broadly in thermal ionization sources. In a single filament source, the evaporation and ionization processes
of the sample are carried out on the same filament surface. Using a double filament source, the sample is placed on one filament used for the evaporation while
the second filament is left free for ionization. In this way, it is possible to set the
sample evaporation rate and ionization temperature independently, thus separating the
evaporation from the ionization process. This is interesting when the vapor pressure
of the studied elements reaches high values before a suitable ionization temperature
can be achieved. A triple filament source can be useful to obtain a direct comparison
of two different samples under the same source conditions.

Positive and negative ions can be obtained by the thermal ionization source.
High yields of positive and negative ions are obtained for atoms or molecules with
low ionization potential and with high electron affinity, respectively. Owing to the
ionization process and the physical parameters of the ions, metals can be analyzed
in positive ion mode whereas many non-metals, semi-metals and transition metals
or their oxides are able to form negative thermal ions.

In thermal ionization sources, the most abundant ions are usually the singly
charged atomic ions. No multiply charged ions can be observed under normal ionization conditions. Cluster ions occur very seldom. However, some metal compounds
lead to abundant metal oxide ions.

For thermal ionization filament sources, the ionization efficiency varies from less
than 1% to more than 10%, depending on the analyzed elements. A significant
improvement of ionization efficiency can be observed with the use of a thermal
ionization cavity source.[67] In this type of thermal source, as displayed schematically
in Figure 1.31, a refractory metal tube is used instead of a filament to evaporate and

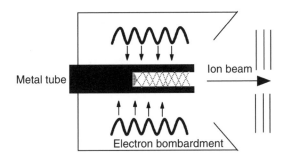

Figure 1.31
Schematic representation of a thermal ionization
cavity source

ionize the sample. High ionization temperatures are achieved by using high-energy electron bombardment to heat the tube. As the sample evaporates, the gaseous atoms interact with each other and with the inner wall of the cavity to produce ions. Compared with filament sources, the cavity sources can provide orders of magnitude enhancement of the ionization efficiency.

1.10.2 Spark source

In spark sources, electrical discharges are used to desorb and ionize the analytes from solid samples.[68] As shown in Figure 1.32, this source consists of a vacuum chamber in which two electrodes are mounted. A pulsed 1 MHz radiofrequency voltage of several kilovolts is applied in short pulses across a small gap between these two electrodes and produces electrical discharges. If the sample is a metal it can serve as one of the two electrodes, otherwise it can be mixed with graphite and placed in a cup-shaped electrode.

Atomization is accomplished by direct heating of the electrode by the electron component of the discharge current and by the discharge plasma. Then ionization of these atoms, which occurs in the plasma, is due mainly to plasma heating by electrons accelerated by the electric field. Chemical reactions also can take place in the plasma, leading to the formation of clusters.

Various types of positive ions are produced in a spark discharge, such as singly and multiply charged atomic ions, polymer ions and heterogeneous compound ions. A spark source mass spectrum is always characterized by singly and multiply charged ions of the major constituents, with a decrease of their intensities when their charges increase. Also, the singly charged species always are the most intense for minor constituents and usually are the only ones used for analytical purposes. Other abundant ions detected in spark sources are heterogeneous compound ions formed by association of the matrix with hydrogen, carbon, nitrogen and oxygen.

It must be noted that this technique does not provide accurate quantitative analysis.

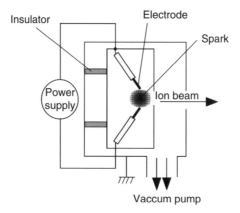

Figure 1.32
Typical spark ion source

1.10.3 Glow discharge source

A glow discharge source is particularly effective at sputtering and ionizing compounds from solid surfaces.[69–71] This source is indicated schematically in Figure 1.33. A glow discharge source consists of one cathode and one anode in a low-pressure gas (0.1–10 Torr), usually one of the noble gases. Argon is the most commonly used gas because of its low cost and its high sputtering efficiency. The sample is introduced in this source as the cathode. Application of an electric current across the electrodes causes breakdown of the gas and the acceleration of electrons and positive ions towards the oppositely charged electrodes. Argon ions from the resulting plasma attack the analyte at the surface of the cathode by bombardment. Collisions of these energetic particles onto the surface transfer their kinetic energy. The species near the surface can receive sufficient energy to overcome the lattice binding and be ejected mainly as neutral atoms. Sample atoms liberated then are carried into the negative glow region of the discharge where they are ionized mainly by electron impact and Penning ionization (Figure 1.33).

A simple DC power supply at a voltage of 500–1000 V and a current of 1–5 mA suffices to obtain glow discharge. However, pulsed DC discharges allow higher peak voltages and currents to be used whereas a radiofrequency (RF) discharge allows the analysis of poorly conducting samples directly.

The principal application of glow discharge is in bulk metal analysis. The conducting solid samples can be made into an electrode, whereas non-conducting materials are compacted with graphite into an electrode for analysis. Solution samples can be analyzed by drying the sample on a graphite electrode. Other applications such as the examination of thin films or solution residues have been described. Glow

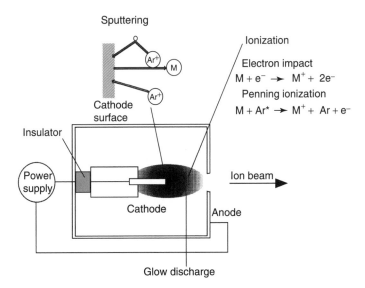

Figure 1.33
Schematic diagram of a glow discharge source with sputtering at the cathode surface and ionization in the discharge

discharge ion sources can be used in conjunction with laser ablation. The pulse of ablated material enters directly into the plasma of the glow discharge where it undergoes ionization.

Owing to the ionization process, only positive ions can be obtained by the glow discharge ionization source. Consequently, the spectra obtained with a glow discharge source are characterized by singly charged positive ions of the sputtered cathode atoms. Almost no doubly charged ions from the sample are observed. Some diatomic cluster ions are formed but normally are observed only for major constituents. The noble gas used (usually argon) and residual gases such as nitrogen, oxygen and water vapor are always observed.

Only a low net ionization in the discharge is produced. The ionization efficiency is estimated to be 1% or less. However, glow discharge produces an atomic vapor representative of the cathode constituents and the discharge ionization processes also are relatively non-selective. Furthermore, because most elements are sputtered and ionized with almost the same efficiency in the source, quantitative analysis without a standard is possible.

It must be noted that glow discharge requires several minutes for the extracted ions to reach equilibrium with elemental concentrations in the analyzed sample. Thus, the sample throughput with this technique is relatively low.

1.10.4 Inductively coupled plasma source

An inductively coupled plasma source is made up of a hot flame produced by inductive coupling in which a solution of the sample is introduced as a spray.[72-74] This source consists of three concentric quartz tubes through which streams of argon flow. As shown in Figure 1.34, a cooled induction coil surrounds the top of the largest tube. This coil is powered by an RF generator that produces between 1.5 and 2.5 kW at 27 or 40 MHz typically. The gas at atmospheric pressure that sustains the plasma is initially made electrically conductive by Tesla sparks, which lead to ionization

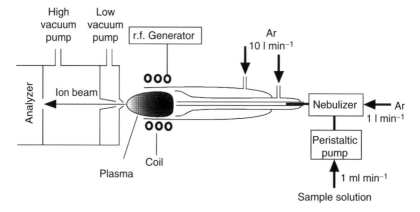

Figure 1.34
Schematic diagram of an inductively coupled plasma source

of the flowing argon. Then, the resulting ions and electrons that are present in the discharge interact with the high-frequency oscillating inductive field created by the RF current in the coil. They are consequently accelerated, collide with argon atoms and ionize them. The products released by this ionization then undergo the same events until the argon ionization process is balanced by the opposing process corresponding to ion–electron recombination. These colliding species cause heating of the plasma to a temperature of about 10 000 K. This temperature requires thermal isolation from the outer quartz tubes by introducing a high-velocity flow of argon at about 10 l min^{-1} tangentially along the walls of these tubes.

The sample is carried into the hot plasma as a thermally generated vapor or a finely divided aerosol of droplets or microparticulates by argon flowing at about 1 l min^{-1} through the central tube. The high temperatures rapidly desolvate, vaporize and largely atomize the sample. Furthermore, this plasma at high temperature and at atmospheric pressure is a very efficient excitation source. The resulting atoms may spend several milliseconds in a region at a temperature between 5000 and 10 000 K. Under these conditions, most elements are ionized to singly charged positive ions with an ionization efficiency close to 100%. Some elements can be ionized to higher charged states but with very low abundance. Ions are extracted from the plasma and introduced in the mass analyzer through a two-stage vacuum-pumped interface containing two cooled metal skimmers.

The most common introduction of the samples in this source consists of a pneumatic nebulizer that is driven by the same flow of argon that carries the resulting droplets in the plasma. An ultrasonic nebulizer or a heated desolvation tube also is used because both allow a better droplet size distribution, which increases the load of the sample into the plasma. Generally, the sample solutions are introduced continuously into the nebulizer at a rate of about 1 ml min^{-1} with the help of a peristaltic pump. However, this is not acceptable with small sample solutions. Therefore, an alternative method using the flow injection technique is employed to introduce a small sample of about 100 μl. The sample solution is injected into a reference blank flow so that the sample is transported in the nebulizer and a transitory signal is observed.

Other alternative methods of sample introduction have been applied. A very small volume of liquid (<10 μl) or solid sample may be introduced in the plasma as a vapor produced by electrothermal vaporization, which vaporizes the sample by flash evaporation from a filament heated by a current pulse. In this case, the sample vapor is produced in a short time of a few seconds and is transferred to the plasma as a gas. Solid samples at atmospheric pressure also can be introduced by laser ablation. This technique also allows spatial analysis of the surface of a sample.

It must be noted that handling samples in solution allows automation and high sample throughput. Indeed, the sample throughput for an inductively coupled plasma mass spectrometry (ICPMS) instrument is typically 20–30 elemental determinations in a few minutes, depending on such factors as the concentration levels and precision required. Handling solution samples includes other advantages such as recording and subtracting a true blank spectrum, adjustment of the dynamic range by dilution, the simplicity of adding internal standards, etc.

1.10.5 Practical considerations

When compared with optical spectrometric techniques of elemental analysis, the techniques based on mass spectrometry provide an increase in sensitivity and analytical working range of some orders of magnitude. For instance, the detection limits with ICPMS are three orders of magnitude better than inductively coupled plasma optical emission spectrometry. Figure 1.35 shows the maximum sensitivity obtained for the different elements using an ICPMS coupling with a quadrupole.

These techniques are relatively interference free but there are, nevertheless, two major types of interferences. A first type are the matrix interferences, which induce suppression or enhancement of the analyte signal. Such interferences are due to the identity and composition of the sample itself. Because they cause a difference between samples and standards for a particular element concentration, they lead to quantitative errors. Matrix effects generally are more serious with these techniques than with optical spectrometric techniques.

However, the most common interferences are the spectral interferences, also called 'isobaric interferences'. They are due to overlapping peaks, which can mask the analyte of interest and give erroneous results. Such interferences may occur from ions of other elements within the sample matrix, elemental combination, oxide formation, doubly charged ions, etc.

A solution to the overlapping peaks problem consists of identifying the interfering species and applying corrections for their contribution to the signal from the analyte. Corrections are based on identifying an isotope of the suspected interfering element that does not itself suffer from a spectral interference and that can be measured with

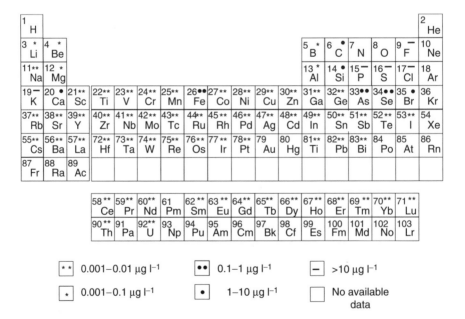

Figure 1.35
Analytical sensitivity of the elements by ICPMS

a sufficiently accurate signal. Then, knowing the relative natural abundance of the isotopes, the contribution of this interfering element to the signal at the mass of interest can be calculated. This correction is not appropriate when the interference is many times more abundant than the analyte because the error on measurement can be too large.

Another very straightforward solution to the overlapping peaks problem is to increase the resolution power of the mass spectrometer. High-resolution mass spectrometers have the ability to resolve many isobaric interferences from the analyte and thus allow unambiguous quantitative analysis to be carried out. However, it should be noted that the sensitivity decreases when the resolution increases.

1.11 Gas-phase Ion–Molecule Reactions

Although electron ionization does not imply any ion–molecule reactions, the latter provide the whole basis of chemical ionization. Other ionization methods give rise to ion–molecule reactions as secondary processes. We will now emphasize some characteristics of these gas-phase reactions and compare them with condensed-phase reactions.

Figure 1.36 displays the energy characteristics of both the gas phase and aqueous solution phase for the substitution reaction:[75]

$$Cl^- + CH_3Br \longrightarrow CH_3Cl + Br^-$$

In the condensed phase, molecules and *a fortiori* ions are strongly solvated. This solvation cage must be at least partially destroyed in order to form the activated complex. This requires a lot of energy and high activation energies thus are observed. Furthermore, each species mingles with the surrounding molecules and continuously exchanges energy.

In the gas phase the opposite occurs: the naked ion interaction with the molecule is exothermic, which leads to the formation of an ion–molecule complex. Some activation energy thus becomes necessary in order to transform the $Cl^-\cdot CH_3Br$ complex into the activated complex and finally into $Br^-\cdot CH_3Cl$. As can be seen in Figure 1.36, this activation energy is generally lower than the energy produced by the first step. Thus, if compared with the starting reactants, the actual activation energy is negative. Another very important difference must be noticed: mass spectrometry literature refers to the gas phase but this association with conventional gases is incorrect. Indeed, the reacting species generally considered in mass spectrometry do not interact with each other, due to the low pressure, and thus do not continuously exchange energy with surrounding molecules. They are never present as an equilibrium phase. Hence the energy produced by the association of the ion with the molecule remains in the complex, thereby allowing it to overcome the activation barrier. Deactivation through radiation is a relatively slow process. This energy further allows the separation of the products into free Br^- ion and CH_3Cl molecules.

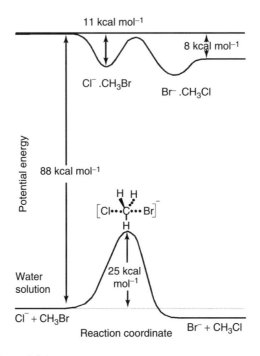

Figure 1.36
Potential energy diagram for a substitution reaction in
the gas phase and in solution in water[75]

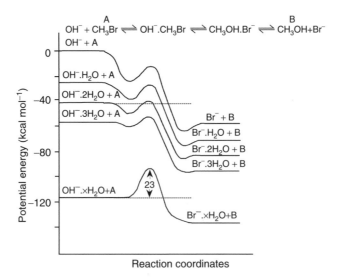

Figure 1.37
(A) CH_3Br; (B) CH_3OH. Potential energy profile for differently
solvated ions reacting with molecules. (Reproduced (modified)
from Ref. 76 with permission)

Figure 1.37 displays the results obtained by measuring the kinetic and thermodynamic parameters for the following reaction:[76]

$$OH^- + CH_3Br \rightleftharpoons Br^- + CH_3OH$$

In this case, the authors succeeded in measuring separately the reaction characteristics both for the naked OH^- ion and for the ion solvated by one to three water molecules. As the number of water molecules increases, the shape of the curves approaches that for the reaction in aqueous solution. The observed rate constants are given in Table 1.3. The rate constant for the gas-phase ion–molecule reaction is 10^{16} times greater than the rate observed in solution. This experimentally observed factor goes up to 10^{20} in the case of other reactions.

Let us now address another problem: are all the ion–molecule reactions possible? Figure 1.38 displays experimental results of proton transfer under chemical ionization conditions. Exergonic reactions, i.e. $\Delta G° < 0$, are highly efficient, with almost every collision giving rise to a proton transfer. However, the efficiency decreases sharply when the process becomes endergonic.[77]

The $\Delta G°$ values for some acids and bases can be found in Appendix 6 of this book. Extensive values can be found in Ref. 78 or at http://webbook.nist.gov.

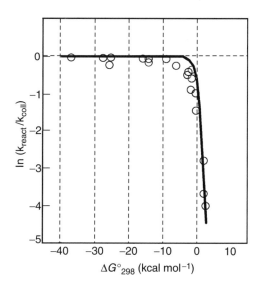

Figure 1.38
The natural logarithm of the number of reactions per collision ratio indicates that the proton transfer is almost 100% efficient when the process is exergonic. When it becomes endergonic, the efficiency drops sharply. (Reproduced (modified) from Ref. 77 with permission)

Table 1.3 Rate constants for the reaction

$$CH_3Br + OH^- \rightleftharpoons CH_3OH + Br^-$$

for naked and solvated ions in the gas phase and in aqueous solution

Hydroxide ion	Rate constant ($cm^3\ mol^{-1}\ s^{-1}$)
OH^-	$(1.0 \pm 0.2) \times 10^{-9}$
$OH^- \cdot H_2O$	$(6.3 \pm 2.5) \times 10^{-10}$
$OH^- \cdot (H_2O)_2$	$(2 \pm 1) \times 10^{-12}$
$OH^- \cdot (H_2O)_3$	$<2 \times 10^{-13}$
Aq. sol. $OH^- \cdot (H_2O)_x$	2.3×10^{-25}

As an example, let us examine whether or not the proton transfer from protonated ammonia to neutral aniline is efficient:

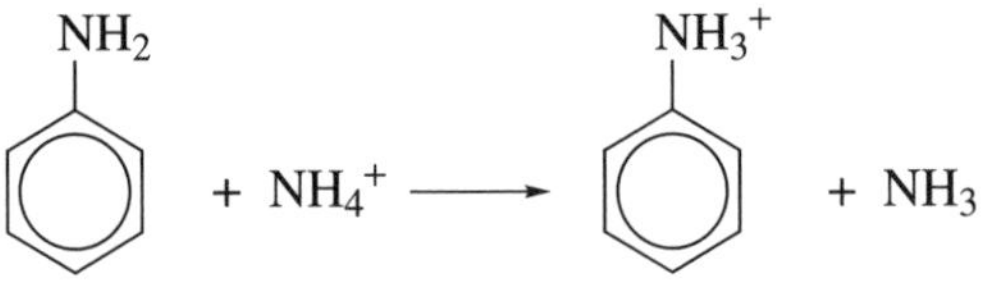

From the individual $\Delta G°$ values, the $\Delta G°$ value for this reaction is calculated:

$$NH_4^+ \longrightarrow NH_3 + H^+ \qquad\qquad \Delta G° = -(-196.4)\ \text{kcal mol}^{-1}$$

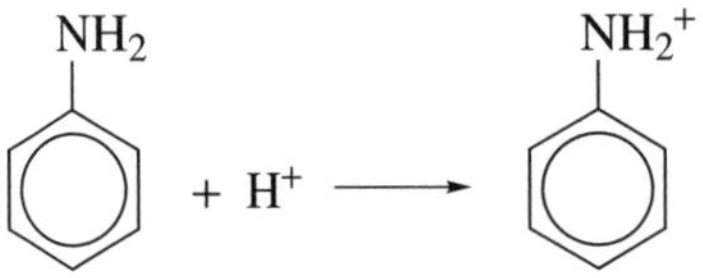

$$\Delta G° = -203.1\ \text{kcal mol}^{-1}$$

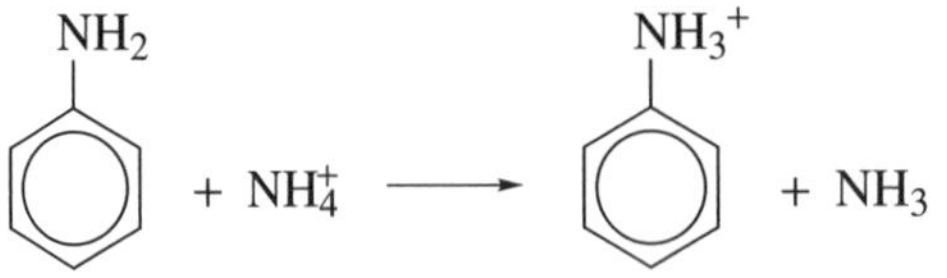

$$\Delta G° = -6.7\ \text{kcal mol}^{-1}$$

In a standard source, with the reaction being exergonic, the proton transfer from ammonium to aniline will be very efficient. Note that in the source we are dealing with the efficiency at each collision and not with the equilibrium. Under the high vacuum conditions, equilibrium is not established. This example was selected because it shows that, in the gas phase, aniline is actually a stronger base than ammonia. The importance of solvation is thus emphasized once again. On the other hand, methylamine is more basic than aniline:

$$CH_3-NH_2 + H^+ \longrightarrow CH_3-NH_3^+ \quad \Delta G° = -205.7 \text{ kcal mol}^{-1}$$

$$\underset{NH_3^+}{\bigcirc} \longrightarrow \underset{NH_3}{\bigcirc} + H^+ \quad \Delta G° = -(203.1) \text{ kcal mol}^{-1}$$

$$\underset{NH_3^+}{\bigcirc} + CH_3-NH_2 \longrightarrow \underset{NH_2}{\bigcirc} + CH_3-NH_3^+ \quad \Delta G° = -2.6 \text{ kcal mol}^{-1}$$

1.12 References

1. Bleakney W., *Phys. Rev.*, **34**, 157 (1929).
2. Nier A.O., *Rev. Sci. Instrum.*, **18**, 415 (1947).
3. Bentley T.W. and Johnstone R.A.W., in *Advances in Physical Organic Chemistry*, ed. by Vol. 8, V. Gold, Academic Press, London, 1970, pp. 151–269.
4. Kienitz H., *Massenspektrometrie*, Verlag Chemie, Weinheim, 1968, p. 45.
5. Harrison A.G., *Chemical Ionization Mass Spectrometry*, CRC Press, Boca Raton, FL, 1983.
6. Lias S.G., Bartmess J.E., Liebman J.F., *et al.*, *J. Phys. Chem. Ref. Data*, **17**, Suppl. 1 (1988). Also available on laser disk at NIST (National Institute of Standards and Technology), Washington, DC, USA.
7. Baldwin M.A. and McLafferty F.W., *Org. Mass Spectrom.*, **7**, 1353 (1973).
8. Van Vaeck L., Adriaens A. and Gijbels R., *Mass Spectrom. Rev.*, **18**(1), 1 (1999).
9. Adriaens A., Van Vaeck L. and Adams F., *Mass Spectrom. Rev.*, **18**(1), 48 (1999).
10. Benninghoven A. and Sichtermann W.K., *Anal. Chem.*, **50**, 1180 (1978).
11. Barber M., Bardoli R.S., Sedgwick R.D., *et al.*, *J. Chem. Soc., Chem. Commun.*, **325** (1981).
12. Aberth W., Straub K.M. and Burlingame A.L., *Anal. Chem.*, **54**, 2029 (1982).
13. Beckey H.D., *Principles of Field Ionization and Field Desorption in Mass Spectrometry*, Pergamon Press, Oxford, 1977.
14. Mcfarlane R.D. and Torgesson T.F., *Science*, **191**, 970 (1976).
15. McNeal C.J. and McFarlane R.D., *J. Am. Chem. Soc.*, **103**, 1609 (1981).
16. Karas M. and Hillenkamp F.H., *Anal. Chem.*, **60**, 2229 (1988).
17. Hillenkamp F., Karas M., Ingeldoh A., *et al.*, in *Biological Mass Spectrometry*, ed. by A.L. Burlingame and J.A. McCloskey, Elsevier, Amsterdam, 1990, p. 49.
18. Hillenkamp F.H., Karas M., Barh U., *et al.*, in *Ion Formation from Organic Solids V*, ed. by A. Benninghoven, Wiley, Chichester, 1990, p. 111.
19. Chen X.J., Carroll J.A. and Beavis R.C., *J. Am. Soc. Mass Spectrom.*, **9**(9), 885 (1998).
20. Gluckmann M. and Karas M., *J. Mass Spectrom.*, **34**(5), 467 (1999).
21. Zenobi R. and Knochenmuss R., *Mass Spectrom. Rev.*, **17**(5), 337 (1998).
22. Karas M. and Hillenkamp F.H., in *Ion Formation from Organic Solids IV*, ed. by A. Benninghoven, Wiley, New York, 1988, p. 103.
23. Spengler B. and Cotter R.J., *Anal. Chem.*, **62**, 793 (1990).
24. Hillenkamp F.H. and Karas M., in *Methods in Enzymology*, Vol. 193, ed. by J.A. McCloskey, Academic Press, New York, 1990, pp. 280–295.

25. Demirev P., Westman A., Reimann C.T., *et al.*, *Rapid Comm. Mass Spectrom.*, **6**, 187 (1992).
26. http://prowl.rockefeller.edu/cgi-bin/websql/websql.dir/prowl/matrixdepot.hts.
27. Vorm O., Roepstorff P. and Mann M., *Anal. Chem.*, **66**, 3281 (1994).
28. Xiang F. and Beavis R.C., *Rapid Commun. Mass Spectrom.*, **8**, 199 (1994).
29. http://www.proteometrics.com/recipes/contents.htm.
30. Vorm O., Chait B.T. and Roepstorff P., *Proc. 41st ASMS Conference*, San Francisco, CA, 1993, pp. 621a–621b.
31. Vestal M.L., *Mass Spectrom. Rev.*, **2**, 447 (1983).
32. Blakney C.R. and Vestal M.L., *Anal. Chem.*, **55**, 750 (1983).
33. Huang E.C., Wachs T., Conboy J.J., *et al.*, *Anal. Chem.*, **62**, 713A (1990).
34. Bruins A.P., *Mass Spectrom. Rev.*, **10**, 53 (1991).
35. Fenn J.B., Mann M., Meng C.K., *et al.*, *Science*, **246**, 64 (1989).
36. Snyder A.P., *Biochemical and Biotechnological Applications of Electrospray Ionization Mass Spectrometry*, ACS Symposium Series 619, American Chemical Society, Washington, DC, 1996.
37. Cole R.B., *Electrospray Ionisation Mass Spectrometry*, Wiley, Chichester, 1997.
38. de la Mora J.F., Van Berkel G.J., Enke C.G., *et al.*, *J. Mass. Spectrom.*, **35**, 939 (2000).
39. Yamashita M. and Fenn J.B., *Phys. Chem.* **88**, 4451 (1988).
40. Yamashita M. and Fenn J.B., *Phys. Chem.* **88**, 4671 (1988).
41. Loo J.A., Udseth H.R. and Smith R.D., *Anal. Biochem.*, **179**, 404 (1989).
42. Ikonomou M.G., Blades A.T. and Kebarle P., *Anal. Chem.*, **62**, 957 (1990).
43. Smith R.D., Loo J.A., Edmons C.G., *et al.*, *Anal. Chem.*, **62**, 882 (1990).
44. Mann M., *Org. Mass Spectrom.*, **25**, 575 (1990).
45. Kebarle P. and Tang L., *Anal. Chem.*, **65**, 972A (1993).
46. Gomez A. and Tang K., *Phys. Fluids*, **6**, 404 (1994).
47. Andersen J., Molina H., Moertz E., *et al.*, *46th Conference on Mass Spectrometry and Allied Topics*, Orlando, Florida, 1998, p. 978.
48. van Berkel W.J.H., van den Heuvel R.H.H., Versluis C., *et al.*, *Protein Sci.*, **9**, 435 (2000).
49. Kelly M.A., Vestling M.M., Fenselau C., *et al.*, *Org. Mass Spectrom.*, **27**, 1143 (1992).
50. Covey T., in *Biochemical and Biotechnological Applications of Electrospray Ionization Mass Spectrometry*, ed. by A.P. Snyder, ACS Symposium Series 619, American Chemical Society, Washington, DC, 1996, pp. 21–59.
51. Emmett M.R. and Caprioli R.M., *J. Am. Soc. Mass Spectrom.*, **5**, 605 (1994).
52. Wilm M. and Mann M., *Proce. 42nd ASMS Conference*, Chicago, IL, 1994, p. 770.
53. Wilm M., Shevchenko A., Houthaeve T., *et al.*, *Nature*, **379**, 466 (1996).
54. Fenn J.B., Rosell J., Nohmi T., *et al.*, in *Biochemical and Biotechnological Applications of Electrospray Ionization Mass Spectrometry*, ed. by A.P. Snyder, ACS Symposium Series 619, American Chemical Society, Washington, DC, 1996, pp. 21–59.
55. Carroll D.I., Dzidic I., Stillwell R.N., *et al.*, *Anal. Chem.*, **47**, 2369 (1975).
56. French J.B. and Reid N.M., *Dynamic Mass Spectrometry*, Vol. 6, Heyden, London, 1980, pp. 220–233.
57. Bruins A.P., *Mass Spectrom. Rev.*, **10**, 53 (1991).
58. Cambier V., Hance T. and de Hoffmann E., *Phytochem. Anal.*, **10**, 119 (1999).
59. Adams F., Gijbels R. and Van Grieken R., *Inorganic Mass Spectrometry (Chemical Analysis: a Series of Monographs on Analytical Chemistry and Its Applications)*, Vol. 95, Wiley, Chichester, 1988.
60. Gijbels R., Van Straaten M. and Bogaerts A., *Adv. Mass Spectrom.*, **13**, 241 (1995).
61. Becker J.S. and Dietze H.J., *Spectrochim. Acta Part B*, **53**(11), 1475 (1998).
62. Steward I.I. and Horlick G., *Trends Anal. Chem.*, **15**(2), 80 (1996).
63. Traeger J.C. and Colton R., *Adv. Mass Spectrom.*, **14**, 637 (1998).
64. Colton R., D'Agostino A. and Traeger J.C., *Mass Spectrom. Rev.*, **14**(2), 79 (1999).

65. Inghram M.G. and Chupka W.A., *Rev. Sci. Instrum.*, **24**, 518 (1953).
66. Heumann K.G., Eisenhut S., Gallus S., *et al.*, *Analyst*, **120**(5), 1291 (1995).
67. Johnson P.G., Bolson A. and Henderson C.M., *Nucl. Instrum. Methods*, **106**, 83 (1973).
68. Dempster A.J., *Rev. Sci. Instrum.*, **7**, 46 (1936).
69. Harrison W.W. and Hang W., *Fresenius J. Anal. Chem.*, **355**(78), 803 (1996).
70. Bogaerts A. and Gijbels R., *Spectrochim. Acta Part B*, **53**(1), 1 (1998).
71. Hang W., Yan X.M., Wayne D.M., *et al.*, *Anal. Chem.*, **71**(15), 3231 (1999).
72. Gray A.L., *Spectrochim. Acta Part B*, **40**, 1525 (1985).
73. Zoorob G.K., McKiernan J.W. and Caruso J.A., *Mikrochim. Acta*, **128**(34), 145 (1998).
74. Hill S.J., *ICP Spectrometry and Its Applications*, Sheffield Academic Press, Sheffield, 1999.
75. McIver R.T., *Sci. Am.*, **243**, 148 (1980).
76. Bohme D.K. and Mackay G.I., *J. Am. Chem. Soc.*, **103**, 978 (1981).
77. Bohme D., Mackay G.I. and Schiff H.I., *J. Chem. Phys.*, **73**, 4976 (1980).
78. Lias S.G., Bartmess J.E., Liebman J.F., *et al.*, *J. Phys. Chem. Ref. Data*, **17**, Suppl. 1 (1988).

2

Mass Analyzers

Once the ions have been produced, they need to be separated according to their masses, which must be determined. There are many different analyzers, just as there are a great variety of sources.

Scanning analyzers transmit ions of different masses successively along a time scale. These are the most frequently encountered in organic analysis. They are either magnetic sector instruments with a flight tube in the magnetic field, allowing only the ions of a given mass-to-charge ratio to go through at a given time, or quadrupole instruments.

Although mass analyzers are generally scanning devices, some allow the simultaneous transmission of all ions, such as the dispersive magnetic analyzers, the time-of-flight mass analyzer, the ion trap or the ion cyclotron resonance instruments. The time-of-flight mass analyzer requires the ions to be produced in bundles and thus is especially well suited for pulsed laser sources. The ion trap, previously used as a gas chromatography/mass spectrometry (GC/MS) analyzer, is now (end of the 1990s) used as a high-performance mass spectrometer. Ion cyclotron resonance mass spectrometers using Fourier transform have shown impressive possibilities at high masses and high resolution.

Instruments combining several analyzers in sequence, such as for tandem mass spectrometry (MS/MS), are becoming increasingly common. They allow one to obtain a mass spectrum resulting from the decomposition of an ion selected in the first analyzer. The time-dependent decomposition of a selected ion can also be observed in cyclotron or ion trap instruments.

The three main characteristics of an analyzer are the upper mass limit, the transmission and the resolution. The mass limit determines the highest value of the m/z ratio that can be measured. It is expressed in thomson, or in atomic mass units (u) for an ion carrying an elementary charge, i.e. $z = 1$. The transmission is the ratio between the number of ions reaching the detector and the number of ions produced in the source. The resolving power is the ability to yield distinct signals for two ions with a small mass difference (Figure 2.1).

Two peaks are considered to be resolved if the valley between the two peaks is equal to 10% of the weaker peak intensity when using magnetic or ion cyclotron resonance instruments and 50% when using quadrupoles.

If δm is the smallest mass difference for which two peaks with masses m and $m + \delta m$ are resolved, the definition of the resolution R is $R = m/\delta m$.

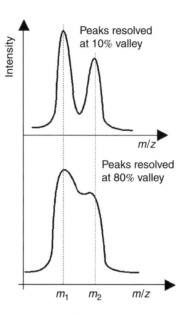

Figure 2.1
Diagram showing the concepts of
peak resolution and valley

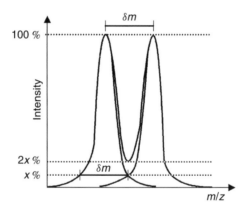

Figure 2.2
Relationship between the two definitions of
resolution

   The resolution of an isolated peak can be defined also as being the peak width
δm at $x\%$ of the peak height. Often x is taken to be 50% and δm is designated as
the full width at half-maximum (FWHM).
   The relationship between the two definitions is obvious for two peaks with equal
intensities. The resolution at $x\%$ of the maximum is equal to the resolution at $2x\%$
for the valley (Figure 2.2).

2.1 Quadrupolar Analyzers

The quadrupole is a device that uses the stability of the trajectories in oscillating electric fields to separate ions according to their m/z ratio. The quistor (quadrupole ion store) or ion trap is based on the same principle.

2.1.1 Description

Quadrupole analyzers[1] are made up of four rods with circular or, ideally, hyperbolic section (Figures 2.3 and 2.4). The rods must be perfectly parallel.

A positive ion entering the space between the rods will be drawn towards a negative rod. If the potential changes sign before it discharges itself on this rod, the ion will change direction.

The principle of the quadrupole was described by Paul and Steinwegen,[3] at Bonn University in 1953. They started from research work on strong focusing of ions carried out in 1951 in Athens by the electrical engineer Christophilos. The quadrupoles have since been developed to commercially available instruments by the work of Shoulders, Finnigan[4] and Story.

Ions traveling along the z-axis are subjected to the influence of a total electric field made up of a quadrupolar alternative field superposed on a constant field resulting from the application of the potentials upon the rods:

$$\Phi_0 = +(U - V \cos \omega t) \text{ and } -\Phi_0 = -(U - V \cos \omega t)$$

In this equation, Φ_0 represents the potential applied to the rods, ω is the angular frequency (in rad s^{-1}) $= 2\pi v$, where v is the frequency of the radio frequency (RF) field, U is the direct potential and V is the 'zero to peak' amplitude of the RF voltage. Typically, U will vary from 500 to 2000 V and V from 0 to 3000 V (from -3000 to $+3000$ V peak to peak).

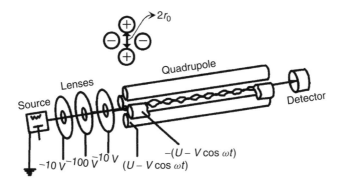

Figure 2.3
Quadrupole instrument made up of the source, the focusing lenses, the quadrupole cylindrical rods and the detector. Ideally, the rods should be hyperbolic. (Reproduced (modified) from Ref. 2 with permission)

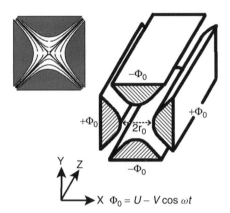

Figure 2.4
Quadrupole with hyperbolic rods and applied potentials.
The equipotential lines are represented on the left

2.1.2 Equations of motion

The ions accelerated along the z-axis enter the space between the quadrupole rods and maintain their velocity along this axis. However, they are submitted to accelerations along x and y that result from the forces induced by the electric fields (Figure 2.4):

$$F_x = m\frac{d^2x}{dt^2} = -ze\frac{\partial \Phi}{\partial x}$$

$$F_y = m\frac{d^2y}{dt^2} = -ze\frac{\partial \Phi}{\partial y}$$

where Φ is a function of Φ_0:

$$\Phi_{(x,y)} = \Phi_0(x^2 - y^2)/r_0^2 = (x^2 - y^2)(U - V\cos\omega t)/r_0^2$$

Derivatizing and rearranging the terms leads to the following equations of motion:

$$\frac{d^2x}{dt^2} + \frac{2ze}{mr_0^2}(U - V\cos\omega t)x = 0$$

$$\frac{d^2y}{dt^2} - \frac{2ze}{mr_0^2}(U - V\cos\omega t)y = 0$$

The trajectory of an ion will be stable if the value of x and y never reach r_0, i.e. if it never hits the rods. To obtain the values of either x or y during the time, these equations need to be integrated. The following equation was established in 1866 by the physicist Mathieu in order to describe the propagation of waves in membranes:

$$\frac{d^2u}{d\xi^2} + (a_u - 2q_u\cos 2\xi)u = 0$$

where u stands for either x or y. Comparing the preceding equations with this one, and taking into account that the potential along y has an opposite sign to the one along x, the following change of variables allow the equations of motion to be depicted in the form of the Mathieu equation:

$$\xi = \frac{\omega t}{2}, \quad a_u = a_x = -a_y = \frac{8zeU}{m\omega^2 r_0^2} \text{ and } q_u = q_x = -q_y = \frac{4zeV}{m\omega^2 r_0^2}$$

We will not attempt to integrate these equations.[5] We only need to recognize that they establish a relationship between the coordinates of an ion and time. As long as x and y (which determine the position of an ion from the center of the rods) both remain less than r_0, the ion will be able to pass the quadrupole without touching the rods. Otherwise, the ion discharges itself against a rod and is not detected. Figure 2.5 shows stable and unstable trajectories in a quadrupole.[6]

For a given quadrupole, r_0 is constant, $\omega = 2\pi\nu$ is maintained constant and U and V are the variables. For an ion of any mass, x and y can be determined during a time span as a function of U and V. Stability areas can be represented in an a_u, q_u-diagram. In these areas the values of U and V are such that x and y do not reach

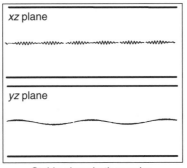

Stable along both x and y

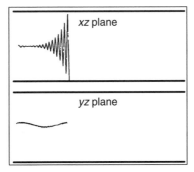

Stable along y, unstable along x

Figure 2.5
Stable and unstable trajectories of ions in a quadrupole.
(Reproduced (modified) from Ref. 6 with permission)

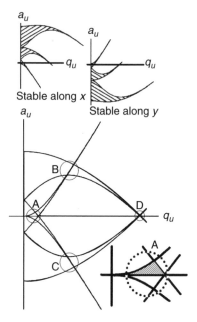

Figure 2.6
Stability areas for an ion along x or y (above) and along
x and y (below); u represents either x or y. The four
stability areas are labeled A to D and are circled. The
area A is that used commonly in mass spectrometers
and is blown up. The direct potential part is shown for
positive U (shaded) or negative U. From now on we will
consider the positive area. (Reproduced (modified) from
Ref. 6 with permission)

values above or equal to r_0. Figure 2.6 represents these stability areas. The upper part
of this figure represents the stability areas along the x- and y-axes. The overlay of
these two diagrams is represented in the lower part of the figure and allows regions
to be identified where the ions will have stable trajectories according to the x- and
y-axes simultaneously. We will consider later the area A with a positive U. The
area A is that currently used, with U either positive or negative. What is true for a
positive U symmetrically holds true for a negative U.

Considering the equations

$$a_u = \frac{8zeU}{m\omega^2 r_0^2} \text{ and } q_u = \frac{4zeV}{m\omega^2 r_0^2}$$

we can deduce:

$$U = a_u \frac{m}{z} \frac{\omega^2 r_0^2}{8e} \text{ and } V = q_u \frac{m}{z} \frac{\omega^2 r_0^2}{4e}$$

The last terms of both the U and V equations are constant for a given quadrupole instrument because they operate at constant ω. We see that switching from one m/z (ratio of the mass to the number of charges) to another results in a proportional multiplication of a_u and q_u, which means changing the scale of the drawing in U, V coordinates; thus the triangular shaped area A will change from one mass to another, like proportional triangles. Figure 2.7 represents in a U, V diagram the A areas obtained with different masses.

We see in this diagram that scanning along a line maintaining the U/V ratio constant allows successive detection of the different masses. As long as the line keeps going through stability areas, the higher the slope and the better will be the resolution.

If $U = 0$ there is no direct current and the resolution becomes zero. However, the value of V imposes a minimum on stable masses. Thus, if V is increased from zero to V so as to reach slightly beyond the stability area m_1, all of the ions with masses equal to or lower than m_1 will have an unstable trajectory, and all of those with masses above m_1 will have a stable trajectory.

In practice, the highest detectable m/z ratio is about 4000 Th and the resolution hovers at around 3000. Thus, beyond 3000 u the isotope clusters are no longer clearly resolved. As is shown in Figure 2.7, adjusting the slope of the operating line allows us to increase the resolution. The resolution normally obtained is not sufficient to deduce the elementary analysis. Usually, quadrupole mass spectrometers are operated at 'unit resolution', i.e. a resolution that is sufficient to separate two peaks one mass unit apart. Quadrupoles are low-resolution instruments.

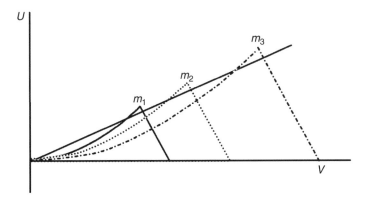

Figure 2.7
Stability areas as a function of U and V for ions with different masses ($m_1 < m_2 < m_3$). Changing U linearly as a function of V, we obtain a straight operating line that allows us to observe those ions successively. A line with a higher slope would give us a higher resolution, as long as it goes through the stability areas. Keeping $U = 0$ (no direct potential) we obtain zero resolution. All of the ions have a stable trajectory as long as V is within the limits of their stability area. (Reproduced (modified) from Ref. 6 with permission)

Operating at constant δm, they require the scanning to be carried out at a uniform velocity throughout the entire mass range, as opposed to the magnetic instruments, which require an exponential scanning.

The quadrupole is a real mass-to-charge ratio analyzer. It does not depend on the kinetic energy of the ions when they leave the source. The only requirements are that the time for crossing the analyzer be short compared with the time necessary to switch from one mass to the other, and that the ions remain long enough between the rods for a few oscillations of the alternative potential to occur. This means that the kinetic energy at the source exit must range from one to a few hundred electronvolts. The weak potentials in the source allow a relatively large tolerance on the pressure.

Because the scan speed easily can reach 1000 Th s^{-1} and more, this analyzer is well suited to chromatographic coupling.

Quadrupoles also have the property of focusing the trajectory of the ions towards the center of the quadrupole. Consider the diagram in Figure 2.8. The potential energy of the positive ion located at the center of the rods of a quadrupole increases if it comes nearer to a positive rod. Conversely, it decreases if it comes nearer to a negative rod. However, the alternative field actually spins the potentials, alternating

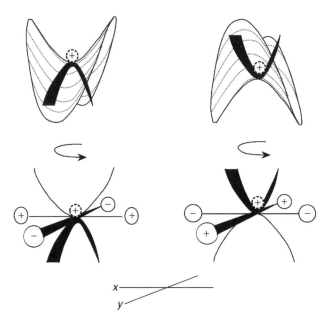

Figure 2.8
A positive ion, represented within a dotted circle, is at the center of quadrupole rods, the potential signs of which are indicated. It goes down the potential 'valley' with respect to the negative rods and acquires some kinetic energy in that direction. However, the potentials quickly change so that the kinetic energy is converted into potential energy and the ion goes back to the center of the rods, as would happen for a ball on a horse saddle that is turned quickly. The name 'saddle field' is an allusion to this phenomenon

the 'wells' and the 'hills'. If the frequency is sufficient, an ion that starts coming down the slope towards a negative barrier is caught in the positive potential well and is thus brought back to the center of the quadrupole rods.

We saw that ions of all m/z ratios higher than a particular lower limit have a stable trajectory when there is no direct potential (quadrupole with RF only) as long as V is within the limit of their stability area. When U is equal to zero (quadrupole with RF only) all of the ions with a mass higher than a certain limit selected by adjusting the value of V have a stable trajectory. Such a quadrupole causes all of the ions to be brought back systematically to the center of the rods, even if they were deflected by a collision. This focusing effect is important to increase the transmission of ions after collisions.

2.1.3 Spectrometers with several quadrupoles in tandem

Figure 2.9 shows the general diagram of an instrument with three quadrupoles.[7] Quadrupole mass spectrometers are symbolized by 'Q' and RF-only quadrupoles by 'q'. A collision gas can be introduced into the central quadrupole at a pressure such that an ion entering the quadrupole undergoes one or several collisions.

When the gas is inert, internal energy is transferred to the ion by converting a fraction of the kinetic energy into internal energy. The ion then fragments and the products are analyzed by quadrupole Q3. The kinetic energy for internal energy transfers is governed by the laws concerning collisions of a mobile species (the ion) and a static target (the collision gas). The collisions in these instruments occur at low energy. The conversion of kinetic to internal energy will be discussed in Chapter 3.

When the gas is reactive, ion–molecule reactions can be induced. The reaction products then are analyzed by quadrupole Q3.

These instruments with several analyzers in series can be scanned in several ways, the most important of which are displayed in Figure 2.10.

The first scanning mode consists of selecting an ion with a chosen mass-to-charge ratio using the first spectrometer. This ion collides inside the central quadrupole and reacts or fragments. The reaction products are analyzed by the second mass spectrometer. This is a 'fragment ion scan' or 'product ion scan'. This method used to be called a 'daughter scan'.

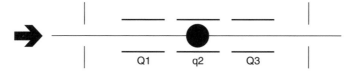

Figure 2.9
Diagram of a triple quadrupole instrument. The first and the last (Q1 and Q3) are mass spectrometers. The center quadrupole, q2, is a collision cell made up of a quadrupole using RF only. The quadrupole mass spectrometers are symbolized by 'Q' and the RF-only quadrupoles by 'q'

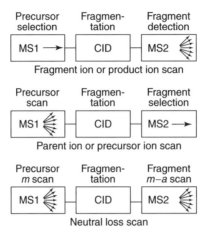

Figure 2.10
Different scan modes for a tandem
mass spectrometer; CID = collision-
induced dissociation

The second possibility consists of focusing the second spectrometer (Q3) on a
selected ion while scanning the masses using the first spectrometer (Q1). All of the
ions that produce the ion with the selected mass through reaction or fragmentation
are thus detected. This method is called 'precursor scan' because the 'precursor ions'
are identified. It used to be called 'parent scan'.

In the third common scan mode, both mass spectrometers are scanned together,
but with a constant mass offset between the two. Thus, for a mass difference a,
when an ion of mass m goes through the first mass spectrometer detection occurs
if this ion has yielded a fragment ion of mass $(m - a)$ when it leaves the collision
cell. This is a 'neutral loss scan', the neutral having the mass a. For example, in
chemical ionization the alcohol molecular ion loses a water molecule. Alcohols thus
are detected by scanning a neutral loss of 18 mass units. On the other hand, a given
mass increase can be detected if a reactive gas is introduced within the collision cell.

2.2 The Quadrupole Ion Trap or Quistor

2.2.1 General principle

Paul and Steinwedel, the former being the inventor of the quadrupole analyzer,
described an 'ion trap'[8,9] in 1960. It has been modified to a useful mass spectrometer
by Stafford et al.[10] of the Finnigan company. It is made up of a circular electrode,
with two ellipsoid caps on the top and the bottom (Figure 2.11).

Conceptually, an ion trap can be imagined as a quadrupole bent on itself in order
to form a closed loop. The inner rod is reduced to a point at the center of the trap,
the outer rod is the circular electrode and the top and bottom rods make up the caps.

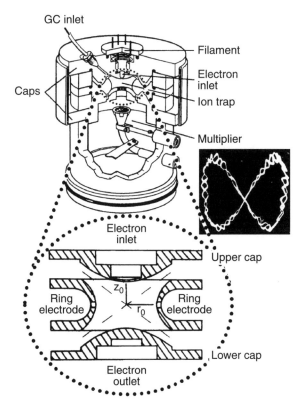

Figure 2.11
(*Top*) A complete ion trap mass spectrometer (Finnigan MAT).
(*Bottom*) Detail of the trap. (*Right*) Figure obtained when injecting
a fine dust of aluminum in a quadrupolar field such as is encountered
in an ion trap. The particles rotate altogether along r and oscillate
along z, yielding a figure-of-eight-shaped trajectory. (Reproduced
(modified) from Finnigan MAT documentation, with permission)

The overlapping of a direct potential with an alternative one gives a kind of
'three-dimensional quadrupole' in which the ions of all masses are trapped on a
three-dimensional trajectory (Figure 2.11). The inventors proposed the use of this
ion trap as a mass spectrometer by applying a resonant frequency along z to expel
the ions of a given mass.

In quadrupole instruments, the potentials are adjusted so that only ions with a
selected mass go through the rods. The principle is different in this case. Ions with
different masses are present together inside the trap and are expelled according to
their masses so as to obtain the spectrum.

As the ions repell each other in the trap, their trajectories expand as a function
of time. To avoid ion losses by this expansion, care has to be taken to reduce the
trajectory. This is accomplished by maintaining in the trap a pressure of helium gas

that removes excess energy from the ions by collision. This pressure hovers around 10^{-3} Torr (0.13 Pa). A single high-vacuum pump with a flow of about 40 1 s^{-1} is sufficient to maintain such a vacuum, as compared with the 250 1 s^{-1} needed for other mass spectrometers. The instrument is very simple and relatively inexpensive.

As is the case with quadrupole instruments, a potential Φ_0 (the sum of a direct and an alternative potential) could be applied to the caps and $-\Phi_0$ to the circular electrode. However, in most commercial instruments Φ_0 is applied only to the ring electrode. In both cases the resulting field must be seen in three dimensions.

Here, again, the mathematical analysis using the Mathieu equations allows us to locate areas wherein ions of given masses have a stable trajectory. These areas may be displayed in a diagram as a function of U (direct potential) and V (amplitude of the alternative potential). The areas in which the ions are stable are those in which the trajectories never exceed the dimensions of the trap, z_0 and r_0. Such a diagram appears in Figure 2.17.

2.2.2 Theory of the ion trap

The mathematical analysis using the Mathieu equations allows us to locate areas wherein ions of given masses have a stable trajectory. The equations are very similar to those used for the quadrupole. However, in the quadrupole, ion motion resulting from the potentials applied to the rods occurs in two dimensions x and y, the z motion resulting from the kinetic energy of the ions when they enter the quadrupole field. In the trap the motion of the ions under the influence of the applied potentials occurs in three dimensions, x, y and z. However, due to the cylindrical symmetry $x^2 + y^2 = r^2$, it can be expressed also using the z, r coordinates (Figure 2.12).

The equations of the movement inside the trap are similar to those encountered for the quadrupole analyzer (section 2.1.2):

$$\frac{d^2z}{dt^2} - \frac{4ze}{m(r_0^2 + 2z_0^2)}(U - V\cos\omega t)z = 0$$

$$\frac{d^2r}{dt^2} + \frac{2ze}{m(r_0^2 + 2z_0^2)}(U - V\cos\omega t)r = 0$$

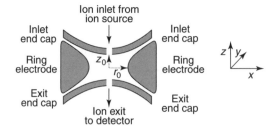

Figure 2.12
Schematic view of an ion trap, and direction of the x, y and z coordinates

In these equations, an arial z is used for the number of charges, to avoid confusion with the z coordinate. The general form of the Mathieu equation, whose solutions are known, is:

$$\frac{d^2u}{d\xi^2} + (a_u - 2q_u \cos 2\xi)u = 0$$

where u stands for either z or r. Thus, the following change of variables allows the equations of motion to be given in the form of the Mathieu equation, where u stands for either r or z:

$$\xi = \frac{\omega t}{2}, \quad a_u = a_z = -2a_r = \frac{-16zeU}{m(r_0^2 + 2z_0^2)\omega^2},$$

$$q_u = q_z = -2q_r = \frac{8zeV}{m(r_0^2 + 2z_0^2)\omega^2}$$

To have a stable trajectory, the movement of the ions must be such that during the time the coordinates never reach or exceed r_0 (r-stable) and z_0 (z-stable). The complete integration of the Mathieu equation by the method of Floquet and Fourier requires the use of a function $e^{(\alpha+i\beta)}$. Real solutions correspond to a continuously increasing, and thus unstable, trajectory. Only purely imaginary solutions correspond to stable trajectories. This requires both $\alpha = 0$ and $0 < \beta_u < 1$ (Figure 2.13).

The parameter β_u can be calculated from the q_u and a_u parameters of the Mathieu equation. Exact values require the use of long series of terms. The following equation allows approximate values to be obtained:[11]

$$\beta_u = \left[a_u - \frac{(a_u-1)q_u^2}{2(a_u-1)^2 - q_u^2} - \frac{(5a_u+7)q_u^4}{32(a_u-1)^3(a_u-4)} - \frac{(9a_u^2+58a_u+29)q_u^6}{64(a_u-1)^5(a_u-4)(a_u-9)}\right]^{1/2}$$

A simpler approximate equation holds for q_u values lower than 0.4:

$$\beta_u = \left[a_u + (q_u^2/2)\right]^{1/2}$$

Figure 2.13
(*Left*) A type of stable trajectory corresponding to a purely imaginary solution of the Mathieu equation, requiring both $\alpha = 0$ and $0 < \beta < 1$. (*Right*) A continuously amplified movement resulting from a not purely imaginary solution. If this occurs along r, the ions will discharge on the wall; along z, the ions will be expelled and 50% will reach the detector, the others being expelled in the opposite direction

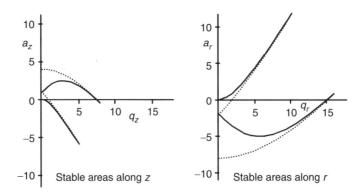

Figure 2.14
Stability diagram along r and z, respectively, for an ion trap. The iso-β lines for $\beta_u = 0$ (continuous lines) and $\beta_u = 1$ (dotted lines) are drawn. The areas inside these limits correspond to stable trajectories for the considered coordinate. They correspond to imaginary solutions of the Mathieu equation

Figure 2.14 displays the iso-β lines for $\beta_u = 0$ and $\beta_u = 1$, respectively. One diagram refers to the z coordinate and the other to the r coordinate. Remember that $a_u = a_z = -2a_r$ and $q_u = q_z = -2q_r$. The areas inside these limit values for β_u correspond to a,q values for a stable trajectory. However, for an ion to be stable in the ion trap it must have a stable trajectory along both z and r, which correspond to the overlap of the stability areas of the two diagrams.

This overlap, as displayed in Figure 2.15, shows common areas at different places. Commercial ion traps use the first stability region, circled and enlarged in the figure.

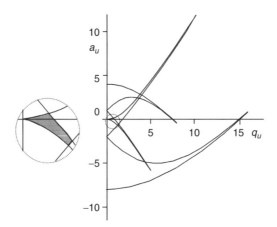

Figure 2.15
Stability areas along both r and z in an ion trap. The common r and z stability area used in commercial ion traps is enlarged. This is displayed again in Figure 2.17. (Redrawn and modified from Ref. 6)

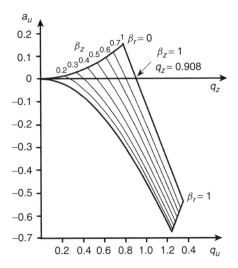

Figure 2.16
Typical stability diagram for a quadrupole ion trap. The
value at $\beta_z = 1$ along the q_z axis is $q_z = 0.908$. At the
upper apex, $a_z = 0.149998$ and $q_z = 0.780909$. (Data
from Ref.12)

This stability region is displayed again in Figure 2.16; iso-β curves are drawn for
intermediate values between 0 and 1.

These areas also may be displayed in a diagram as a function of U (direct potential)
and V (amplitude of the alternative potential). The areas in which the ions are stable
are those in which the trajectories never exceed the dimensions of the trap, z_0 and
r_0. Such a diagram appears in Figure 2.17.

Two other important parameters depending on q_z or β will be considered. First,
if an RF voltage with an angular velocity $\omega = 2\pi v$ is applied to the trap, the ions
will not oscillate at this same 'fundamental' v frequency because of their inertia,
which causes them to oscillate at a 'secular' frequency f lower than v that decreases
with increasing masses. It should be noted that a_u and q_u, and thus β, are inversely
proportional to the m/z ratio. The relation between v and f along the z-axis is:

$$f_z = \beta_z v/2$$

Because the maximum value of β for a stable trajectory is $\beta = 1$, the maximum
secular frequency f_z of an ion will be half the fundamental frequency v. We will
see later on that this is important for ion excitation or for resonant expulsion.

The second important parameter, which is a function of q_z, is the Dehmelt pseudo-
potential well. The trapping efficiency of ions injected in the trap can be described
using the pseudo-potential well given by the following equation:

$$\overline{D_z} = q_z \frac{V}{8} = \frac{eV^2}{m(r_0{}^2 + 2z_0{}^2)\omega^2}$$

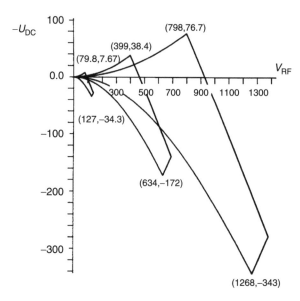

Figure 2.17
Stability areas[13] for ions simultaneously stable along r and z. The mass-
to-charge ratios of the ions displayed are 10, 50 and 100 Th. The trap
being considered has a diameter of 1 cm and operates at a frequency of
1.1 MHz.[13] Under such conditions, all the ions have a stable trajectory for
$U = 0$ if their mass is lower than 0.108V. For instance, a 1 Th ion has a
stable trajectory up to $V = 1/0.108 = 9.26$ V. As shown, a 50 Th ion is
unstable starting from $V = 463$ V. Increasing V destabilizes the trajectories
of ions having an increasingly high mass-to-charge ratio. (Reproduced from
Ref. 13 with permission)

2.2.3 Injection or production of ions in the trap

The first commercially available ion traps were coupled to a gas chromatograph,
which provided helium for ion cooling. Ionization occurred in the trap by injection of electrons through an end cap. Some GC/MS^n ion traps still use this internal
ionization. Most of today's (1999) instruments have an external source. Figure 2.18
shows a popular ion trap instrument with an external electrospray ionization (ESI)
source, the Finnigan liquid chromatography quadrupole (LCQ). Ions produced in
the source are focused through a skimmer and two RF-only octopoles. Differential
pumping ensures a vacuum in the trap while the source is at atmospheric pressure.
A gating lens limits the number of injected ions to an acceptable limit. Indeed,
too many ions in the trap causes a loss of resolution, and too few a loss of sensitivity. For positive ions a positive potential is applied, except during injection of
ions, which occurs when a negative potential is applied. The following sequence of
events is used. First, a negative potential is applied for a fixed time, and the number
of ions in the trap is measured by expelling them together to the detector. From
the value obtained a gating time is calculated so as to inject the selected number
of ions.

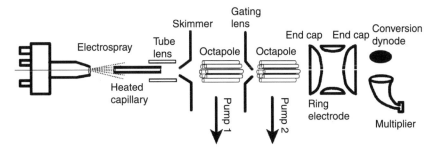

Figure 2.18
An ion trap with an external electrospray ionization (ESI) source (Finnigan LCQ).
Ions produced by the ESI source are focused through a skimmer and octapole
lenses to the ion trap. The gating lens is used to limit the number of ions injected
in the trap, to avoid the space charge effect

2.2.4 Mass analysis of ions in a trap

After ion injection, ions of different masses are stored together in the trap. They must
be analyzed now according to their mass. Most commercial traps operate by applying
a fundamental RF voltage on the ring. Its frequency is constant, but its amplitude V
can be varied. Additional RF voltages of selected frequencies and amplitudes can be
applied to the end caps. Figure 2.19 schematically represents the wiring of such an

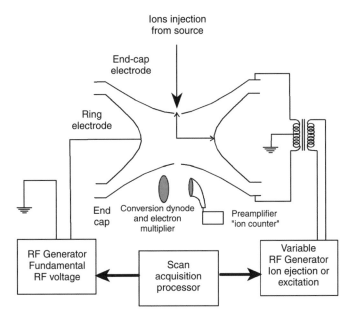

Figure 2.19
Ion trap with an RF voltage applied to the ring electrode, providing the
fundamental ν frequency and its associated variable amplitude V. Instead
of injecting ions, electrons may be injected for internal ionization. Variable
RF voltage can be applied to the end caps for ion excitation or ion ejection

ion trap. In this example, no direct current (DC) voltage is applied. Other ion traps allow the application of DC voltages too. To explain the principle of ion analysis, we will consider an ion trap wired as in Figure 2.19.

2.2.5 Mass analysis by ion ejection at the stability limit

Because no DC voltage is applied, the trap will be operated along the q_u axis because in the absence of a DC voltage $a_u = 0$. As explained already, q_z is given by the following equation:

$$q_z = \frac{8zeV}{m(r_0{}^2 + 2z_0{}^2)\omega^2}$$

Because the frequency is fixed, ω is a fixed value. We will consider only monocharged ions ($z = 1$). The elementary charge e, is a constant and r_0 and z_0 are constant for a given trap, q_z will increase if V increases and will decrease if m increases. This is

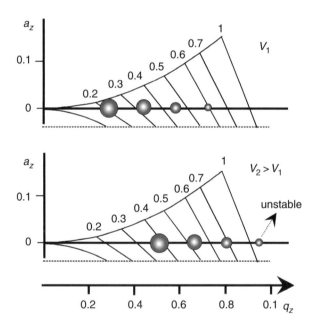

Figure 2.20
At a fixed value of the RF potential V applied to the ring electrode, heavier ions will have lower β_z values and thus lower secular frequencies. If V is increased, β_z values increase for all the ions, as do the secular frequencies. In the example given, the lightest ion now has a β_z value larger than unity and is thus expelled from the trap. The highest mass that can be analyzed depends on the limit V value that can be applied: around 7000–8000 V from zero to peak. For a trap having $r_0 = 1$ cm and operating at a v frequency of 1.1 MHz, the highest detectable mass-to-charge ratio is about 650 Th

represented in Figure 2.20, where higher mass ions are represented by larger balls. If V is increased, all the ions will have a higher q_z value. If this value is equal to 0.908, β is equal to unity and the ion has reached its stability limit. A slight increase of V will cause this ion to have an unstable trajectory, and will be expelled from the trap in the z direction; 50% of the expelled ions will reach the detector. This allows the ions present in the trap to be analyzed.

If we consider again the equation for q_z, we can write an equation with these limit values giving the maximum observable mass m_{MAX}:

$$q_z = \frac{8zeV}{m(r_0{}^2 + 2z_0{}^2)\omega^2} \qquad m_{MAX} = \frac{8ze8000}{0.908(r_0{}^2 + 2z_0{}^2)(2\pi v)^2}$$

Example of calculation

$$q_z = \frac{8zeV}{m(r_0{}^2 + 2z_0{}^2)\omega^2}$$

Let us refer to the first-generation LCQs for which $r_0 = 7.07$ mm and $z_0 = 7.83$ mm. In this equation must be in meters and thus:

$$(r_0{}^2 + 2z_0{}^2) = 1.726 \times 10^{-4} \text{ m}^2$$

The RF voltage 'zero to peak' ($V_{0\rightarrow P}$, expressed in volts) is variable up to a maximum of 8500 V. The fundamental RF v is equal to 0.76 MHz or 7.6×10^5 Hz and $\omega = 2\pi v$:

$$\omega^2 = (6.28 \times 7.6 \times 10^5)^2 = 2.278 \times 10^{13} \text{ (rad s}^{-1})^2$$

The number of charges, z is an entire number and is unity for a monocharged ion. The elementary charge e (absolute value of the charge of one electron) is 1.602×10^{-19} C.

The equation now can be written with all the constant terms:

$$q_z = \frac{8zeV}{m(r_0{}^2 + 2z_0{}^2)\omega^2} \qquad q_z = \frac{8(1.602 \times 10^{-19})}{(1.726 \times 10^{-4})(2.278 \times 10^{13})} \times \frac{zV}{m}$$

In this equation, m is expressed in kilograms per ion. If we want the classical value in atomic mass units (u = Da = 1.66022×10^{-27} kg), denoted m', we must multiply the equation by 1000 to go from kilograms to grams and by 6.02×10^{23} for one mole of ions rather than one ion. We then get the following equation, with the constant 0.1963 valid only, of course, for the first-series LCQs:

$$q_z = \frac{8(1.602 \times 10^{-19})(1000)(6.02 \times 10^{23})}{(1.726 \times 10^{-4})(2.278 \times 10^{13})} \times \frac{zV}{m'} = 0.1963 \times \frac{zV}{m'}$$

Continued on page 82

Continued from page 81

What is the value of q_z for an ion of mass $m' = 1500$ u, monocharged ($z = 1$) at 5000 V RF amplitude?

$$q_z = 0.1963 \times \frac{(1)(5000)}{1500} = 0.6543$$

Because this value of q_z is lower than 0.908 (the $\beta = 1$ stability limit), this ion will have a stable trajectory in this trap operated at 5000 V amplitude of the fundamental RF.

Thus, besides trying to increase V at higher values without arcing, the maximum observable mass can be increased by reducing the size of the trap or by using a lower RF v. For example, if both r_0 and z_0 are reduced to 7 mm and the RF frequency to 600 kHz, and the maximum value for V is 8000 V, the mass limit will become about 3260 Th.

2.2.6 Mass analysis by resonant ejection

If we remember that the secular frequency at which an ion oscillates in the trap is given by:

$$f_z = \beta_z v / 2$$

and remembering that β_z increases if q_z increases (Figure 2.16), it is possible to calculate the value of V to be applied for ions of a given mass to oscillate at a selected frequency f_z.

If an RF voltage at frequency f_z is applied to the end caps, i.e. along the z-axis, the ion will come in resonance and the amplitude of its oscillations will increase. If the applied amplitude is sufficient, the oscillations of the ion will be so large that it will be destabilized and ejected from the trap in the z direction (see Figure 2.21).

If this frequency is near to $v/2$, then β_z will be near to unity and it is at the stability limit (q_z=0.908). If, however, the applied frequency corresponds to a lower q_z, the ions will be ejected at a lower V value. Thus, for the same maximum V value, the highest ejectable mass will be higher. This is another way to increase the mass range. For example, the Bruker Esquire instrument allows ions to be ejected at $\beta_z = 2/11$, corresponding to $q_z = 0.25$, which allows the mass range to be extended up to 6000 Th.

It should be noted that if an ion fragments during the analysis, it is possible that its mass-to-charge ratio is such that its q_z value is higher than the resonant ejection value. If, later on, by increasing V it reaches the stability limit ($q_z = 0.908$) and is ejected, it will be detected then at the wrong m/z because the data system expects it to be expelled by resonance. Its apparent m/z will be higher than the true one. These 'ghost peaks' will occur more if the resonance frequency corresponds to a lower q_z value.

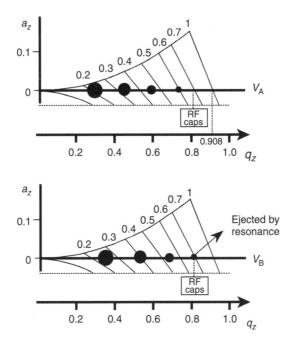

Figure 2.21
Principle of resonant ejection. (*Upper*) Ions are stored in the trap at a voltage V_A of the fundamental RF. An additional RF is applied to the end caps corresponding to $q_z = 0.8$. (*Lower*) Increasing V ions are moved to higher q_z values. The smallest ion has reached $q_z = 0.8$ and is ejected by resonance. This ejection occurs at a lower value of V than that needed to eject by instability at $q_z = 0.908$

Figure 2.22 displays the sequence of events for a resonance ejection analysis in a pictorial way.[14] Note that in this figure the supplementary applied voltage on the end caps is designated as 'AC' (alternating current).

2.2.7 Tandem mass spectrometry in the ion trap

There are several ways to perform tandem mass spectrometry in an ion trap. *Time-dependent* rather than space-dependent tandem mass spectrometry occurs in the trap. The general sequence of operations is as follows:

1. Select ions of one mass-to-charge ratio by expelling all the others from the ion trap. This can be performed either by selecting the precursor ion at the apex of the stability diagram or by resonant expulsion of all ions except for the selected precursor.

2. Let these ions fragment. Energy is provided by collisions with the helium gas, which is always present. This fragmentation can be improved by excitation of the selected ions by irradiation at their secular frequency.

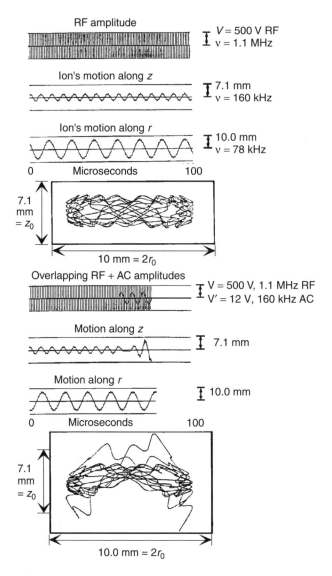

Figure 2.22

Resonant ejection of a 100 Th ion. Ions of a given mass-to-charge ratio have a characteristic oscillation frequency along both r and z. Superimposing on the fundamental RF potential an adjustable AC potential, with a frequency equal to that of the characteristic ion oscillation, transfers energy to the ion through resonance. In this example, 160 kHz is chosen. The ion trajectory along z is thus destabilized. This resonant mode allows the ejection of high-mass ions. The boxes display computer simulations of the ion trajectories. (Reproduced (modified) from Ref. 14, with permission)

3. Analyze the ions by one of the described scanning methods: stability limit or resonant ejection.

4. Alternatively, select a fragment in the trap and let it fragment further. This step can be repeated to provide MS^n spectra.

As a first example, let us look at an MS/MS/MS or MS^3 experiment performed by selection of ions at the stability apex. Figure 2.23 shows how correctly selecting values of U and V allows only ions with a given m/z ratio to be trapped. These ions fragment in time within the trap, which then contains all of the product ions. The latter then can be analyzed by selective expulsion. *Time-dependent*, instead of space-dependent, MS/MS is thus carried out. The process can be repeated several times, selecting successive fragment ions.

Figure 2.24 shows a typical sequence of operations used to observe third-generation ions.

A second example describes the use of resonant ejection of ions by selected waveform inverse Fourier transform (SWIFT). Figure 2.25 describes an MS/MS experiment with an instrument using RF voltages applied to the caps but no DC voltage. In this example, the final analysis of the fragments is performed by the stability limit method.

To fragment the ions by resonance excitation, an RF voltage must be applied to adjust the q_z value for the selected m/z to the frequency of the RF generator used. This means, if we remember that increasing V causes ejection of the lower mass ions, that the lower observable m/z limit will be the higher if the applied voltage V is increased. The lowest m/z fragments are lost for this reason. In general, fragments with m/z values lower than about 20% of the precursor ion's m/z will be lost. Thus there is an advantage to work at lower V, which means lower q_z, to have a larger mass limit. But at higher V the Dehmelt potential is higher because it is proportional to V^2:[15]

$$\overline{D_z} = q_z \frac{V}{8} = \frac{zeV^2}{m(r_0{}^2 + 2z_0{}^2)\omega^2}$$

Figure 2.26 displays this Dehmelt potential graphically as a function of q_z, itself a function of V.

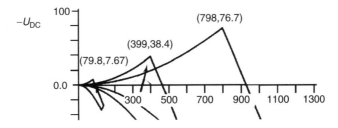

Figure 2.23
Fragment of Figure 2.17. The arrow indicates a point where only 50 Th ions have a stable trajectory. (Reproduced (modified) from Ref. 6 with permission)

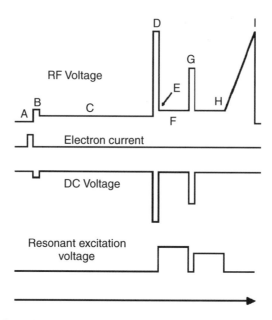

Figure 2.24
Computer-controlled sequence of operations used to carry out a typical MS/MS/MS experiment in an ion trap. (A) Electron impact ionization of the chemical ionization gas. (B) Correct tuning of U (DC voltage) and V (RF voltage) allows the selection of ions with a given m/z for chemical ionization. (C) Protonation of the analyte by the chosen ion for 200 ms. (D) The parent ion is selected. (E) Voltage V is increased in order to select a mass range to observe the fragments, while keeping the surviving precursors in the trap. (F) Resonant excitation is used to fragment the selected precursor ions. (G) One of the fragments is selected. (H) Resonant excitation is used to fragment this fragment. (I) Mass scanning to observe the second-generation fragments

Actually, when an ion produces fragments they need to be trapped efficiently, which depends on the value of the Dehmelt potential. Otherwise, some of the fragments are lost by lack of trapping efficiency. There is thus a compromise to find between loss of sensitivity and loss of lower m/z fragment ions.

The gas used to slow the ions within the trap is normally helium, which has a 4 Da mass. The ion kinetic energy amounts to only a few electronvolts. The maximum fraction of kinetic energy convertible into internal energy is given by the equation discussed in Chapter 3:

$$E_t = E_c \frac{m_2}{(m_1 + m_2)}$$

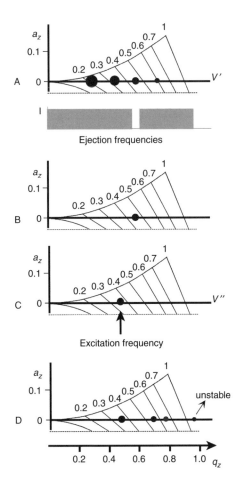

Figure 2.25
One possible sequence of events to produce an MS/MS spectrum. (A) Ions of one mass-to-charge ratio are selected by expelling all the others at their resonance frequency applied to the caps. (B) Only ions of the selected m/z are present in the trap. (C) Voltage V is adjusted so as to bring the ion in resonance with the excitation frequency applied to the caps. (D) Ions are analyzed by ejection at the stability limit

Such a weak energy is often insufficient to induce ion fragmentation within a short time span. The kinetic energy of the ions can be increased by superimposing their resonant frequency on the radiofrequency with an amplitude that is weak enough not to expel them from the trap.

In a quadrupole collision cell, the ions undergo multiple collisions. The fragments, as soon as they are formed, are reactivated by collision and can fragment further. In the ion trap, if excitation occurs by irradiation at the secular frequency of the

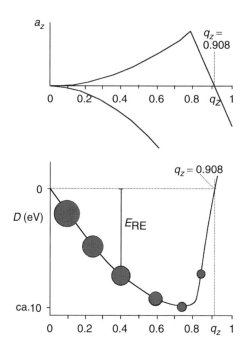

Figure 2.26
Dehmelt potential D (in eV, lower graph) as a function of q_z. Heavier ions, for a given V value, have lower Dehmelt trapping energies. An increase in V will displace all the ions along the curve to higher q_z values. If q_z passes the 0.908 value, the ion will be expelled. The E_{RE} line represents the energy that has to be provided by resonant excitation in order to expel the selected ion from the trap. The top graph is a reminder of the stability diagram. The ordinate is not the same for the two diagrams

precursor, only this ion is excited, and the product ion may be too cool to fragment further. Figure 2.27 gives an example of this behavior.[16]

In the ion trap, instead of selectively irradiating the precursor ion, broadband excitation can be used: irradiation occurs over a large range of frequencies, so that all the ions in the trap are excited and also the fragments formed.

The ion trap allows MS^n experiments to be performed. For this purpose, an ion produced in the source is selected and fragmented. One of the fragments is selected and fragmented again. This process can be repeated. As an example, Figure 2.28 displays the MS^2 and MS^3 spectra obtained from permethylated oligosaccharides LNT (Gal($\beta1-3$) GlcNac($\beta1-3$) Gal($\beta1-3$) Glc) and LNnT, which differs by the 1–4 position of the first galactose residue with the N-acetylglucosamine (Gal($\beta1-4$) GlcNac($\beta1-3$) Gal($\beta1-3$) Glc). The spectra show that the two different terminal disaccharides can be distinguished easily.[17] Here it is shown also that these same disaccharides originating from larger oligosaccharides yield the same fragmentation

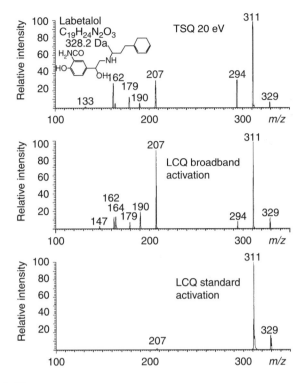

Figure 2.27

(*Top*) Fragment ion obtained from ESI protonated Labetalol molecular ion in a triple quadrupole instrument. (*Center*) Same spectrum obtained in an ion trap with broadband excitation. All the ions (precursor or fragment) are activated by resonant irradiation. (*Bottom*) Same spectrum as obtained with an ion trap when only the precursor ion is activated. (Redrawn from Ref. 16 with permission)

spectra even if several fragmentation steps are used — MS^4, MS^5 or more — provided that the collision energy in the last step is the same.

2.2.8 Space charge effect

When too many ions are introduced in an ion trap, those located at the outside will act as a shield. The field acting on the ions located at the inside will be modified and the shape of the stability diagram will be modified, as displayed in Figure 2.29.

2.3 Time-of-flight Analyzers

The concept of the linear time-of-flight (TOF) analyzer was described by Stephens in 1946.[18] Wiley and McLaren in 1955 published the design of a linear TOF mass

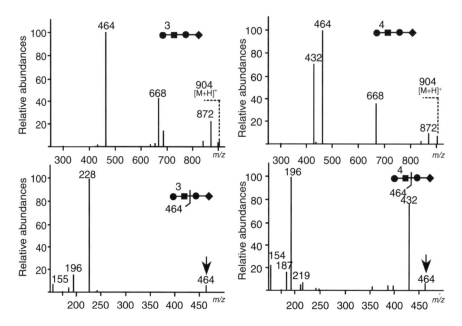

Figure 2.28
The MS^2 and MS^3 mass spectra from two identical oligosaccharides except for the position of the first galactose residue on the N-acetylglucosamine: (●) galactose; (■) N-acetylglucosamine; (♦) glucose. These same disaccharides originating from more complex oligosaccharides produce the same fragmentations. (Reproduced from Ref. 17 (Figure 1, modified) with permission)

spectrometer that later became the first commercial instrument.[19] There has been renewed interest in these instruments since the end of the 1980s. First, progress in electronics simplifies the handling of the high data flow. Second, the TOF analyzer is well suited to the pulsed nature of laser desorption ionization. The development of matrix-assisted laser desorption/ionization (MALDI)/TOF has paved the way for new applications, not only for biomolecules but also for synthetic polymers and polymer/biomolecule conjugates. Time-of-flight analyzers were reviewed by Cotter,[20] Mann,[21] Weickhardt[22] and Wollnik.[23] Books on TOF mass spectrometers were published in 1994 and 1997.[24-26]

2.3.1 Linear time-of-flight mass spectrometer

Figure 2.30 displays the scheme of a linear TOF instrument. Ions are expelled from the source in bundles that are either produced by an intermittent process such as plasma or laser desorption or expelled by a transient application of the required potentials on the source-focusing lenses. They are accelerated by a potential V_s and fly a distance d before reaching the detector.

Mass-to-charge ratios are determined by measuring the time that ions take to move through a field-free region between the source and the detector. Indeed, as it leaves

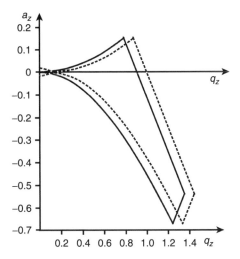

Figure 2.29
The dotted line represents the shift of the stability diagram resulting from the space charge effect. To reach the stability limit $\beta = 1$, q_z and thus V must have higher values. This could lead to an error on the mass if proper caution is not taken

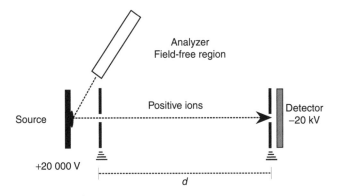

Figure 2.30
Principle of a linear TOF instrument tuned to analyze positive ions produced by focusing a laser beam on the sample

the source, an ion with mass m and total charge $q = ze$ has a kinetic energy

$$\frac{mv^2}{2} = qV_S = zeV_S = E_k$$

The time needed to fly the distance d is given by

$$t = \frac{d}{v}$$

Replacing *v* by its value in the previous equation gives

$$t^2 = \frac{m}{z}\left(\frac{d^2}{2V_s e}\right)$$

This equation shows that m/z can be calculated from a measurement of t^2, the terms in parentheses being constant. This equation also shows that, all other factors being equal, the lower the mass of an ion, the faster it will reach the detector.

In principle, the upper mass range of a TOF instrument has no limit, which makes it especially suitable for soft ionization techniques. For example, samples with masses above 300 kDa have been observed by MALDI/TOF.[27,28] Another advantage of these instruments is their high transmission efficiency, which leads to very high sensitivity. For example, the spectrum from 10^{-15} mol of gramicidin[29] and the detection of 100–200 amol amounts of various proteins (cytochrome *c*, ribonuclease A, lysozyme and myoglobin)[30] have been obtained with TOF analyzers. All the ions are produced in a short time span and temporal separation of these ions allows all of them to be directed towards the detector. Therefore, all the ions formed are, in principle, analyzed.

The most important drawback of the linear TOF analyzers is their poor mass resolution. Mass resolution is affected by factors that create a distribution in flight times among ions with the same m/z ratio. These factors are the length of the ion formation pulse (time distribution), the size of the volume where the ions are formed (space distribution), the variation of the initial kinetic energy of the ions (kinetic energy distribution), etc. Electronics and, more particularly, the digitizers also can affect the resolution and precision of the time measurement.

Because the mass resolution is proportional to the flight time, one solution to increase the resolution of these analyzers is to lengthen the flight tube. It is possible also to increase the flight time by lowering the acceleration voltage. But lowering this voltage reduces the sensitivity, so the only way to have both high resolution and high sensitivity is to use a long flight tube of 1–2 m for a higher resolution and an acceleration voltage of at least 20 kV for keeping the sensitivity high.

2.3.2 Delayed pulsed extraction

To reduce the kinetic energy spread among ions with the same m/z ratio leaving the source, a time lag or delay between ion formation and extraction can be introduced. The ions first are allowed to expand into a field-free region in the source and after a certain delay (hundreds of nanoseconds to several microseconds) a voltage pulse is applied to extract the ions outside the source. This mode of operation is referred to as delayed pulsed extraction, to differentiate it from continuous extraction used in conventional instruments. Delayed pulsed extraction, also known as pulsed ion extraction, pulsed extraction or dynamic extraction, is a revival of time-lag focusing, which was developed initially by Wiley and McLaren in the 1950s, shortly after apparition of the first commercial TOF instrument.

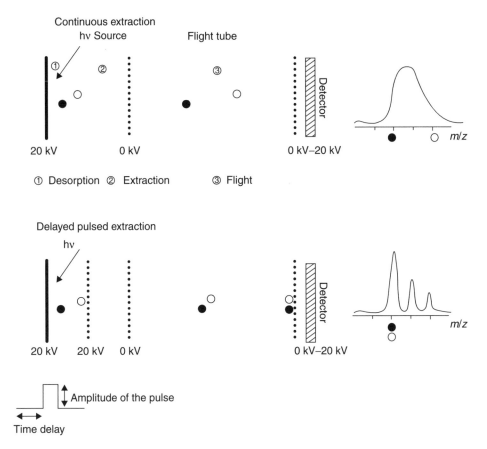

Figure 2.31
Schematic description of a continuous extraction mode and a delayed pulsed extraction mode in a linear TOF mass analyzer: (○) ions of a given mass with correct kinetic energy; (●) ions of the same mass but with a kinetic energy that is too low. Delayed pulsed extraction corrects the energy dispersion of the ions leaving the source with the same m/z ratio

As shown in Figure 2.31, the ions formed in the source using the continuous extraction mode are extracted immediately by a continuously applied voltage. The ions with the same m/z ratio but with different kinetic energy reach the detector at slightly different times, resulting in peak broadening. In the delayed pulsed extraction mode, the ions initially are allowed to separate according to their kinetic energy in the field-free region. For ions of the same m/z ratio, those with more energy move further towards the detector than the initially less energetic ions. The extraction pulse applied after a certain delay transmits more energy to the ions that remained for a longer time in the source. Thus, the initially less energetic ions receive more kinetic energy and join the initially more energetic ions at the detector.

Energy focusing can be accomplished by adjusting the amplitude of the pulse and the time delay between ion formation and extraction. For optimal focusing, both the pulse and the delay can be adjusted separately. It should be noted that the optimal pulse voltage and delay are mass dependent. Lower pulse voltages or shorter delays are required to focus ions of lower m/z ratio.

2.3.3 Reflectrons

Another way to improve mass resolution is to use an electrostatic reflector, also called a reflectron. The reflectron, defined usually by a series of grids and ring electrodes, was proposed for the first time by Mamyrin.[31] It creates a retarding field that acts as an ion mirror by deflecting the ions and sending them back through the flight tube. The term reflectron TOF is used to differentiate it from linear TOF.

The reflectron corrects the energy dispersion of the ions leaving the source with the same m/z ratio, as shown in Figure 2.32. Indeed, ions with more kinetic energy will penetrate the reflectron more deeply and will spend more time in the reflectron. Thus, they reach the detector at the same time as slower ions of the same m/z. However, the reflectron increases the mass resolution at the expense of sensitivity and introduces a mass range limitation.

If the electric field in the reflectron is E, an ion of charge q with a kinetic energy E_k will enter with a velocity v_{ix} and penetrate in the reflectron at a depth x such that

$$x = \frac{E_k}{qE}$$

Its speed v_x along the x-axis will then be zero and its mean velocity into the reflectron will be equal to $v_{ix}/2$. The time needed to penetrate at a distance x is thus

$$t_0 = \frac{x}{v_{ix}/2}$$

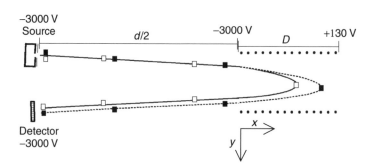

Figure 2.32
Schematic description of a TOF instrument equipped with a reflectron; (■) ions of a given mass with correct kinetic energy; (□) ions of the same mass but with a kinetic energy that is too low. The latter reach the reflectron later, but come out with the same kinetic energy as before (see text). With properly chosen voltages, path lengths and fields, both kinds of ions reach the detector simultaneously

Then the ion will be repelled symmetrically outside of the reflectron, so that its kinetic energy will be restored at the same absolute value as before but the velocity will be in the opposite direction. The total flight length in the reflectron will be $2x$ and the total time t_r in the reflectron will be

$$t_r = 2t_0 = \frac{2x}{v_{ix}/2} = \frac{4x}{v_{ix}}$$

Let us now consider two ions sharing the same mass m, one coming from the source with the correct kinetic energy E_k and the other with a kinetic energy E_k'. We define a^2 as

$$E_k'/E_k = a^2$$

The speed along the x-axis during the field-free flight will be:

$$E_k = \frac{mv_{ix}^2}{2}; \quad v_{ix} = (2E_k/m)^{1/2}; \quad v_{ix}' = (2E_k a^2/m)^{1/2}; \quad v_{ix}' = av_{ix}$$

The time of field-free flight out of the reflectron for a path length d will be:

$$t = d/v_x; \quad t' = d/v_x' = d/av_x; \quad t' = t/a$$

In the reflectron, the ions will penetrate at a depth x or x':

$$x = E_k/(qE); \quad x' = E_k'/qE = a^2 E_k/qE; \quad x' = a^2 x$$

The time spent in the reflectron will be:

$$t_r = 4x/v_{ix}; \quad t_r' = 4x'/v_{ix}' \text{ or } t_r' = 4a^2 x/av_{ix} = 4ax/v_{ix} \text{ hence } t_r' = at_r$$

The total flight time for these ions sharing the same mass but having different kinetic energies will be the sum of the flight time in and out of the reflectron:

$$t_{TOT} = t + t_r \text{ and } t'_{TOT} = t/a + at_r$$

This means that if $a > 1$, then the ion with an excess kinetic energy will have a shorter flight time out of the reflectron (t/a) but a longer one in the reflectron (at_r). The opposite holds for $a < 1$. The variations of the flight times thus compensate each other. A correct compensation yielding the same total flight time for all ions sharing the same mass but different kinetic energies requires choosing the proper values for E, V_s and d. Furthermore, a complete treatment would take into account the displacement along the y-axis. However, in practice, the reflection angle is typically less than 2°, and this error will be small. In the figures, much larger angles have been used for clarity of the drawings.

2.3.4 Tandem mass spectrometry with time-of-flight analyzer

Tandem mass spectrometry can be performed with reflectron TOF instruments. This is best achieved by combining a linear with a reflectron spectrometer, as shown in

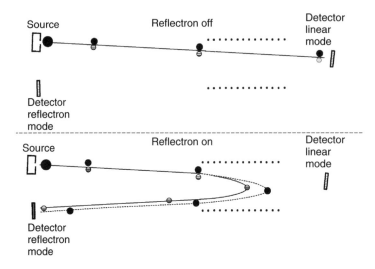

Figure 2.33
Tandem mass spectrometry with combined linear and reflectron TOF
instrument: (●) bundle of ions with one given mass leaving the source;
during the flight, a fraction of these ions fragments (•) survivor ions;
(⊝) fragment ions. Those ions fragmenting between the source and the
reflectron are called post-source decay (PSD) ions

Figure 2.33. When an ion fragments after acceleration and before the entrance in
the reflectron, its product ions and the neutral product have the same velocity as
the precursor, and thus the same flight time in the absence of any field. However,
they will have different kinetic energies. If m_p is the mass of the precursor and m_f
the fragment mass, because they have the same velocity v_{ix} their respective kinetic
energies E_{kp} and E_{kf} are:

$$E_{kp} = \frac{m_p v_{ix}^2}{2} \text{ and } E_{kf} = \frac{m_f v_{ix}^2}{2}$$

thus

$$E_{kp}/E_{kf} = m_p/m_f \text{ or } E_{kf} = E_{kp}\frac{m_f}{m_p}$$

The penetration depth x in the reflectron is given by $x = E/qE$. Hence, for the
precursor and the fragment, the relevant equations are:

$$x_p = E_{kp}/qE \text{ and } x_f = E_{kf}/qE = \frac{E_{kp}(m_f/m_p)}{qE}$$

hence

$$x_f = x_p\frac{m_f}{m_p}$$

The respective flight times in the reflectron are given by:

$$t_{rp} = \frac{4x_p}{v_{ix}} \text{ and } t_{rf} = \frac{4x_f}{v_{ix}} = \frac{4x_p(m_f/m_p)}{v_{ix}} = t_{rp}(m_f/m_p)$$

This demonstrates that the time spent in the reflectron will be shorter for the fragment than for the precursor, the ratio of the respective times being equal to the mass ratio. Thus, as shown in Figure 2.33, the time of flight will be the same for the precursors and fragments in the linear mode but will differ in the reflectron mode. When comparing the two spectra, ions observed in the reflectron spectrum but not in the linear mode result from post-source decay (PSD) fragmentations having occurred between the source and the reflectron device. Measurement of the flight times of fragment ions correlated with a particular precursor ion allows the fragmentation of this ion to be described.

An alternative method consists of detection of the neutrals that are not reflected with a linear detector while the fragment ion from the decomposition is reflected by the reflectron to a second detector.[32] The flight time of the neutral at the detector behind the reflectron therefore identifies the precursor ion. However, this method requires acceleration of the ions at several tens of kilo-electrovolts, typically 30 keV, in order to have enough kinetic energy in the neutrals to be detectable. This is illustrated in Figure 2.34.

If the instrument is fitted with electrodes after the source on the ion path, a precursor ion can be selected to analyze its fragmentation by PSD. As shown in Figure 2.35, a potential is applied to eliminate the ions. Cutting off this potential for a short time, corresponding to the passage of the selected ion, allows this ion to be selected. The resolution on modern (1995) instruments is about 50. This gives a 20 Th window at m/z 1000 Th. The mass of the passing ions can be determined in linear flight mode, whereas the fragments are observed in reflectron mode.[33]

Cornish and Cotter[34] have developed a compact laser desorption tandem TOF mass spectrometer. The instrument incorporates two dual-stage reflectron analyzers

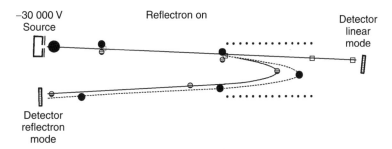

Figure 2.34
($\bullet$) Parent ions or survivors ions. ($\ominus$) Fragment ions. ($\square$) Neutral fragments. The last are not deflected and reach the linear detector. They will be detected provided that the parent ions have been accelerated at a sufficient kinetic energy, typically 30 keV

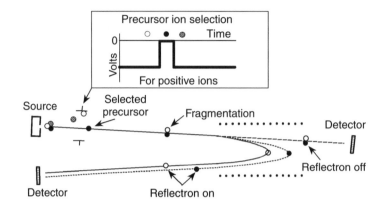

Figure 2.35
Time-of-flight spectrometer fitted with a deflection electrode for precursor
ion selection. The instrument can be operated in either linear or reflectron
mode. The selection resolution is about 50, which, for example, gives a 20
Th window at mass 1000 Th

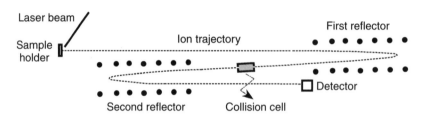

Figure 2.36
Schematic representation of a tandem TOF mass spectrometer comprising two
reflectron analyzers with a collision cell in-between.[26]

separated by a collision region for producing ions by collision-induced dissociation
(Figure 2.36). The first reflectron provides energy focusing whereas the second
provides both energy focusing and time dispersion of ions having different masses.

Cotter also has introduced a curved field reflectron[35] to allow the recording of
product ion spectra without stepping or scanning of the reflectron potential. Indeed,
with a linear field reflectron, optimal focusing in energy is obtained when $L_1 + L_2 =
4x$, where L_1 and L_2 are the distances covered before and after the reflectron and
x is the penetration depth in the reflectron. The result is a great dispersion of the
focal points ($L_2 = 4x - L_1$) for ions of different masses, and thus a loss of mass
resolution. For example, the theobromine $m_p = 137$ u produces, by loss of CO in
PSD, a fragment ion $m_f = 109$ u. If the acceleration voltage $V_s = -3$ kV, $L_1 = 1$ m,
the electric field in the reflectron $E = 3120$ V and its length $D = 0.522$ m, then
$L_{2p} = 1$ m and $L_{2f} = 0.59$ m. To correct this situation, it is necessary to adapt the
potential of the reflectron to each mass: $E_{2R} = (m_2/m_1)E_{1R}$. Thus, the focal points
and times of flight are the same for the two ions. A better solution is the introduction

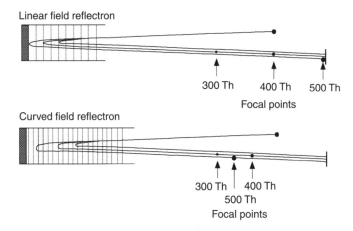

Figure 2.37
Trajectories for collision-induced dissociation (CID) fragmentation of ions of 500 Th. The focal points in energy of the product ions of 400 and 300 Th are indicated for a linear field reflectron and for a curved field reflectron. (Reproduced (modified) from Ref. 26 with permission)

of a curved field reflectron, which decreases the penetration distance for the heavy ions, as indicated in Figure 2.37.

2.3.5 Coupling time-of-flight analyzer with continuous ionization

Time-of-flight analyzers are directly compatible with pulsed ionization techniques such as plasma or laser desorption because they provide short, precisely defined ionization times and a small ionization region. Continuous ionization techniques also can be compatible with TOF analyzers but require some adaptations to transform a continuous ion beam into a pulsed process. For instance, the coupling of an electrospray (or any other atmospheric pressure ionization) source with a TOF mass spectrometer is difficult, because an electrospray yields a continuous ion beam whereas the TOF system works on a pulsed process.

The classical coupling technique consists of waving the ion beam coming out of the electrospray over a slit to the ion source, thereby yielding a pulsed inlet in the source. However, most of the ions are lost with this method. A better technique is described in Figure 2.38.

Ions from the electrospray are carried to an ion trapping device by low-potential lenses and are stored there by a decelerating field. They are extracted from the storage device and then pass into the flight tube, which is perpendicular to the incident beam direction. This occurs through 1 μs pulses that can be repeated at a frequency as high as 1 kHz. The perpendicular orientation of the source with respect to the flight tube reduces the dispersion in kinetic energy in the direction of the flight tube. This gives a mass resolution of 25 000 in the case of high masses.[36]

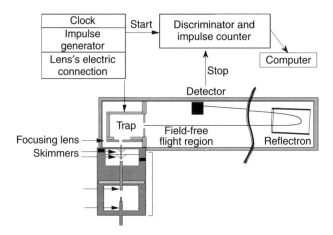

Figure 2.38
Coupling of a continuous ion source, here an electrospray, with a TOF instrument using an ion trapping device. (Reproduced (modified) from Ref. 36 with permission)

2.4 Magnetic and Electromagnetic Analyzers

Consider an ion with mass m and charge q, accelerated in the source by a potential difference V_s. At the source outlet, its kinetic energy is

$$E_k = \frac{mv^2}{2} = qV_s \tag{2.1}$$

2.4.1 Action of the magnetic field

If the magnetic field has a direction that is perpendicular to the velocity of the ion, the latter is submitted to a force F_M as described in Figure 2.39. Its magnitude is given by:

$$F_M = qvB$$

where B is the magnetic force.
The ion follows a circular trajectory with a radius r so that the centrifugal force equilibrates the magnetic force

$$qvB = \frac{mv^2}{r} \quad \text{or} \quad mv = qBr \tag{2.2}$$

For every value of B, ions with the same charge and the same momentum (mv) have a circular trajectory with a characteristic r value. Thus, the magnetic analyzer selects ions according to their momentum. However, taking into account the kinetic energy of the ions at the source outlet leads to

$$mv^2 = 2qV_s$$

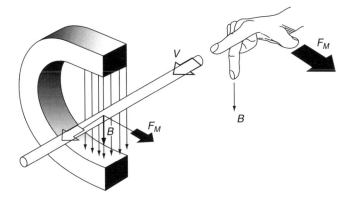

Figure 2.39
Orientation of the magnetic force B upon a moving ion

hence

$$\frac{m}{q} = \frac{r^2 B^2}{2V_s}$$

If the radius r is imposed by the presence of a flight tube with a fixed radius r, for a given value of B only ions with the corresponding value of m/q go through the analyzer. Changing B as a function of time allows successive observations of ions with various values of m/q. If $q = 1$ for all of the ions, the magnetic analyzer selects the ions according to their mass, provided that they all have the same kinetic energy. Thus the magnetic analyzer (which is fundamentally a momentum analyzer) can be used as a mass analyzer provided that the kinetic energy of the ions or at least their velocity is known.

Instead of positioning a guide tube and detecting the ions successively while scanning the magnetic field, it is possible also to use the characteristic that ions with the same kinetic energy but different m/q ratios have trajectories with different r values. Such ions emerge from the magnetic field at different positions. These instruments are said to be dispersive (Figure 2.45 shows some samples). Furthermore, Equations (2.1) and (2.2) yield

$$r = \frac{\sqrt{2m E_k}}{q B}$$

The result is that ions with identical charge and mass are dispersed by a magnetic field according to their kinetic energy. In order to avoid this dispersion, which alters the mass resolution, the kinetic energy dispersion must be controlled. This is achieved with an electrostatic analyzer.

2.4.2 Electrostatic field

Suppose a radial electrostatic field is produced by a cylindrical condenser. The trajectory then is circular and the velocity is constantly perpendicular to the field. Thus, the centrifugal force equilibrates the electrostatic force according to the following equations, where E stands for the intensity of the electric field

$$q E = \frac{m v^2}{r}$$

Introducing the entrance kinetic energy

$$r = \frac{2 E_k}{q E}$$

with the trajectory being independent of the mass, the electric field is not a mass analyzer but rather a kinetic energy analyzer, just as the magnetic field is a momentum analyzer. Note that the electric sector separates the ions according to their kinetic energy.

2.4.3 Dispersion and resolution

The resolving power of an analyzer was defined earlier. We have seen at the beginning of this chapter how the resolution depends inversely on the dispersion at the analyzer outlet. Three factors favor the dispersion, and thus the loss of resolution:

1. If the ions entering the field do not have the same kinetic energy, they follow different trajectories through the field. This is called energy dispersion (Figure 2.40).

2. If the ions entering the field follow different trajectories, this divergence may increase during the trip through the field. This is called angular dispersion (Figure 2.40).

3. The incoming ions do not originate from one point; they are issued from a slit. The magnetic or electric field can only yield, at best, a picture of that slit. The picture width depends on the width of the slit and on the magnifying effect of the analyzer.

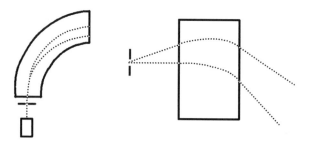

Figure 2.40
(*Left*) Energy dispersion. (*Right*) Angular dispersion

2.4.3.1 Direction focusing

An ion entering into the magnetic field along a trajectory perpendicular to the field edge follows a circular trajectory as was described earlier. An ion entering at an angle α with respect to the previous perpendicular trajectory follows a circular trajectory with an identical radius and thus converges with the previous ion when emerging from the sector (Figure 2.41). Choosing correctly the geometry of the magnetic field (sector field) thus allows focusing of the incoming beam.

An ion entering the electric sector perpendicular to the field edge follows a curved trajectory, as was described earlier. However, if the ion trajectory at the inlet is not perpendicular to the edge, its trajectory is longer if it enters the sector closer to the outside and shorter if it enters the sector closer to the inside (Figure 2.42). Here, again, a suitably chosen geometry results in a direction focusing.

2.4.3.2 Energy focusing

As can be observed in Figure 2.43, when a beam of ions with different kinetic energies is issued from the source, the electric and the magnetic sector produce an energy dispersion and a direction focusing. If two sectors with the same energy dispersion are oriented as is shown in Figure 2.44, the first sector dispersion energy is corrected by the second sector convergence. Double-focusing instruments use this principle and a few examples appear in Figure 2.45. A rudimentary ion optics of the mass spectrograph has been described in a simple yet usable manner by Burgoyne and Hieftje.[37]

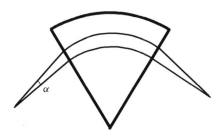

Figure 2.41
Direction focusing in a magnetic sector

Figure 2.42
Direction focusing in an electric sector

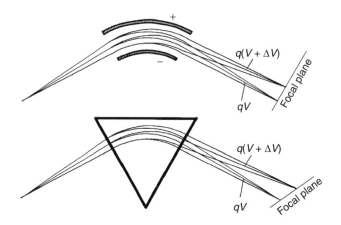

Figure 2.43
Energy dispersion in an electric sector (*top*) and in a magnetic
sector (*bottom*)

Figure 2.44
Combination of the two sectors, electrical and magnetic, shown in
Figure 2.43; the magnetic sector is turned over to obtain double focusing

2.4.4 Practical considerations

The magnetic instrument's source must function with a potential V_s of about 10 kV.
The vacuum in the source thus must be very high to avoid arcing.

Classical magnets were not well suited to fast scanning, which is necessary for gas
chromatography (GC) coupling, for instance, because of the hysteresis phenomenon
and the magnet heating by Foucauld currents induced by rapidly changing magnetic
fields. Lamellar magnets avoid such inconveniences; they have been well developed
and now are widely used.

It can be shown that magnetic instruments function at constant resolution $R =
m/\delta m$. As a result, δm varies proportionally with m. In the low-mass range δm
is small whereas in the high-mass range it is large. As an example, suppose that
the resolution is adjusted to 1000. For an ion with a mass of exactly 100.0000,
$\delta m = 100/1000 = 0.1$. An ion with a mass of 1000 yields $\delta m = 1000/1000 = 1$.
The ion of mass 100 is observed while the instrument is scanning from mass 99.95

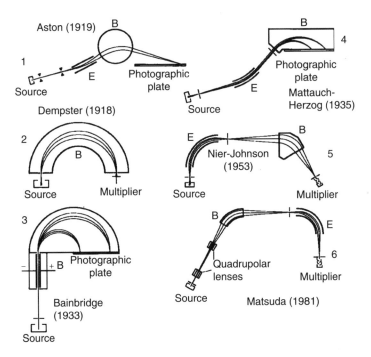

Figure 2.45
Six types of magnetic mass spectrometers are displayed. Instruments 4, 5 and 6, with double focusing, are most frequently used. Instrument 4 is dispersive whereas 5 and 6 are scanning types. The photographic plate in 4 can be replaced by an array detector for limited mass ranges. B, Magnetic sector; E, electric sector. (Reproduced from Finnigan MAT documentation, with permission)

to 100.05. At mass 1000, this ranges from 999.5 to 1000.5. If the scanning is carried out so as to cover a mass unit within a time t, the ion of mass 100 is observed during a time equal to $0.1t$, whereas that with mass 1000 is observed during $1t$. So if the number of ions produced in the source during this time t is the same at mass 100 as at mass 1000, then the total number of ions measured at the detector is ten times smaller at mass 100 than at 1000: the number of detected ions does not correspond to the number of ions that are produced. The scanning thus is made exponential in order to correct this error and the time spent scanning at mass 100 is ten times longer than that at mass 1000. The number of detected ions then is proportional to the number that is produced.

An alternative technique allows one to increase artificially the mass precision of magnetic spectrometers at a given resolution; this is called peak matching. The technique consists of comparing the masses of two compounds that are ionized simultaneously in the spectrometer source: one is unknown and its exact mass is sought; the other is a reference and its mass is known with accuracy. This comparison is achieved by a very rapid alternative modification of the acceleration voltage so as to focus the two ions, the intensities of the magnetic and the electric field being kept constant. The match is perfect when the two mass profiles exactly overlap. If the acceleration voltages necessary for focusing the two ions are known with accuracy, the mass of the unknown compound can be determined with accuracy.

Since sources were developed that produce high-mass ions, researchers have sought to extend the mass range of mass spectrometers. In the case of magnetic instruments, the basic equation shows how to act on the mass range:

$$\frac{m}{q} = \frac{r^2 B^2}{2V_s}$$

where V_s, the acceleration voltage in the source, can be reduced. This is a classical approach but it entails a loss of resolution and sensitivity.

Increasing B allows one to increase the mass range while keeping V_s constant. It is not possible to use superconducting magnets because the magnetic field cannot be scanned over a wide range, as is required for these instruments. Classical magnets allow one to reach a maximum field that depends on the nature of the alloy being used; the highest value is obtained with Permadur, a steel containing cobalt. The maximum then is 2.4 T. Finally, another technique consists of increasing the radius r. This causes a direct increase in the instrument size, especially because a greater radius corresponds to a greater focal distance.

Recent instruments have attempted to overcome this inconvenience by modifying the ion trajectories and thus the focal distances. This is achieved by using two factors: the entrance and outlet angles of the ions can be different from the normal perpendicular to the field edge; and non-homogeneous fields can be used, which modify the ion trajectories. For example, on a Kratos MS50RF instrument the focal distance was reduced by 60% using a non-homogeneous field.

2.4.5 Tandem mass spectrometry in electromagnetic analyzers

We saw that some ions decomposed during the time of flight inside the analyzer. These metastable or collision-induced fragmentations can occur in different parts of the instrument. The fragmentations that occur inside the analyzer normally are not observable. However, those that occur in the field-free regions can be observed under proper experimental conditions and yield interesting information. The first region located between the source and the first analyzer is called the 'first field-free region'. The 'second field-free region' is located between the first and the second analyzers, and so on. They can have dimensions ranging from a few centimeters to 1 m. The region that is most appropriate to the type of study being carried out is chosen: ion structure, reaction mechanism, thermochemical determinations, ion–molecule reactions, etc.

Instruments with combined magnetic and electric analyzers can be assembled according to either of two configurations. The electric sector is located either in front of the magnetic sector, which is the most frequent case, or behind it. The magnetic sector is labeled B and the electric sector is labeled E. The first configuration is called EB (or also Nier–Johnson).

The geometry is less frequently a BE one. It is sometimes called a 'reverse Nier–Johnson' or 'reverse geometry'. This geometry allows easier analysis of the ion kinetic energy.

Tandem mass spectrometry is possible with a double-sector instrument even though it cannot be considered as made up of exactly two mass spectrometers hooked in series. A technique called linked scan is used. It consists of a simultaneous scan of the E and B sectors according to a mathematical relationship dependent on the system geometry, on the region under study and on the type of information that is looked for. These different types of scans are listed in Table 2.1.

We will now examine the most common types of scans.

2.4.5.1 Detection of metastable ions in normal scan

All of the ions have the same kinetic energy at the source outlet. If an ion dissociates in a field-free region, the fragments have approximately the same velocity as the precursor ion and thus have different kinetic energies from that of the precursor ion. Because the electric sector sorts the ions according to their kinetic energy, all of the metastable ions formed before the electric sector are not observed in the normal spectrum. Thus the metastable ions produced in the first field-free region of an EB spectrometer or in BE instruments are not detected in a normal scan. Only EB instruments produce spectra that show the metastable ions formed between the two sectors. At the electric sector outlet the ion with a mass m_p has a velocity

$$v = \sqrt{\frac{2q\,V_s}{m_p}}$$

The fragment ion with mass m_f has the same velocity, and thus a momentum

$$m_f v = m_f \sqrt{\frac{2q\,V_s}{m_p}}$$

Table 2.1 Types of scans for electromagnetic analyzers

EB Configuration		
Product ion spectrum	FFR1	B/E constant
Precursor ion spectrum	FFR1	B^2/E constant
Neutral loss spectrum	FFR1	$(B^2/E^2)(1 - E)$ constant
BE Configuration		
Product ion spectrum	FFR1	B/E constant
	FFR2	E scan (MIKES)
Precursor ion spectrum	FFR1	B^2/E constant
	FFR2	$B^2 E$ constant
Neutral loss spectrum	FFR1	$(B^2/E^2)(1 - E)$ constant
	FFR2	$B^2(1 - E)$ constant

FFR1, first field-free region; FFR2, second field-free region; MIKE, mass-analyzed ion kinetic energy spectroscopy.

with the focusing condition in the magnetic sector being $Bq = mv/r$, we have, for the precursor with mass m_p:

$$B_p = \sqrt{\frac{m_p}{q}}\frac{1}{r}\sqrt{2V_s}$$

and for the fragment with mass m_f:

$$B_f = \frac{m_f}{qr}\sqrt{\frac{2qV_s}{m_p}}$$

which corresponds to an apparent measured mass m^*:

$$B_f = \sqrt{\frac{m^*}{qr}}\sqrt{2V_s} = \frac{m_f}{qr}\sqrt{\frac{2qV_s}{m_p}} \quad \text{which gives} \quad m^* = \frac{m_f^2}{m_p}$$

The ion issued from the metastable fragmentation shows up at the apparent mass m^*, linked to m_p and m_f through the relation $m^* = m_f^2/m_p$. The kinetic energy released during the fragmentation brings up a dispersion in velocity and thus an alteration in the resolution.

Figure 2.46 shows, as an example, the spectrum of theobromine and a scheme of the metastable fragmentation.

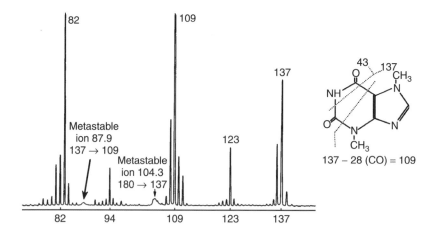

Figure 2.46
Example of a theobromine spectrum obtained with a magnetic instrument using an EB configuration. The signal detected at an apparent m/z 87.9 comes from metastable fragmentation of the ion at 137 Da, which loses 28 Da and yields the 109 Da fragment. Similarly, the signal at apparent m/z 104.3 comes from the metastable fragmentation 180 → 137 Da. (Reproduced from Kratos documentation, with permission)

2.4.5.2 BE Instruments and mass-analyzed ion kinetic energy spectroscopy

Mass-analyzed ion kinetic energy spectroscopy (MIKES) is the simplest observation method for metastable or collision-induced ions. It requires an instrument with a reverse BE geometry.

At the source outlet, every ion, if we consider only the singly charged ions, has a kinetic energy expressed by the following equation:

$$eV_s = \frac{mv^2}{2}$$

This kinetic energy is constant throughout the flight.

If an ion dissociates between the magnetic and electric sectors, the fragments have, as a first approximation, the same velocity as the precursor ion.

Consider an ion M with mass m_p yielding a fragment F with mass m_f. The fragment has the velocity v_p of the ion M. The kinetic energies are, respectively

$$E_{kp} = eV_s = \frac{m_p v_p^2}{2}$$

$$E_{kf} = \frac{m_f v_p^2}{2}$$

We therefore deduce

$$\frac{E_{kp}}{E_{kf}} = \frac{m_p}{m_f}$$

The ions going through the electric sector must obey the condition

$$eE = \frac{mv^2}{r}$$

Normally, the value of E is the same for all the primary ions, because they all have the same kinetic energy. The metastable ions then are eliminated: they are not observed in the normal spectrum taken with a BE instrument. However, if we modify the value of the field E so that

$$eE_p = \frac{m_p v_p^2}{r} \quad \text{and} \quad eE_f = \frac{m_f v_p^2}{r}$$

then the value E_f of the metastable ion field is observed. Thus

$$\frac{E_p}{E_f} = \frac{m_p}{m_f}$$

Knowing the ratio E_p/E_f and the value of m_p allows us to determine the mass m_f of the fragment that is observed. The result is the mass-analyzed ion kinetic energy.

Normally, it is sufficient to isolate a beam of precursor ions by the magnetic sector (constant B value) and to scan the electric sector. As opposed to the other linked scans, the MIKES experiments exploit the independent use of magnetic and electric sectors.

A first important use for MIKES is the determination of the fragment filiations that are observed in the spectrum. Such information is, of course, important for establishing fragmentation mechanisms.

The analysis that we carried out supposes that the fragment ion retained the velocity of the precursor. This is true only if the fragmentation occurs without any release of kinetic energy. Usually the fragmentation goes along with a conversion of part of the internal energy into kinetic energy of the fragments. If the reaction releases kinetic energy E_f, the fragment then has a kinetic energy ranging from $(E_{kf} + E_f)$ to $(E_{kf} - E_f)$, depending on whether the orientation of the precursor ion induced an increase or a decrease in the velocity. This variation results in a widening of the peak compared with the width due to the normal dispersion of velocities and to the instrument resolution.

The MIKES technique thus allows direct measurement of the kinetic energy released during fragmentation between the magnetic and electric sectors in a BE configuration instrument by scanning the electric sector. The kinetic energy released during the fragmentation of a metastable ion also can be measured by scanning the acceleration potential (high voltage, HV) scan. Both of these methods are shown in Figure 2.47.[38]

2.4.5.3 The B/E = constant scan

Consider an ion that fragments between the source and the first analyzer. The fragment always has the velocity of the precursor. If the precursor is focused in

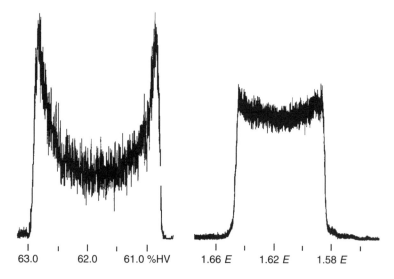

Figure 2.47
Widening of metastable peaks analyzed by high voltage (HV) scan (*left*) and electric sector (MIKES) scan (*right*) resulting from the kinetic energy that is released during the fragmentation. (Reproduced from Ref. 38 with permission)

the magnetic sector for a field B_p such that

$$e B_p = \frac{m_p v_p}{r}$$

the fragment ion is focused for

$$e B_f = \frac{m_f v_p}{r}$$

Hence

$$\frac{B_p}{B_f} = \frac{m_p}{m_f}$$

For the electric sector, these conditions are

$$e E_p = \frac{m_p v_p{}^2}{r'}$$

$$e E_f = \frac{m_f v_p{}^2}{r'}$$

$$\frac{E_p}{E_f} = \frac{m_p}{m_f} = \frac{B_p}{B_f}$$

or

$$\frac{B_p}{E_p} = \frac{B_f}{E_f} = \text{constant}$$

The measurement principle is as follows: both of the instrument sectors first are focused on the precursor normal focusing values. The magnetic and electric fields then are scanned simultaneously to respect the condition that $B/E = \text{constant}$ by reducing the values of B and E with respect to their initial values. All of the fragments from the selected precursor thus are detected successively. This is called the B/E linked scan. Note that this technique is useful both for BE and for EB geometry instruments. The mass resolution is better than for the MIKES technique because double focusing is still available. However, because the energy dispersion of the ions remains relatively high, especially if a collision gas is used, only a weak resolution is possible in order to have sufficient sensitivity. The effective resolution is about 350, which allows one to reach up to a 350 Da mass with a 'unit' or better resolution. This type of scan does not yield information on the kinetic energy that is released, because ions with equal masses but with different kinetic energies are focused upon one point.

2.4.5.4 The $B^2/E = \text{constant}$ scan

The fragment and precursor ions produced in the source carry the same kinetic energy $(m_p v_p{}^2 = m_f v_f{}^2)$. However, the fragment ions formed in the first field-free region have a velocity that is identical with that of their precursors issued from the source ($v_p = v_f'$).

Given E_f and B_f (the respective values of the electric and magnetic fields allowing the transmission of fragment ions produced in the source)

$$eE_f = \frac{m_f v_f^2}{r'} \text{ and } eB_f = \frac{m_f v_f}{r}$$

and E_f' and B_f', (the values of the electric and the magnetic fields allowing the transmission of the fragment ions produced in the first field-free region from the same precursors)

$$eE_f' = \frac{m_f v_f'^2}{r'} \text{ and } eB_f' = \frac{m_f v_f'}{r}$$

the following relationships can be established:

$$\frac{E_f}{E_f'} = \frac{m_f v_f^2}{m_f v_f'^2} = \frac{m_p v_p^2}{m_f v_p^2} = \frac{m_p}{m_f}$$

$$\frac{B_f}{B_f'} = \frac{m_f v_f}{m_f v_f'} = \frac{v_f}{v_p} = \sqrt{\frac{m_p}{m_f}}$$

Hence

$$\frac{B_f^2}{E_f} = \frac{B_f'^2}{E_f'} = \text{constant}$$

In fact, the initial values of the magnetic and electric fields are those allowing detection of selected fragment ions issued from the source. Then the two sectors are scanned simultaneously, while keeping B^2/E constant.

Another scanning technique, the defocusing voltage scan, allows the determination of precursor ions from a given fragment.

2.4.5.5 Accelerating voltage scan or defocusing

If an ion fragments between the source outlet and the first analyzer, the product ion has the same velocity as the precursor, instead of its own velocity, and thus is not detected. However, if the precursor is accelerated to a velocity equal to that needed by the product ion to be focused, then the product ion is detected

$$eV_s = \frac{m_p v_p^2}{2} = \frac{m_f v_f^2}{2}$$

In order for the product ion of mass m_f issued from the precursor of mass m_p to be focused, the precursor ion has to be accelerated, before its dissociation, by a potential difference V' such that its velocity is v_f. We thus have

$$eV_s' = \frac{m_p v_f^2}{2}$$

Dividing the last two equations, we obtain:

$$\frac{V_s'}{V_s} = \frac{m_p}{m_f}$$

In order to exploit the method, we proceed as follows: first, the instrument is focused entirely on an ion issued from precursors that we want to identify, with a source potential V_s. The potential V_s then is increased progressively. The fragment ion is 'defocused' if it is produced inside the source, but detected again every time the potential V_s' corresponds to a precursor of this ion. The mass m_p of the precursor is calculated easily:

$$m_p = \left(\frac{V_s'}{V_s}\right) m_f$$

It is thus possible to detect and analyze the mass of all of the precursors of a given fragment, as long as they yield a metastable or collision-induced fragmentation.

2.4.5.6 Constant neutral loss scan: $B^2(1 - E)/E^2 = constant$ scan

Some groups give rise to the loss of typical neutrals. Consider the alcohols as an example, because they easily lose H_2O. The detection, in a spectrum, of all the fragments that lose 18 Da allows the detection of those with a high probability of containing one of the hydroxyl groups of the starting molecule.

During a chromatographic analysis, the detection of compounds that give rise to losses of 18 Da leads to highly selective detection of the chromatographic peaks of compounds containing a hydroxyl group. This technique is called neutral loss scanning.

Given m_n as the mass of a neutral loss, then starting from a precursor of mass m_p the fragments of mass m_f equal to $m_p - m_n$ have to be focused.

The focusing condition in the electric sector is written in the same way as for analysis of the product ions:

$$\frac{m_f}{m_p} = \frac{E_f}{E_p} = E' = 1 - \frac{m_n}{m_p}$$

$$m_p = \frac{m_n}{(1 - E')}$$

The focusing condition in the magnetic sector, again, is that the B value corresponds to an ion with an apparent mass m^* such that

$$m^* = \frac{m_f^2}{m_p} = m_f E' = (m_p - m_n)E'$$

$$= \left(\frac{m_n}{(1 - E') - m_n}\right) E' = \frac{m_n E'^2}{(1 - E')}$$

$$\frac{m^*}{q} = \frac{m_n E'^2}{(1 - E')q} = \frac{r^2 B_f^2}{2V_s}$$

Grouping the constant terms on the right, we obtain

$$\frac{B_f^2(1 - E')}{E'^2} = \frac{2m_n V_s}{qr^2} = \text{constant}$$

If the scan is carried out while respecting the condition resulting from this equation, the fragment ion formed between the source and the analyzer is passed on by the two sectors only if it differs by a constant mass m_n from its precursor.

2.4.5.7 The metastable map $B = f(E)$

A map of the detected ions can be drawn as a plot of B as a function of E. In order to obtain experimental points, E is kept constant and B is scanned. Then E is increased and a new scan of B is carried out. Repeating this process allows one to cover all of the possible combinations of E and B within a chosen range. Such a map contains simultaneously any function of B and of E, as is shown in Figure 2.48.

2.4.5.8 Instruments with more than two sectors

We can imagine combining more than two sectors, thereby obtaining an increasing number of possibilities for MS/MS analyses. Instruments with three and four sectors in some configurations are commercially available.

The combination of two electric sectors and two magnetic sectors within an EBEB configuration, for example, allows, in theory, high-resolution analysis of both precursors and fragments. However, the ions produced during fragmentation have the velocity of the corresponding precursors. If a collision chamber is located between the two mass spectrometers and if the ions are not slowed, then the situation at the inlet of the second analyzer is the same as that prevailing for the fragments

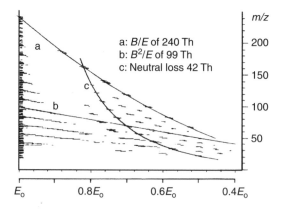

Figure 2.48
Metastable map obtained by scanning B for different values of E using a heptadecane sample. (Reproduced from Kratos documentation, with permission)

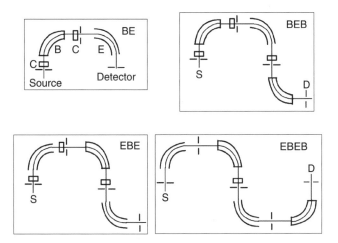

Figure 2.49
Common combinations of electric (E) and magnetic (B) sectors
and collision cells (C)

formed between the source and the analyzer in single-analyzer instruments. The focusing condition was shown earlier. It calls for scanning while keeping the B/E ratio constant and equal to the value that this ratio normally has for the precursor ion.

The second instrument thus must scan according to a function that depends on the precursor selected in the first analyzer. In practice, this can be carried out only by automatic computer control.

The second possible solution consists of slowing the ions issued from the first spectrometer, achieving low-energy collisions and then in re-accelerating the ions so that they all have the same kinetic energy. The second analyzer then functions as a normal spectrometer. However, the greater energy dispersion brings about a resolution loss of the second spectrometer. Figure 2.49 shows examples of such configurations.

2.5 Ion Cyclotron Resonance and Fourier Transform Mass Spectrometry

2.5.1 General principle

We saw how ion trajectories are curved in a magnetic field. If the ion velocity is low and if the field is intense, the radius of the trajectory becomes small. The ion thus can be 'trapped' on a circular trajectory in the magnetic field: this is the principle of the ion cyclotron or Penning trap.

Suppose that an ion is injected into a magnetic field B with a velocity v. The equations are:

$$\text{Centripetal force: } F = qvB$$

$$\text{Centrifugal force: } F' = \frac{mv^2}{r}$$

The ion stabilizes on a trajectory resulting from the balance of these two forces:

$$qvB = \frac{mv^2}{r} \text{ or } qB = \frac{mv}{r}$$

The ion completes a $2\pi r$ circular trajectory with a frequency

$$v = \frac{v}{2\pi r}$$

Thus the angular velocity ω is equal to

$$\omega_c = 2\pi v = \frac{v}{r} = \frac{q}{m}B$$

As a result of this equation, the frequency and the angular velocity depend on the ratio $(q/m)B$, and are thus independent of the velocity. However, the radius of the trajectory increases, for a given ion, proportionally to the velocity. If the radius becomes larger than that of the cell, the ion is expelled.

In practice, the ions are injected into a box (Figure 2.50) a few centimeters along its side, located in a magnetic field of 3–9.4 T produced by a superconducting magnet. For a 3 T field, the cyclotron frequency is 1.65 MHz at 28 Th and 11.5 kHz at 4000 Th. The frequency range is thus very large. Currently (2000) magnets giving a 9.4 T maximum field are used. Magnets of 24 T have been tested at the National High Magnetic Field Laboratory.[39]

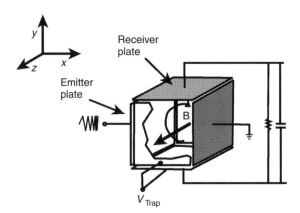

Figure 2.50
Diagram of an ion cyclotron resonance instrument. The magnetic field is oriented along the z-axis. Ions are injected in the trap along the z-axis. They are trapped along this axis by a trapping voltage, typically 1 V, applied to the front and back plates. In the x,y plane, they rotate around the z-axis due to the cyclotronic motion and then go back along the z-axis between the electrostatic trapping plates. The sense of rotation indicated is for positive ions. Negative ions will orbit in the opposite direction

The relationship between the frequency and the mass shows that determining the mass in this case consists of determining the frequency. The latter can be measured according to several methods which are classified in either of two categories: those based on observing isolated frequencies and those using complex waves and Fourier transforms.

2.5.2 Ion cyclotron resonance

The abbreviation ICR is normally used to refer to ion cyclotron resonance. The first application of ICR to mass spectrometry is due to Sommer.[40]

Irradiating with an electromagnetic wave that has the same frequency as an ion in the cyclotron allows resonance absorption of this wave. The energy that is thus transferred to the ion increases its kinetic energy, which causes an increase in the trajectory radius. The 'image current' that is induced by the ions circulating in the cell wall perpendicular to the ions' trajectory can be measured. In this case an ion excitation phase, targeting only ions with a given mass so as to have them flying close to the wall, alternates with a detection phase.

To be detected, ions of a given mass must circulate as tight packets on their orbits. Ions of the same mass excited to the same energy will be on the same orbit and rotate with the same frequency. If, however, they are located anywhere on the orbit when one ion passes close to one of the detecting plates, statistically there will be another ion of the same mass passing close to the opposite detecting plate. The resulting induced current will be null. To avoid this, ions have to be excited in a very short time, so that they are all grouped together on the orbit and thus in phase.

2.5.3 Fourier transform mass spectrometry

Fourier transform mass spectrometry (FTMS) was first described by Comisarow and Marshall in 1974,[41,42] has been reviewed by I.J. Amster[43] in 1996 and by Marshall *et al.*[44] in 1998. This technique consists of simultaneously exciting all of the ions present in the cyclotron by a rapid scan of a large frequency range within a time span of about 1 μs. This induces a trajectory that comes close to the wall perpendicular to the orbit and also puts the ions in phase. This allows one to transform the complex wave detected as a time-dependent function into a frequency-dependent intensity function through a Fourier transform, as shown in Figures 2.51 and 2.52.

Ions of each mass have their characteristic cyclotronic frequency. It can be demonstrated that ions excited by an AC irradiation at their own frequency and with the same energy, thus the same V_0 potential, applied over the same time T_{exc} will have an orbit with the same radius, and thus with an appropriate radius will all pass close to the detection plate:

$$r = \frac{V_0 T_{exc}}{B_0}$$

This equation, demonstrated in Ref. 44, is indeed independent of the m/z ratio. Thus, broadband excitation will bring all the ions on the same radius but at frequencies

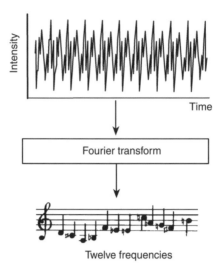

Figure 2.51
Principle of the Fourier transform: a sound signal whose
intensity is measured as a time-dependent function is
made up of many frequencies superposed one over the
other, each with its own intensity. The Fourier transform
allows one to find the individual frequencies and their
intensities. (Reproduced (modified) from Finnigan MAT
documentation, with permission)

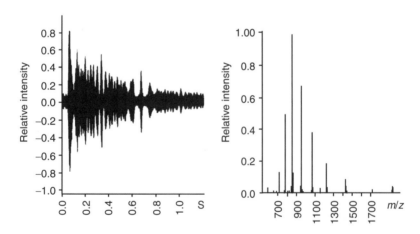

Figure 2.52
Signal intensity as a time function is transformed, through a Fourier transform,
into intensity as a function of frequency and hence into an intensity to *m/z*
relationship. (Redrawn from Ref. 43 with permission)

dependent on their mass-to-charge ratio provided that the voltage is the same at each frequency. This can be performed best by applying a waveform calculated by inverse Fourier transform, i.e. selected waveform inverse Fourier transform (SWIFT).[45]

As usual for a technique based on Fourier transform, the resolution depends on the observation time, which is linked to the disappearance of the detected signal (relaxation time). Here, the disappearance of the signal mainly results from the ions being slowed by residual ion–molecule collisions. In order to achieve high resolution, a very high cell vacuum is necessary (about 10^{-5} Pa), which is an important limitation of this technique.

The Fourier transform requires a considerable amount of calculations, because the sampling velocity must be at least twice as great as the highest frequency that has to be measured (Nyquist theorem). The data flow to be treated is very large and needs appropriate computers. This inconvenience at present (1999) is impossible to overcome in the case of coupling with a chromatographic technique, unless we want only the spectra of a very limited number of components.

Comparing FTMS with Fourier transform nuclear magnetic resonance (FTNMR), we first notice how the frequency range to be covered here is very large. Second, relaxation in NMR is invariably linked with interaction among liquid-phase or solid-phase molecules. In the gas phase, relaxation depends on the vacuum and on the stability of the ions being observed. If the vacuum is not sufficient, collisions slow the ions and their movement becomes incoherent. The observation of an ion also is limited to its lifetime.

The use of FTMS allows one to achieve time spans of about 1 s per spectrum, such as in other mass spectrometric methods.

These instruments are sensitive enough to detect about ten ions in the cell. However, the number of ions in the cell cannot exceed 10^6, because the repulsion between the ions scatters them considerably. The dynamic range is thus limited to about 10^5.

Techniques based on cyclotron resonance also are interesting because they allow the observation of ions over long time spans. This allows the study of slow fragmentations that are not observable in classical mass spectrometry and also the equilibria between ionic species and ion–molecule reactions.

The possibility of selectively eliminating ions from the cell through intense irradiation at resonance frequencies, and of eventually keeping only ions of a single mass within the cell, offers possibilities for the high-resolution study of ion reactions. For example, a gas-phase acidity scale was obtained with this method.[46]

Using an electrospray ionization (ESI) source coupled to an ICR/FTMS system, Smith et al.[47] were able to observe an ion with a mass of 5×10^6 Da, with 2610 charge units. One ion was isolated in the cell and allowed to discharge with a collision gas. Figure 2.53 shows that the ion discharges in a quantified way. The number of charges can be deduced from the masses shifts that are observed.

2.5.3.1 Additional features of modern (2000) FTMS

In modern Fourier transform ion cyclotron resonance (FTICR) instruments, an ion of mass m and charge q is submitted to an axial homogeneous magnetic field B and

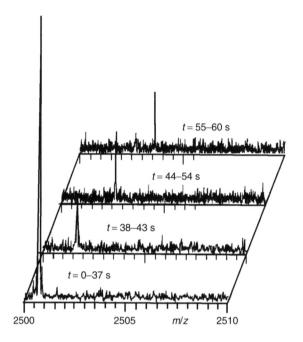

Figure 2.53
One multiply charged ion, produced in an electrospray source, is isolated in an ICR/FTMS cell. During this time it discharges by collision with a neutral gas in a quantified way, proving that it is indeed an isolated ion. From the observed masses, the number of charges can be determined, as explained for the electrospray source. (Reproduced from Ref. 47 with permission)

a quadrupolar electric field E derived from a potential $V = V_0 (z^2 - \rho^2/2)/2d^2$, with $d^2 = (z_0^2 + \rho_0^2/2)/2$, where ρ_0 is the polar coordinate in the x, y plane.

The motion of this ion results[48] from the superposition of an axial oscillation due to the trapping voltage V_0 on the end plates separated by a distance d with frequency $\omega_z = (qV_0/md^2)^{1/2}$, a cyclotron motion with frequency $\omega_c = qB/m$ and a magnetron motion with frequency $\omega_m = \omega_z^2/2\omega_c$ It may be shown that:

$$\omega_m \ll \omega_z \ll \omega_c$$

so that the magnetron motion (Figure 2.54) has a radius much larger than the cyclotron motion.

Collisional damping (reducing the energy of the ion by collision with an inert gas) reduces the amplitude of the axial and the cyclotron motions while it increases the radius of the magnetron motion and hence the ions are lost on the wall of the cell. In order to prevent this inconvenience, methods have been devised for axialization of the motion. One efficient method is the superposition of an azimuthal electric quadrupolar field of frequency ω_c that converts the magnetron motion into

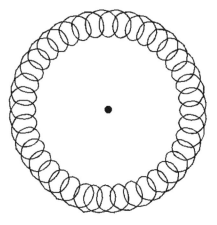

Figure 2.54
Cyclotron and magnetron motion. The ions turn on an orbit known as the cyclotron movement. But that orbit itself turns around a center (represented by a black dot) in a magnetron movement, responsible for a loss of resolution. The magnetron motion is the more important the larger the ion, and thus causes mainly loss of resolution at higher masses. Application of a quadrupolar radio frequency field allows this magnetron movement to be suppressed

collision-damped cyclotron motion.[49] To achieve the collision damping, helium gas is injected for a short time to raise the pressure to about 10^{-2} Pa. This gives a spectacular increase in resolution, sensitivity and selectivity.[50] A resolution of 1 770 000 has been obtained for leucine enkephalin.[51]

There is another advantage[52] of the high resolution so obtained: for the multiply charged ions produced by ESI, the $^{12}C/^{13}C$ isotopic peaks must exhibit unit mass spacing so that the number of them in a unit mass-to-charge ratio must represent the number of charges. A theoretical intensity distribution may be compared with the experimental distribution. Figure 2.55 displays a spectrum of ubiquitin from McLafferty's laboratory.[53]

2.5.4 MS^n in ICR/FTMS instruments

In an ICR cell, ions exposed to an axial magnetic field intensity B and having a charge ze have a cyclotron motion with frequency:

$$\omega_c = \frac{zeB}{m}$$

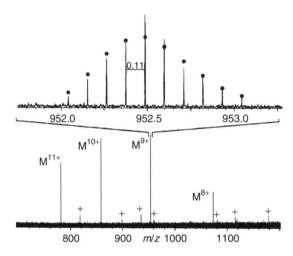

Figure 2.55
High-resolution Fourier transform mass spectrum of
ubiquitin. The upper trace displays the detail of the
peak at 952.5 Th. The distance of 0.11 Th between the
isotopic peaks allows one to deduce that the charge
is 9. This spectrum was obtained with 5 amol of
ubiquitin. (Reproduced from Ref. 53 with permission)

Thus, monocharged ions will have, for a given value of B depending on the
instrument, a frequency inversely related to their mass. They can be detected by
the current they induce in the wall of the cell, in a non-destructive way. This is a
distinct feature of ICR that is important in MS/MS experiments. By applying an AC
electric field perpendicular to the axis, it is possible to excite ions or to expel ions
by discharge on the wall. This allows one to expel all the ions except those having a
selected mass. These ions then are allowed to fragment during the time. Sometimes a
collision gas is introduced for a short time to induce fragmentation by collision, and
excitation by irradiation at the cyclotron frequency also is used. However, sustained
excitation at the resonance frequency results in larger cyclotron radii, and fragment
ions then are produced with non-zero magnetron radii. It is now common practice
to use sustained off-resonance irradiation (SORI).[54,55] This results in ions being
alternatively accelerated and decelerated, limiting the cyclotron radius. The fragments
then are produced close to the center. Fragments are detected while they are formed
in a non-destructive way. They can be detected repetitively, increasing sensitivity and
resolution. The selection and fragmentation process then can be repeated, providing
MS^n capability without reloading ions from the source because the detection is not
destructive. This contrasts with the situation in ion traps, where the detection of ions
empties the trap, which then has to be reloaded from the source.

However, the magnetic field has no focusing properties. Over time, there is an
off-axis displacement of the center of the ion cyclotron orbit, known as magnetron
radial expansion. Re-axialization can be performed using the focusing properties of
a quadrupolar RF electric field. Combined with collision damping, this allows the

magnetron radial expansion to be reverted and greatly improves the resolution and the sensitivity, especially by reducing the loss of ions during MS/MS experiments. High resolution can be obtained on both precursor *and* fragment ions.

2.6 Hybrid Instruments

Different types of mass analyzers can be coupled together, in which case they are called hybrid instruments. Many hybrid instruments have been built over the years. The most common ones result from the coupling of a magnetic instrument with a quadrupole. They are most frequently made up of a magnetic instrument in front of a collision chamber with a quadrupole analyzer, either in the BEqQ or the EBqQ configurations. The ions can be analyzed with high resolution in the magnetic instrument and then with low resolution in the quadrupole part. Two options are available between the two instruments, which respect the requirements of each type of analyzer regarding kinetic energy. Figure 2.56 shows examples of such configurations.

The first option consists of slowing all of the ions at the outlet of the magnetic instrument, which is easy because they all have the same kinetic energy. Ions with a weak energy then are fragmented through collisions in the first quadrupole and analyzed in the second quadrupole.

The second option consists of achieving high-energy collisions at the magnetic sector outlet, before entering the quadrupoles. The fragments then are slowed to a kinetic energy that is compatible with the quadrupolar analyzer. However, in this case, the fragments have the same velocity as their precursors, and thus a kinetic energy equal to a fraction m_1/m_2 of their precursors' constant energy. The voltage applied on the electrode before the quadrupoles is a function of the ratio of the mass

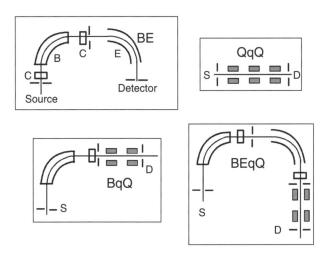

Figure 2.56
Common combinations of electric (E) and magnetic (B) sectors, quadrupoles (Q) and collision cells (C)

m_1 focused at the magnetic instrument outlet and of the mass m_2 analyzed by the quadrupole.

Another recent instrument combines a magnetic mass spectrometer with an orthogonal TOF spectrometer. A TOF analyzer can be advantageous as the second stage of the instrument because its integrating capability leads to a useful increase of sensitivity. As mentioned already, when ions fragment during their flight, with or without collision-induced dissociation, precursor and fragments have the same velocity and will arrive together at the detector. If, however, they are accelerated perpendicular to

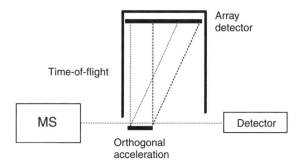

Figure 2.57
Principle of the combination of a mass spectrometer (MS) with an orthogonal TOF spectrometer. Ions coming from the mass spectrometer are directed to the detector. When a pulse voltage is applied on the orthogonal acceleration repeller, the ions are analyzed by the TOF instrument

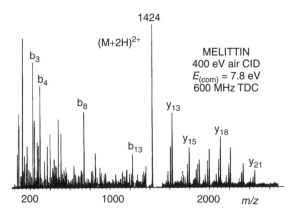

Figure 2.58
Electrospray ionization tandem mass spectrum obtained with an orthogonal TOF mass spectrometer. The doubly charged ion of the molecular species yields monocharged y fragments. (Reproduced from Micromass documentation, with permission)

this velocity and so are pushed into the TOF mass spectrometer, they will arrive at different times and at different places on the array detector and thus will be detected according to their m/z ratio. Figure 2.57 displays a scheme for such an instrument.

An electrospray ionization spectrum of melittin. obtained with a hybrid magnetic orthogonal TOF instrument is displayed in Figure 2.58. For the notation of the fragments, see Biemann's notation in Chapter 7.

2.7 Detectors and Computers

2.7.1 Detectors

The ion beam passes through the mass analyzer and then is detected and transformed into a usable signal by a detector. Different types of detectors exist, which can be classified in either of two categories. The photographic plate and the Faraday cage allow a direct measurement of the charges that reach the detector, whereas electron or photon multiplier detectors and array detectors increase the intensity of the signal.

2.7.1.1 Photographic plates and faraday cylinders

The first mass spectrometers used photographic plates located behind the analyzer as detectors: ions sharing the same m/z ratio all reach the plate at the same place. A calibration scale allows the determination of m/z and m values. The darkness of the spots gives an approximate value of the beam relative intensity.

A Faraday cylinder also can be used. Ions reach the inside of the cylinder where they give up their charge. The discharge current then is amplified and measured. The sensitivity of such detectors is limited by the noise of the amplifiers; these detectors are nevertheless very precise because the charge of the cylinder is independent of the mass and energy of the detected ions. It is necessary to use such detectors in the measurement of highly precise isotopic ratios.

2.7.1.2 Electron multipliers

At present (1996), electron multipliers are used as detectors where a positive or negative ion reaching the plate (conversion dynode) causes the emission of several secondary particles. These secondary particles can include positive ions, negative ions, electrons and neutrals. When positive ions strike the negative high-voltage conversion dynode, the secondary particles of interest are negative ions and electrons. When negative ions strike the positive high-voltage conversion dynode, the secondary particles of interest are positive ions. These secondary particles are accelerated into the continuous-dynode electron multiplier. They strike the cathode with sufficient energy to dislodge electrons as they collide with its curving inner walls. These electrons pass further into the electron multiplier, again striking the walls, causing the emission of more and more electrons as they travel towards the ground potential. Thus a cascade of electrons is created that finally results in a measurable current at the end of the electron multiplier (Figure 2.59). The amplifying power is the product of the conversion factor (number of secondary particles emitted by the conversion

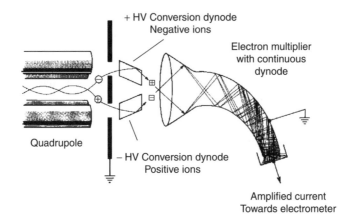

Figure 2.59
Electron multiplier: (⊖) incident ions; (□) secondary particles. (Reproduced (modified) from Finnigan MAT documentation, with permission)

dynode for one incoming ion) and the multiplying factor of the continuous-dynode electron multiplier. It may reach 10^7. Their lifetime is limited to 1 or 2 years because of surface contamination from the ions or from a relatively poor vacuum. The conversion factor depends on the nature (mass, charge and structure) and on the energy of the detected ions, so these detectors are not as precise as Faraday cylinders. However, their sensitivity is such that they allow rapid scanning.

2.7.1.3 Array detectors

An array detector is a plate where parallel cylindrical channels have been drilled. The channel diameter ranges from 4 to 25 μm with a center-to-center distance ranging from 6 to 32 μm (Figure 2.60). The plate input side is kept at a negative potential of about 1 kV compared with the output side.

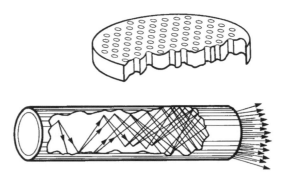

Figure 2.60
Cross-section of an array plate and electron multiplication within a channel. (Reproduced from Galileo documentation, with permission)

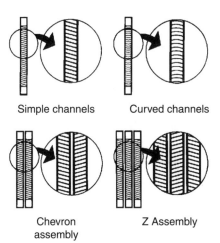

Simple channels   Curved channels

Chevron
assembly   Z Assembly

Figure 2.61
Array plate types and connections.
(Reproduced from Galileo documenta-
tion, with permission)

Electron multiplication is ensured by a semiconductor substance covering each channel and giving off secondary electrons. Curved channels prevent the acceleration of positive ions towards the input side. Two plates also can be connected herringbone-wise or three plates can be connected following a Z shape, as shown in Figure 2.61.

The snowball effect within a channel can multiply the number of electrons by 10^5. Using several plates allows an amplification that can reach 10^8. At every channel exit, a metal anode gathers the stream of secondary electrons and the signal is transferred to the processor. The plate geometry resembles that of a photographic plate: ions with different m/z ratios reach different spots and may be counted at the same time during the analyzer magnetic field scan.

2.7.1.4 Photon multipliers

This type of detector (Figure 2.62) is made up of two conversion dynodes, a phosphorescent screen and a photomultiplier. This device allows the detection of both positive and negative ions. In the positive mode, secondary ions are accelerated towards the dynode that carries the negative potential, whereas in the negative mode, secondary ions are accelerated towards the positive dynode. Secondary electrons that are given off then are accelerated towards the phosphorescent screen, where they are converted into photons. These photons are detected by the photomultiplier. The phosphorescent screen surface is covered with a thin layer of aluminum conductor to avoid the formation of a charge, which would prevent new electrons from reaching it. The amplification ranges from 10^4 to 10^5. However, their lifetime is longer than the lifetime of the electron multipliers.

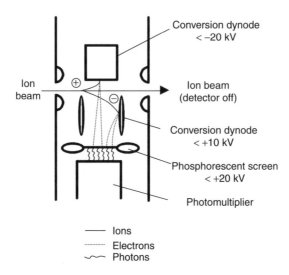

Figure 2.62
Photomultiplier

2.7.2 Computers

This section describes briefly the functioning of a computer dedicated to mass spectrometry. Applications in GC/MS and HPLC/MS will appear in Chapter 4.

2.7.2.1 Functions

A computer dedicated to mass spectrometry registers the data given out by the mass spectrometer and converts them either into values of masses and peak intensities or into total ionic current, temperatures, acceleration potential values, etc.

The computer also can calculate the possible compositions of ions of a given mass, taking into account only the elements in the molecular formula; it can compare the spectra that it observes with a library of spectra. For example, if a compound is known to contain only CHON as elements, a fragment detected at m/z 39 can only have C_2HN or C_3H_3 as elemental compositions. More examples can be found in Appendix 5.

The computer can control the mass spectrometer by introducing the values and variations of different parameters.

Because the computer treats digital data whereas the mass spectrometer produces and receives analog data, an interface is necessary to convert one type of data into another.

The computer calculation power is gauged according to the number and the speed of operations that it allows.

2.7.2.2 Instrumentation

The interface receives, conditions and digitizes the analog signals (those given by a beam, an ionic current, temperatures, voltages) output by the mass spectrometer and sends them to the computer. It is an analog-to-digital converter (ADC).

The dynamic range of analogic values that are measured increases with the number of bits (with 24 bits, it ranges from 0 to $2^{24} - 1$); the sampling velocity determines the number of measures per peak that are possible.

A DAC (digital-to-analog converter) transmits to the mass spectrometer the analogic voltages that are needed for the functioning and that are determined by the computer according to the values given by the operator.

The computer furnishes the calculation capacities and controls the operation sequence required by a program. It transfers the data to the central memory. The software transmits the instructions typed on the keyboard and converts them into machine language.

2.7.2.3 Data acquisition

Ions leaving the analyzer are detected by an electron multiplier that furnishes an analogic signal. The latter is preamplified. The background noise is reduced by a filter that cuts off high frequencies.

Using a time base given by a crystal oscillator, a voltage is obtained at precise time intervals (at least ten measurements for every peak). These voltages are digitized through a comparison with all the possible voltages precisely given out by a DAC:

$$0000\ 0001 = 1\ \text{mV}$$

$$0000\ 0010 = 2\ \text{mV}$$

$$0000\ 0011 = 3\ \text{mV}$$

etc.

$$1111\ 1111 = 255\ \text{mV}$$

using eight bits.

The peak position is calculated by the centroid:

$$t_c = \sum V_n t_n \Big/ \sum V_n$$

The peak intensity is given either by the maximum voltage V_{MAX}, or by the peak area.

Digital thresholding forbids the transmission of the baseline data and thus reduces the quantity of digitized data. A peak recognition routine eliminates the signals that do not produce three consecutive digital values above the threshold.

2.7.2.4 Data conversion

A peak position is given by a time value (or a Hall voltage value in a magnetic analyzer or a voltage value in a quadrupole), which must be converted into a mass value. This supposes a preliminary calibration with known products (PFK, $(CsI)_n$, etc.). The masses furnished by the calibrating product are stored in the computer. A relationship of the type $m = ax + b$ is calibrated with two known peaks and is checked or corrected with known peaks located in other mass areas.

2.7.2.5 Data reduction

The computer can order the printer or the screen to furnish spectra in table or graph form, supposing a normalization to 100 of the base peak intensity or of a fraction of this intensity if the detection of weak peaks is looked for. It can also order the production of time-dependent data (total ion current, temperatures, voltages).

It can print a chromatogram: a variation with time of the intensities registered at a few given masses. It can suggest elementary compositions for the peaks.

2.7.2.6 Library search

The computer can hold a database where the main peaks of known products are stored. The spectra obtained by electron ionization alone are reproducible enough to be useful.

Two types of library search have been developed. The first type, called forward search, compares the new spectrum with the spectra stored in the library and looks for the best match of the spectrum. The second type, called reverse search, checks for the possible presence in the new spectra of a spectrum chosen in the library.

A filtering process to eliminate the majority of the library spectra as being far too different from the new spectrum precedes the search. The search may identify the new spectrum or suggest possible structures. Figure 2.63 shows an example of the result of a library search. For other examples, see Ref. 56 and 57.

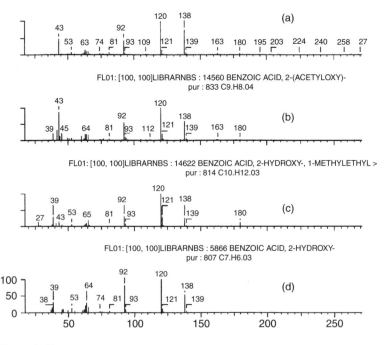

Figure 2.63
Example of a library search: (A) spectrum of an aspirin sample; (B–D) library spectra identified by the computer

Products having isomeric structures most often have almost identical spectra. Moreover, totally different products accidentally may have very similar spectra. Identification mistakes are thus frequent. A systematic recourse to another criterion is necessary, e.g. the retention time.

2.8 References

1. Ferguson R.E., McKulloh K.E. and Rosenstock H.M., *J. Chem. Phys.*, **42**, 100 (1965).
2. Kienitz H., *Massenspektrometrie*, Verlag Chemie, Weinheim, 1968.
3. Paul W. and Steinwedel H.S., *Z. Naturforsch.*, **8a**, 448 (1953).
4. Finnigan R.E., *Anal. Chem.*, **66**, 969A (1994).
5. March R.E. and Hughes R.J., *Quadrupole Storage Mass Spectrometry*, Wiley, New York, 1989; Campbell R., *Théorie Générale de l'Equation de Mathieu*, Masson, Paris, 1955.
6. March R.E. and Hughes R.J., *Quadrupole Storage Mass Spectrometry*, Wiley, New York, 1989.
7. Yost R.A. and Enke C.G., *Anal. Chem.* **51**, 1251A (1979).
8. Paul W. and Steinwedel H.S., *US Patent* 2 939 952 (1960).
9. Todd J.F.J., *Mass Spectrom. Rev.*, **10**, 3 (1991).
10. Stafford G.C., Kelley P.E., Syka J.E., *et al.*, *Int. J. Mass Spectrom. Ion Processes*, **60**, 85 (1984).
11. March R.E. and Hughes R.J., *Quadrupole Storage Mass Spectrometry*, Wiley, New York, 1989, p. 200.
12. March R.E. and Hughes R.J., *Quadrupole Storage Mass Spectrometry*, Chemical Analysis Vol. **102**, Wiley-Interscience, New York, 1989.
13. March R.E. and Hughes R.J., *Quadrupole Storage Mass Spectrometry*, Wiley, New York, 1989, p. 199.
14. Cooks R.G., Glish G.L., McLuckey S.A., *et al.*, *Chem. Eng. News*, **69**, 26 (1991).
15. Major F.G. and Dehmelt H.G., *Phys. Rev.*, **170**, 91 (1968).
16. Senko M.W., Cunniff J.B., Land A.P., 46th ASMS Conference on Mass Spectrometry and Allied Topics, Orlando, FL, 1998, p. 486.
17. Viseux N., de Hoffmann E. and Domon B., *Anal. Chem.*, **69**(16), 3193 (1997).
18. Stephens W., *Phys. Rev.*, **69**, 691 (1946).
19. Wiley W.C. and McLaren J.B., *Rev. Sci. Instrum.*, **16**, 1150 (1955).
20. Cotter R.J., *Anal. Chem.*, **64**, 1027A (1992).
21. Mann M. and Talbo G., *Curr. Opin. Biotechnol.*, **7**, 11 (1996).
22. Weickhardt C., Moritz F. and Grotemeyer J., *Mass Spectrom. Rev.*, **15**(3), 139 (1996).
23. Wollnik H., *Mass Spectrom. Rev.*, **12**, 89 (1993).
24. Cotter R.J., *Time-of-Flight Mass Spectrometry*, ACS Symposium Series No. 549, ACS, Washington, DC, 1994.
25. Schlag E.W., *Time-of-Flight Mass Spectrometry and its Applications*, Elsevier, Amsterdam, 1994.
26. Cotter R.J., *Time-of-Flight Mass Spectrometry: Instrumentation and Applications in Biological Research*, ACS, Washington, DC, 1997.
27. Imrie D.C., Pentney J.M. and Cottrell J.S., *Rapid Commun. Mass Spectrom.*, **9**, 1293 (1995).
28. Moniatte M., vanderGoot F.G., Buckley J.T., *et al.*, *FEBS Lett.*, **384**(3), 269 (1996).
29. Lange W., Greifendorf D., Van Leyen D., *et al.*, *Springer Proc. Phys.*, **9**, 67 (1986).
30. Onnerfjord P., Nilsson J., Wallman L., *et al.*, *Anal. Chem.*, **70**(22), 4755 (1998).
31. Mamyrin B.A., Karataev V.I., Schmikk D.V., *et al.*, *Sov. Phys. JETP*, **37**4 (1973).

32. Standing K.G., in *Proc. 4th Texas Symposium on Mass Spectrometry*, ed. by C.J. McNeal, Wiley, New York, 1988, pp. 267–278.

33. *Communication by Dr Christiaensen, Finnigan MAT User's Meeting*, Chicago, May 1994.

34. Cornish T.J. and Cotter R.J., *Anal. Chem.*, **65**, 1043 (1993).

35. Cornish T.J. and Cotter R.J., *Rapid Commun. Mass Spectrom.*, **7**, 1037 (1993).

36. Boyle J.G. and Whitehouse C.M., *Anal.Chem.*, **64**, 2084 (1992).

37. Burgoyne T.W. and Hieftje G.M., *Mass Spectrom. Rev.*, **15**(4), 241 (1996).

38. Cooks R.G., Beynon J.H., Caprioli R.N., *et al.*, *Metastable Ions*, Elsevier, New York, 1973.

39. Shi S.D.H., Drader J.J., Hendrickson C.L., *et al.*, *J. Am. Soc. Mass Spetrom.*, **10**, 265 (1999).

40. Sommer H., Thomas H.A. and Hipple J.A., *Phys. Rev.*, **76**, 1877 (1949).

41. Comisarow M.B. and Marshall A.G., *Chem. Phys. Lett.*, **25**, 282 (1974).

42. Comisarow M.B. and Marshall A.G., *Chem. Phys. Lett.*, **26**, 489 (1974).

43. Amsetr I.J., *J. Mass Spectrom.*, **31**, 1325 (1996).

44. Marshall A.G., Hendrickson C.L. and Jackson G.S., *Mass Spectrom. Rev.*, **17**, 1 (1998).

45. Marshall A.G., Wang T.C.L. and Ricca T.L., *J. Am. Chem. Soc.* **107**, 7893 (1985).

46. Bowers M.T., *Gas Phase Ion Chemistry*, Vol. 2, Academic Press, New York, 1979.

47. Smith R.D., Cheng X., Bruce J.E., *et al.*, *Nature*, **369**, 137 (1994).

48. Brown L.S. and Gabrielse G., *Rev. Mod. Phys.*, **58**, 233 (1986).

49. Bollen G., Moore R.B., Savarde G., *et al.*, *Appl. Phys.* **681**, 4355 (1990).

50. Guan S., Wahl M.C., Wood T.D., *et al.*, *Anal. Chem.*, **65**, 1753 (1993); Scheikhard L., Guan S. and Marshall A.G., *Int. J. Mass Spectrom. Ion Processes*, **120**, 71 (1992).

51. Guan S. and Marshall A.G., *Rapid Commun. Mass Spectrom.*, **7**, 857 (1993).

52. Beu S.C., Senko M.W., Quiinn J.P., *et al.*, *J. Am. Soc. Mass Spectrom.*, **4**, 190 (1993).

53. Kelleher N.L., Senko M.W., Little D., *et al.*, *J. Am. Soc. Mass Spectrom.*, **6**, 220 (1995).

54. Gauthier J.W., Trautman T.R. and Jacobson D.B., *Anal. Chim. Acta*, **246**, 211 (1991).

55. Amster I.J., *J. Mass Spectrom.*, **31**, 1325 (1996).

56. Debska B., *J. Mol. Struct.*, **294**, 17 (1993).

57. Henneberg D., Weimann B. and Zalfen U., *Org. Mass Spectrom.*, **28**, 198 (1993).

3

Tandem Mass Spectrometry (MS/MS)

Tandem mass spectrometry, abbreviated MS/MS, is any general method involving at least two stages of mass analysis, either in conjunction with a dissociation process or a chemical reaction that causes a change in the mass or charge of an ion.[1-3]

In the most common MS/MS experiment a first analyzer is used to isolate a precursor ion, which then undergoes spontaneously or by some activation a fragmentation to yield product ions and neutral fragments:

$$m_p^+ \longrightarrow m_f^+ + m_n$$

A second spectrometer analyzes the product ions. The principle is illustrated in Figure 3.1. The product ion spectrum will not display isotope peaks if the selected precursor m/z contains only one isotope for each atomic species, which most often will be the case.

It is possible to increase the number of steps: select ions of a first mass, then select ions of a second mass from the fragments, obtained and finally analyze the fragments of these last selected ions. This is labeled as an MS/MS/MS or MS^3 experiment. The number of steps can be increased further her to yield an MS^n experiment (where n refers to the number of generations of ions being analyzed).

3.1 Tandem Mass Spectrometry in Space or in Time

Basically, a tandem mass spectrometer can be conceived in two ways: in space by the coupling of two physically distinct instruments; or in time by performing an appropriate sequence of events in an ion storage device. Thus there are two main categories of instruments that allow MS/MS experiments: tandem mass spectrometers in space or in time.

Common in-space instruments have two mass analyzers, allowing MS/MS experiments to be performed. A common instrument of this type uses quadrupoles as analyzers. The QqQ configuration indicates an instrument with three quadrupoles where the second one, indicated by a lower-case q, is the reaction region. It operates in RF-only mode and thus acts like a lens for all the ions. Other instruments combine electric and magnetic sectors (E and B) or even E, B and qQ, i.e. electric

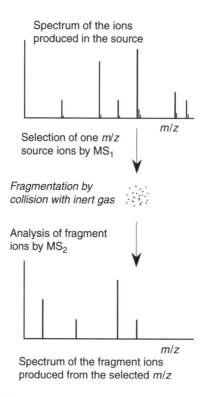

Figure 3.1
Principle of MS/MS: an ion M_1 is selected by the first spectrometer MS_1, fragmented through collision and the fragments are analyzed by the second spectrometer MS_2. Thus ions with a selected m/z value, observed in a standard source spectrum, can be chosen and fragmented to obtain their product ion spectrum

and magnetic sectors and quadrupoles. Time-of-flight instruments with a reflectron, or a combination of a magnetic instrument with a time-of-flight instrument, are also used.

To obtain higher order MS^n spectra requires n analyzers to be combined, increasing the complexity and thus also the cost of the spectrometer. Furthermore, if successive analyzers can be arranged in any number, the practical maximum is three or four analyzers in the case of beam instruments.

Besides this spatial separation method using successive analyzers, tandem mass spectrometry can be achieved also through time separation with a few analyzers such as the quistor (quadrupole ion store) or ion trap and ion cyclotrom resonance (ICR) or Fourier transform mass spectrometry (FTMS), both programmed so that the different steps are carried out successively in the same instrument. This method was described in the case of the ion trap and FTMS in Chapter 2. The practical maximum number of steps for these instruments is seven or eight.

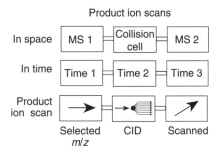

Figure 3.2
Comparison of a product ion scan performed by a space-based and a time-based instrument

A significant difference between the two types of trapping instruments is that in the ion trap mass spectrometer ions are expelled from the trap to be analyzed. Hence, they can be observed only once at the end of the process. In the Fourier transform mass spectrometer they can be observed non-destructively and hence measured at each step in the sequential fragmentation process.

Figure 3.2 represents a schematic comparison of a simple MS/MS product ion scan performed by either a 'space'-based or a 'time'-based instrument.

Nomenclature: the main names and acronyms used in MS/MS[4,5]

Molecular ion: ion formed by the addition or the removal of one or several electrons to or from the sample molecule.

Adduct ion: ion formed through the interaction of two species and containing all the atoms of one of them plus one or several atom(s) of the other.

Ion of the molecular species or pseudomolecular ion: ion originating from the analyte molecule by abstraction of a proton $(M - H)^-$ or by hydride abstraction $(M - H)^+$, or by the formation of an adduct with another ion of the ionizing plasma. These ions allow one to deduce the molecular weight.

Precursor ion: (used to be parent ion) any ion undergoing either a decomposition or a charge change.

Product ion: (used to be daughter ion) ion resulting from the above reaction.

Fragment ion: ion resulting from the fragmentation of a precursor ion.

Neutral loss: fragment lost as a neutral species

Continued on page 136

_ Continued from page 135 _

CA	collisional activation
CAD	dissociation of an ion activated through collision
CID	collision-induced dissociation
DADI	(direct analysis of daughter ion) obsolete name for MIKES
MIKES	mass-analyzed ion kinetic energy spectroscopy
CAR	collision-activated reaction
E	electric sector
B	magnetic sector
Q	quadrupole
q	RF-only quadrupole as collision cell
N	target gas in a collision
NR	neutralization reionization
N_fR	neutral fragment reionization

Source spectrum: spectrum acquired with a single analyzer.

3.2 Tandem Mass Spectrometry Scan Modes

The four main scan modes available using MS/MS are represented in Figure 3.3. Many other MS/MS scan modes are possible. We will focus on fragmentations that occur with an inert collision gas. Tandem mass spectrometry needs a computer system able to control all the experimental factors and to record the results.

1. Product ion scan (daughter scan) consists of selecting a precursor ion (or parent ion) of chosen mass-to-charge ratio and determining all of the product ions (daughter ions) resulting from collision-induced fragmentation (CID). If a reactive gas is used in the collision cell, collision-activated reaction (CAR) products are observed. When only fragment ions are produced, this scan mode also is referred to as 'fragment ion scan'.

2. Precursor ion scan (parent scan) consists of choosing a product ion (or daughter ion) and determining the precursor ions (or parent ions). This method is called 'precursor scan' because the 'precursor ions' are identified. This scan mode cannot be performed with time-based mass spectrometers. The scan requires the focusing of the second spectrometer on a selected ion, while scanning the masses using the first spectrometer. All of the precursor ions that produce ions with the selected mass through reactions or fragmentations thus are detected.

3. Neutral loss scan consists of selecting a neutral fragment and detecting all the fragmentations leading to the loss of that neutral fragment. As in the case of the precursor ion scan, this scan mode is not available with time-based mass spectrometers. The scan requires that both mass spectrometers are scanned together,

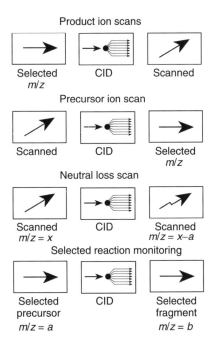

Figure 3.3
Main processes in tandem mass spectrometry (MS/MS);
CID stands for collision-induced dissociation, as occurs
when an inert gas is present in the collision cell

but with a constant mass offset between the two. Thus, for a mass difference
a, when an ion of mass m goes through the first mass spectrometer, detection
occurs if this ion has produced a fragment ion of mass $m - a$ when it leaves the
collision cell.

4. Selected reaction monitoring consists of selecting a fragmentation reaction. For
    this scan, both the first and second analyzers are focused on selected masses.
    There is thus no scan. The method is analogous to 'selected ion monitoring' in
    standard mass spectrometry but here the ions selected by the first mass analyzer
    are only detected if they produce a given fragment by a selected reaction. The
    absence of scanning allows one to focus on the precursor and fragment ions over
    longer times, increasing the sensitivity as for selected ion monitoring, but this
    sensitivity is now associated with a high increase in selectivity.

A symbolism to describe the various scan modes is represented in Figure 3.4.[6]
Note that precursor or neutral loss scans are not possible in time separation
analyzers. The product ion scan is the only one available directly with these mass
spectrometers. However, with these instruments, the process can be repeated easily
over several ion generations. Thus time-based instruments easily allow an MS^n
product ion spectrum to be obtained.

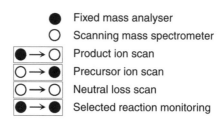

Fixed mass analyser
Scanning mass spectrometer
Product ion scan
Precursor ion scan
Neutral loss scan
Selected reaction monitoring

Figure 3.4
Symbolism proposed by Cooks *et al.*[6] for
the easy representation of various scan
modes

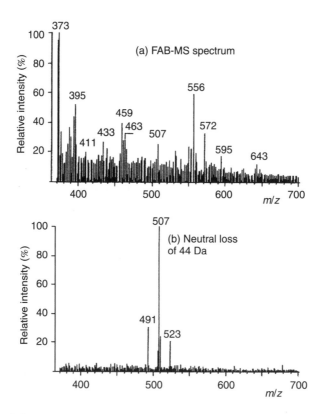

Figure 3.5
(a) Source FAB spectrum of the bile acids in a urine sample. (b) Neutral
loss scan of 44 Da measured immediately after spectrum A. Although
only about 5% of the ions are transmitted, the signal-to-noise ratio is
considerably increased. The ion at m/z 507 is now well detected and the
m/z 491 and 523 ions (not detected in a) are now visible. (Reproduced
(modified) from Ref. 7 with permission)

Because the number of transmitted ions decreases at every step due to the trajectory lengths of the ionic species and the presence of a collision gas, the practical maximum number of steps is three or four in the case of in-space MS/MS and seven or eight for in-time MS/MS. However, the chemical noise may decrease more than the ionic signal, so often the signal-to-noise ratio increases in addition to the sensitivity. This principle is illustrated in Figure 3.5, displaying the analysis of abnormal bile acids in urine.[7] Although only a few percent of the ions are transmitted in the 44 Da neutral loss scan spectrum, the signal-to-noise ratio is strongly increased. This allows very good detection of the m/z 507 ions and displays ions corresponding to two analogous bile acids containing one less hydroxyl group at m/z 491 and one more at m/z 523. These two ions are completely buried in the noise in the source fast atom bombardment (FAB) spectrum.

3.3 Collision-activated or Collission-induced Decomposition (CAD or CID)

Tandem mass spectrometry requires the fragmentation of precursor ions selected by the first analyzer in order to allow the second analyzer to analyze the product ions. According to the Warhaftig diagram (see Chapter 6), the ions leaving the source can be classified into three categories. The first category of ions, with a lifetime greater than 10^{-6} s, reach the detector before any fragmentation has occurred. The second category, with a lifetime smaller than 10^{-7} s, fragments before leaving the source and only the fragments are detected. The third category, called metastable ions, have an intermediate lifetime. These ions are stable enough to be selected by the first analyzer while containing enough excess energy to allow their fragmentation before they reach the second analyzer. The probability associated with this phenomenon is relatively low (1%) because their number is small and they spend a very short time in the reaction region. If precursor ions undergo a collision-induced activation, i.e. an increase in their internal energy that induces their decomposition (CID or CAD), then the situation described above is much improved. This CID technique allows one to increase the number of precursor ions that fragment in the reaction region and also the number of fragmentation paths. The structural analysis thus becomes easier.

An overall view shows the CID process as a sequence of two steps. The first step is very fast (10^{-14}–10^{-16} s) and corresponds to the collision between the ion and the target when a fraction of the ion translational energy is converted into internal energy, bringing the ion into an excited state. The second step is the unimolecular decomposition of the activated ion. The collision yield then depends on the activated precursor ion decomposition probability according to the quasi-equilibrium or RRKM theory (explained elsewhere). Let us recall that it is based on four suppositions:

1. The ion dissociation time is long with respect to the formation time and the excitation time.

2. The dissociation rate is low with respect to the rate of redistribution of the excitation energy among all of the ion internal modes.

3. The ion achieves an internal equilibrium condition where the energy is distributed with an equal probability among all of the internal modes. Considering that an ion with N non-linear atoms has $3N-6$ vibration modes, it is easy to understand how the collision yield decreases in a manner inversely proportional to the ion mass.

4. The dissociation products that are observed result from a series of competitive and consecutive reactions.

Several methods exist that activate the ions through collisions. The most common one consists of colliding the low- or high-energy accelerated ions with gas molecules as immobile targets. To achieve collisional activation in MS/MS instruments with spatially separate analyzers, a collision cell is placed between the two mass analyzers. This cell often corresponds simply to a small chamber with entrance and egress apertures and contains an inert target gas at a pressure sufficient for collisions with ions to occur. In MS/MS instruments based on time-separated mass analysis steps, an inert gas is simply introduced into the ICR or the ion trap. In some elaborations of these experiments, it is introduced only into a certain region or at a particular time in the operating sequence using a pulsed valve. An article described the history of this technique[8] and fundamental aspects are detailed in the literature.[9-11]

3.3.1 Collision energy conversion to internal energy

The kinetic energy for internal energy transfers is governed by the laws concerning collisions of a mobile species (the ion) and a static target (the collision gas).

Consider an ion with mass m_1 hitting a target with mass m_2. If $\mathbf{R}$ represents the position vectors in the laboratory reference frame O, and $\mathbf{r}$ represents the position vectors in the reference frame linked to the center of gravity G, we see on the diagram that:

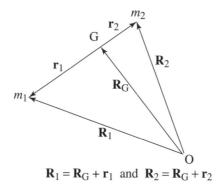

$$\mathbf{R}_1 = \mathbf{R}_G + \mathbf{r}_1 \quad \text{and} \quad \mathbf{R}_2 = \mathbf{R}_G + \mathbf{r}_2$$

Defining the center of gravity: $(m_1 + m_2)\mathbf{R}_G = m_1\mathbf{R}_1 + m_2\mathbf{R}_2$. Thus

$$\mathbf{r}_1 = m_2(\mathbf{R}_1 - \mathbf{R}_2)/(m_1 + m_2) \quad \text{and} \quad \mathbf{r}_2 = -m_1(\mathbf{R}_1 - \mathbf{R}_2)/(m_1 + m_2)$$

Differentiating with respect to t:

$$\mathbf{u}_1 = m_2 \mathbf{g}/(m_1 + m_2) \text{ and } \mathbf{u}_2 = -m_1 \mathbf{g}/(m_1 + m_2)$$

with

$$\mathbf{g} = \mathbf{v}_1 - \mathbf{v}_2 \text{ and } m_1 \mathbf{u}_1 + m_2 \mathbf{u}_2 = 0$$

According to the law of momentum conservation, the kinetic energy of a rapid particle colliding with a static target ($v_2 = 0$) cannot be converted entirely into internal energy. The kinetic energy available for conversion, E_t, termed 'relative kinetic energy', is actually the kinetic energy in the center of mass reference frame. Indeed, in this G frame the momentum is equal to zero. We thus obtain

$$E_t = (m_1 u_1^2 + m_2 u_2^2)/2 = \mu g^2/2$$

with

$$\mu = m_1 m_2/(m_1 + m_2)$$

Thus

$$E_t = E_k m_2/(m_1 + m_2) \text{ where } E_k = m_1 v_1^2/2$$

The kinetic energy is conserved in an *elastic* collision

$$\mu g^2/2 = \mu g'^2/2$$

In an *inelastic* collision, a part Q of the kinetic energy, at a maximum equal to E_t, is converted into internal energy:

$$\mu g^2/2 = Q + \mu g'^2/2$$

Thus, the energy and momentum conservations imply that only a fraction of the translational energy is converted into internal energy under inelastic conditions. This energy fraction is given by the following equation:

$$E_{com} = E_{lab} \frac{M_t}{M_i + M_t}$$

where M_i is the ion mass, M_t is the target gas mass, E_{lab} is the ion kinetic energy in the laboratory frame of reference and E_{cm} is the maximum energy fraction converted into internal energy. Consequently, an increase in the ion kinetic energy or in the target gas mass increases the energy available for the conversion. This energy decreases as a function of $1/M_i$.

For instance, a 100 u ion with a kinetic energy of 10 eV colliding with argon (atomic mass 40) has a maximum increase in its internal energy amounting to

$$10 \left(\frac{40}{40 + 100} \right) = 2.86 \text{ eV}$$

At low collision energy this maximum will be almost reached, but at high collision energy, i.e. in the range of kilo-electronvolts, a fraction of this maximum will be converted to internal energy. Always keep in mind that 1 eV per ion is equal to around 100 kJ mol^{-1}.

A measure of the average amount of kinetic energy converted into internal energy in the collision process was obtained by Harrison and Lin.[12] They observed that fragmentation of the n-butylbenzene molecular ion gives two fragments, $C_7H_7^+$ and $C_7H_8^{\bullet+}$, and that the ratio of their intensities increases with increasing amount of internal energy of the parent ion $C_{10}H_{14}^{\bullet+}$ measured by charge exchange. Nacson and Harrison,[13] using similar techniques, concluded that 87% of the available collision energy is converted into internal energy for collision of 60 eV n-octylbenzene with an N_2 target, whereas the efficiency decreases to 33% for n-butylbenzene with an Ar target.

In practice, two collision regimes should be distinguished: low energy, in the range of 1–100 eV, as occurs in quadrupole or ion trap instruments; and high energy, several thousands of electronvolts, as is common for magnetic instruments.

For theoretically comparable energy exchange, different fragmentation patterns are observed for low and high collision energy. Generally, one might state that high-energy spectra give simpler, more clearcut fragmentations, whereas low-energy spectra lead to more diverse fragmentation pathways, often including more rearrangements.

3.3.2 High-energy collision (keV)

Although the various types of analyzers (electromagnetic, quadrupolar, time-of-flight, ion trap and ICR) allow MS/MS, only electromagnetic or hybrid instruments can function at high energy. The instrument configuration that is most favorable to CID studies at high energy is that which offers the best resolution both of the precursor ions and of the product ions, i.e. EBEB or BEEB. An alternative may be found in the use of triple-sector instruments such as EBE and BEB, or double-sector instruments such as EB or BE. Each of these configurations sacrifices at least the resolution of the precursor or product ions.

The technique consists of introducing a gas into the collision cell as a molecular beam perpendicular to the ion beam. The ion beam has a kinetic energy of a few keV. The molecular beam reaches an outlet where it is pumped out to preserve the vacuum. This collision cell is located at a focal point in order to avoid losses. It is better to float the collision cell electrically to distinguish the metastable from the CID products.

If, for example, an ion of m/z 100 dissociates to yield an ion of m/z 60 with a source at 8 kV and with grounded analyzers, the product ions have an energy of 4.8 keV, whether they are formed through collision or through metastable dissociation. However, if the collision cell is floated at 6 keV, the parent ions undergoing CID have a kinetic energy of 2 keV and the product ions thus have an energy of 1.2 keV plus 6 keV acceleration when leaving the collision cell, i.e. 7.2 keV. The kinetic energy difference allows one to distinguish the ions formed through the two types of reaction because the ions issuing from metastable decomposition have an energy of 4.8 keV.[14]

At high energy, the ion excitation is mostly electronic.[15] The internal kinetic energy conversion occurs most efficiently when the collision interaction time and the internal period of the mode that undergoes the excitation are comparable.[16] Hence, in the case of an ion of mass 1000 Da having an energy of 8 kV, the value for the interaction time with a target a few angstroms wide is close to 10^{-15} s. This time corresponds to the vertical electronic transition. The energy thus acquired is then redistributed in the form of vibrational energy. A bond cleavage may result.

Helium is the most common target gas used in CID studies at high energy. In fact, it minimizes the dissociation reaction, i.e. neutralization of the precursor ion and deviation of the product ions beyond the angle allowed for focalization of ions into the second sector of the spectrometer. However, helium is not very efficient in transferring internal energy, so using a heavier gas such as argon or xenon allows the collision yield to be increased.

The internal energy that is acquired during the ions' stay in the collision cell is, on average, 1–3 eV with an energy dispersion that can reach 15 eV.[17,18] Only a small fraction of the energy available is actually converted. Let us recall that the maximum energy that can be converted in a system where an ion of mass 1000 Da has an energy of 4 keV, and where the collision gas is helium, is 16 eV. In addition, only collisions with an angular deviation of less than 1° are observable through this technique. Mechanisms producing a great angular dispersion of the products thus are rarely observed.

3.3.3 Low-energy collision (between 1 and 100 eV)

Low-energy CID spectra are most often measured using hybrid or triple quadrupole instruments. The collision chamber is most often a quadrupole in RF mode only, which allows one to focus the ions that are dispersed angularly by the collision. The pressure difference between the collision cell and the rest of the analyzer is obtained through differential pumping.

At low energy, the ions' excitation energy is mostly of a vibrational nature[19] because the interaction time between an ion of mass 200 Da at an energy of 30 eV with a target of a few angstroms is about 10^{-14} s, i.e. corresponding approximately to the bonds' vibration period.

The nature of the collision gas is more important than it is for the high-energy collisions. Normally, heavier gases such as argon, xenon or krypton are preferred because they allow the transfer of more energy.

The energy that is deposited is slightly lower than at high energy but has a weaker dispersion. However, the collision yields are extremely high when compared with the energy available. This is due, on the one hand, to the great focusing characteristic of RF-only quadrupoles and, on the other, to the length of the collision chamber that allows multiple collisions. Let us recall that the maximum energy that can be converted in such a system where an ion of m/z 1000 has a kinetic energy of 20 eV, and where the collision gas is xenon, is 2.3 eV.

One of main inconveniences with CID is the limitation of the energy transferred to a molecule and thus the limitation of its degree of fragmentation. In order to avoid this inconvenience, other activation techniques were developed. The most promising investigations seek to replace the gas molecules as targets by electron beams (electron-induced dissociation) or by laser beams (photodissociation). These two techniques are most often used with analyzers such as the ion trap or the ICR because the residence time and the interaction time are longer. The photodissociation has some advantages over the other activation techniques. Indeed, this technique is selective because only the ions that absorb at the wavelength of the incident radiation can be photodissociated and the distribution of the internal energy that is acquired by photodissociation is narrow and well-defined.

Another type of activation was introduced into MS/MS, which uses collisions with a solid surface.[20] This technique is called surface-induced dissociation (SID). In practice, this technique involves typically collisions of precursor ions on a metallic surface at an approximate 45° angle of incidence. The products of SID depend on the energy of the ion and on the nature of both the ion and the surface. In addition to dissociation of the impacting ions, other reactions can be observed. At a collision energy below 100 eV, dissociation is in competition with the formation of ion–surface reaction products, whereas at several hundred electronvolts collision energy the sputtering of materials absorbed on the surface dominates over dissociation. This technique minimizes the pumping that is necessary because no gas is introduced into the collision cell. Moreover, the interaction efficiency is 100% and it is able to transfer large energy (7–100 eV) with an energy distribution that is narrower than that observed in collision activation.

3.4 Reactions Studied in MS/MS

The type of reaction produced by collision with a neutral molecule N that is most widely studied in MS/MS corresponds to a mass change. It is illustrated by the following equations that describe covalent dissociation by ejection of neutral species:

$$m_p^{n+} + N \longrightarrow m_f^{n+} + m_n + N$$

If the target gas is a chemically reactive species, an association reaction may occur, which can yield an adduct ion with a mass greater than that of the precursor:

$$m_p^+ + m_n \longrightarrow (m_p + m_n)^+$$

These two types of reactions also occur with negative ions.

The reaction of a fullerene radical cation with helium, to produce endohedral He@C_{60}, is a particular example of the last type of reaction and is displayed in figure 3.6.[21]

Another type of reaction produced through collision is charge variation:

1. Charge exchange:

$$m_p^{\bullet+} + N \longrightarrow m_p + N^{\bullet+}$$

$$m_p^{\bullet-} + N \longrightarrow m_p + N^{\bullet-}$$

Example:[22] $Ne^{\bullet+} + Pb \longrightarrow Pb^{\bullet+} + Ne$

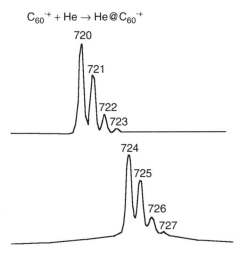

$$C_{60}^{\cdot +} + He \rightarrow He@C_{60}^{\cdot +}$$

720
721
722
723

724
725
726
727

Figure 3.6
Fullerene C_{60} after electron ionization gives the spectrum at the top. The m/z 720–723 ions result from the ^{13}C isotopes. If these fullerene ions are allowed to collide at high energy with ^4He, the spectrum at the bottom is observed, resulting from endohedral addition of He 'in' the fullerene. The endohedral nature has been demonstrated by neutralization-reionization (NRMS) of the m/z 724 ion. (Data from Ref. 21)

2. Partial charge transfer:

$$m_p^{2+} + N \longrightarrow m_p^{\cdot +} + N^{\cdot +}$$

Example:[23] $C_6H_6^{2+} + C_6H_6 \longrightarrow 2C_6H_6^{\cdot +}$

3. Ionization, charge stripping or charge inversion (normally observed at collision energies of about 1 keV):

$$m_p + N \longrightarrow m_p^{\cdot +} + N + e^-$$
$$m_p^{\cdot +} + N \longrightarrow m_p^{2+} + N + e^-$$
$$m_p^- + N \longrightarrow m_p^+ + N + 2e^-$$

Examples:

$$C_6H_5CH_3 + He \longrightarrow C_6H_5CH_3^{\cdot +} + He + e^- \quad \text{(Ref. 24)}$$
$$Ar^{\cdot +} + Ar \longrightarrow Ar^{2+} + Ar^* + e^- \quad \text{(Ref. 25)}$$
$$\text{lactic acid } (M - H)^- + 2SF_6 \longrightarrow (M - H)^+ + 2SF_6^{\cdot -} \quad \text{(Ref. 26)}$$

Because all of these processes, by necessity, can result only from electronic excitation, this suggests that collision activation under these conditions leads at least in part to electronic excitation. However, especially for larger molecules, the mechanism of collisional activation is more likely to involve vibrational excitation. In this case, it may occur through impulsive collision of the target atom or molecule with a selected atom or group of atoms in the region of the collision site.

The charge change can be combined with a mass change, such as in

$$m_p^- + N \longrightarrow m_f^+ + m_n + N + 2e^-$$

which corresponds to dissociative charge stripping, and in

$$m_p^{2+} + N \longrightarrow m_{f1}^+ + m_{f2}^+ + N$$

which corresponds to Coulomb explosion with charge separation.

The type of reaction that is observed depends on the kinetic energy of the precursor ion when it collides with N. When $N = O_2$, the neutral ionization reaction occurs readily:

$$m_n + O_2 \longrightarrow m_n^{\bullet+} + O_2^{\bullet-}$$

This allows the ionization of neutrals with kilo-electronvolt kinetic energy that can be formed either through neutralization of an ion by a charge exchange reaction or

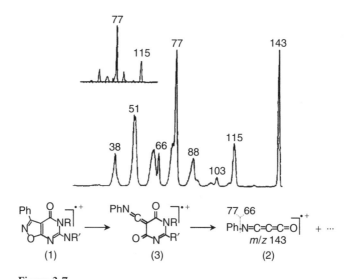

Figure 3.7
An NR reaction for the fragment at m/z 143 from compound (1). The insert shows the CID spectrum. The NR spectrum is the same as the CID spectrum except for m/z 66. This shows that the m/z 66 ion is produced during the neutralization process and is reionized. Both spectra confirm structure (2) for the m/z 143 fragment. It is possible that structure (3) is an intermediate in this process. (Reproduced (modified) from Ref. 30 with permission)

by fragmentation. The acronyms NR are used for 'neutralization reionization' and $N_f R$ for 'neutral fragment reionization'.[27-29]

Positive ion neutralization is performed using collisions with molecules whose ionization energy is low. Ammonia is often used for this purpose:

$$m_n^{\bullet+} + NH_3 \longrightarrow m_n + NH_3^{\bullet+}$$

A typical NR experiment combines the following steps: collision with ammonia to neutralize the ions, elimination of all charged species by passing between the plates of a condenser and reionization by collision with oxygen. Figure 3.7 gives an example of such an NR reaction.[30]

3.5 Tandem Mass Spectrometry Applications

Tandem mass spectrometry applications are plentiful, e.g. in elucidation of structure, determination of fragmentation mechanisms, determination of elementary compositions, applications to high-selectivity and high-sensitivity analysis and observation of ion–molecule reactions. We will examine them in detail through a few examples.

3.5.1 Structure elucidation

Tandem mass spectrometry allows more structural information to be obtained on a particular ionic species, either because the ionization method used yields relatively few structurally diagnostic fragments or because its fragmentation is obscured by the presence of other compounds in the mixture introduced in the source or other ions generated from the matrix in the course of ionization.

An example of structure elucidation by MS/MS is displayed in Figure 3.8: the FAB mass spectrum of an isolated fraction containing several nodulation factors appearing at different m/z values.[31] One ion at m/z 1244 is selected and fragmented using a B/E linked scan on a magnetic instrument. The observed fragment ions suggest the structure shown, corresponding to features known for this class of compounds, i.e. an oligosaccharide bearing at one end an alkyl chain bound to an amino-sugar.

The analogy to a chromatographic separation coupled with a mass spectrometer should be mentioned. In this case, the FAB mass spectrum is analogous to a chromatographic trace displaying compounds of the mixture. The B/E linked scan yields the mass spectrum of one of these compounds. Although chromatography gives a separation requiring a long time, separation according to the mass is almost instantaneous, providing a much shorter analysis time.

Tandem mass spectrometry also allows the determination of isomers and diastereoisomers. As an example, glycosylmonophosphopolyisoprenols[32,33] display abundant $(M - H)^-$ ions in the negative ion FAB mode. When these ions are fragmented under low-energy collisions, an abundant fragment corresponding to the phospholipid anion is observed, together with fragments across the sugar ring. This fragmentation is very sensitive to the stereochemistry, as illustrated in Figure 3.9.

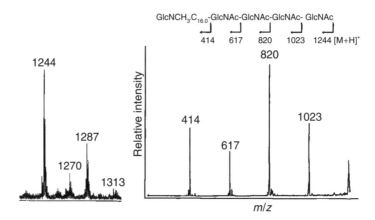

Figure 3.8
Structure elucidation by MS/MS of a nodulation factor. (*Left*) The
m/z values observed by FAB from a chromatographic fraction of
nodulation factors. (*Right*) The *m/z* 1244 ion is selected as precursor
and fragmented by *B/E* linked scan in a magnetic instrument. The
clearcut fragmentation at the glycosylic bonds allows straightforward sequence assignment, as shown by the fragmentation scheme.
(Reproduced (modified) from Ref. 31 with permission)

Decaprenylphospho-D-ribose displays the decaprenylphosphate anion as almost the
sole fragment, whereas the analogous D-arabinose derivatives also produce fragments
resulting from cleavage across the sugar ring.

From a study of several compounds, it appears that a 2-hydroxyl *trans* to the phosphate strongly favors the formation of the phospholipid anion. The ribose derivative
differs from the arabinose derivative only by the *trans* configuration of the 2-hydroxy
group with respect to the phosphate.

An example of elucidation of fragmentation mechanisms by MS/MS is displayed
in Figure 3.10, which shows the mass spectrum of *p-t*-butylphenol obtained
through electron ionization. An important fragment is observed at *m/z* 107, which
corresponds to a loss of 43 Da. We can consider either the loss of a methyl
radical followed by the elimination of CO or the loss of a C_3H_7 fragment after a
rearrangement in one or several steps. Spectrum A gives a partial answer: selecting
the ion of *m/z* 135 as a precursor, we observe that it produces the fragment of
107 Th. Thus the first step is the loss of a methyl that produces the ion of 135 Th.
The subsequent loss of 28 Da may result from a neutral CO or C_2H_4. In order to
determine this, we select 137 Th as a precursor, an ion corresponding to 135 Th with
an ^{18}O isotope or with two ^{13}C. Spectrum B shows that the ion produced is displaced
from 107 to 109 Th: the oxygen atom is preserved in the fragment. The fragments
that are observed at 108 and 107 Th in this last spectrum are due to the contribution
of ^{13}C. A complete calculation[34] shows that the observed masses correspond to the
loss of a neutral with composition C_2H_4. We can thus conclude that the molecular
ion loses a methyl, rearranges and then loses C_2H_4.

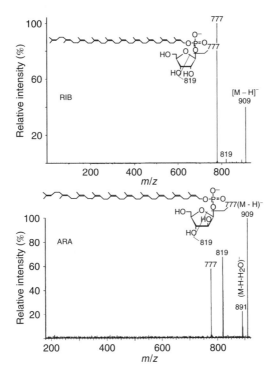

Figure 3.9
Fast atom bombardment CID at low collision energy of the $(M - H)^-$ anion from the decaprenylphosphates of β-D-ribose (*top*) and β-D-arabinose (*bottom*), respectively. The *trans* configuration of the 2-OH and phosphate groups in β-D-ribose favors the formation of the phospholipid fragment. (Reproduced (modified) from Ref. 33 with permission)

3.5.2 Selective detection of target compound class

The following example explains the development of a rapid selective analysis method for the components of a complex mixture based on MS/MS. The aim is to find an easy manner of detecting a class of compounds, the carnitines, in biological fluids. Such compounds participate in carrying fatty acids of various chain lengths through cell membranes. It also contributes to eliminating excess fatty acids in urines, and is therefore of interest to detect them in order to arrive at a diagnosis.

Elaborating a selective detection method starts with the study of fragmentation reactions. Ions that are typical of carnitine are looked for, independently from the nature of the fatty acid. Figure 3.11 shows the product ion spectrum of the molecular ion of octanoylcarnitine.

The fragmentation schemes deduced from this spectrum, and confirmed by the CID spectra of the fragments and the product ion spectra of other carnitines, are

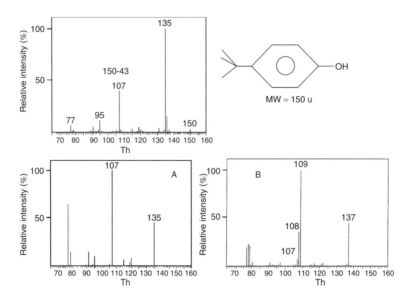

Figure 3.10
(*Top*) Electron ionization spectrum of *p-t*-butylphenol. Spectrum A: fragments of the precursor ion 135 Th. This spectrum shows that the ion of *m/z* 107 results from a loss of 28 Da after the loss of a methyl by the molecular ion. Spectrum B proves that the oxygen atom is not lost during this fragmentation: the loss of 28 Da is thus a loss of C_2H_4, not of CO

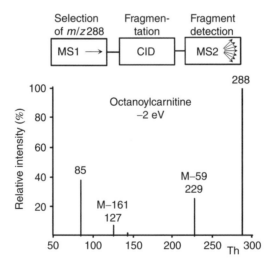

Figure 3.11
Spectrum of the ions produced by CID of the pseudomolecular ion of octanoylcarnitine. This spectrum is obtained by selecting the 288 Th ion by means of the first analyzer, fragmenting it through collision in the collision cell and analyzing the fragments using the second spectrometer

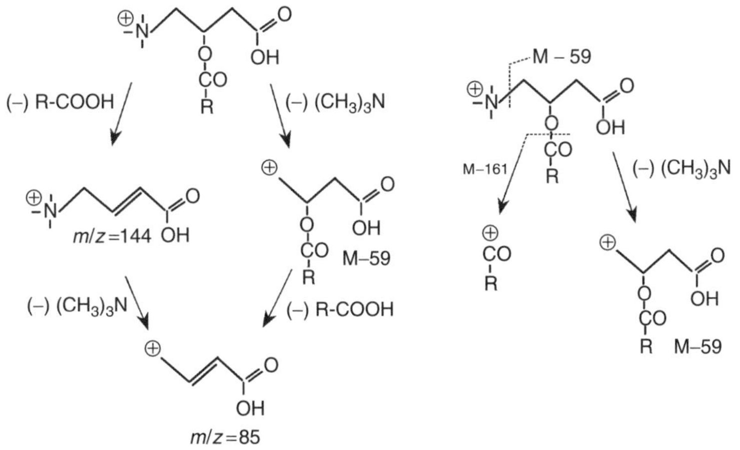

Figure 3.12
Fragmentation diagram of carnitines leading to 144 and 85 Th fragment
ions and to ions resulting from losses of the 59 and 161 Da neutrals.
These fragmentations are typical of all the carnitines, whatever the nature
of the fatty acid chain R

shown in Figure 3.12. Referring to these fragmentation schemes, it can be seen that
precursor scans of m/z 144 and 85 are good candidates. Indeed, these fragments
do not contain parts of the fatty acids and thus should not be sensitive to their
nature. The fragment ions at m/z 229 (M − 59) and 127 (M − 161) (acylium ion
obtained after loss of the carnitine moiety) contain the fatty acid chain and thus are
not good candidates for precursor ion scans, which would detect all the carnitine
conjugates. However, the neutrals lost in these fragmentations contain part (59 Da)
or the complete (161 Da) carnitine moiety. A neutral loss scan of either 59 or 161 Da
thus should allow selective detection of all the carnitine conjugates.

We conclude that we could selectively detect the carnitine conjugates by looking
for the precursors of the 85 and 144 Th fragments or by detecting the losses of 59
or 161 Da neutrals.

An example of an application is given in Figure 3.13. The first spectrum is the
FAB spectrum, sometimes called the 'source spectrum' because it detects the ions
formed in the source. The second spectrum is that obtained by selectively detecting
the precursors of m/z 85. We see that the carnitine conjugates clearly dominate.

This first sample came from the urine of a patient suffering from a short-chain
fatty acid metabolism defect (short-chain acylCoA dehydrogenase deficiency). As
a comparison, the spectrum of a urine sample from a patient suffering from a
medium-chain fatty acid metabolism defect (medium-chain acylCoA dehydrogenase
deficiency) is displayed in Figure 3.14. Other examples of selectivity obtained by
MS/MS are given in Chapter 4.

This general methodology for the selective detection of compounds or compound
classes can be combined with a chromatographic separation, allowing selective detection of one compound in a complex mixture. Furthermore, as discussed already at

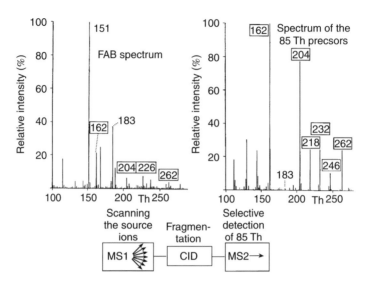

Figure 3.13
(*Left*) The FAB spectrum of a crude biological sample containing carnitine conjugates. (*Right*) Spectrum of the same sample obtained by detecting precursors of the 85 Th fragment. At 162 Th, carnitine; at 204 Th, acetylcarnitine; at 218 Th, propionyl, etc

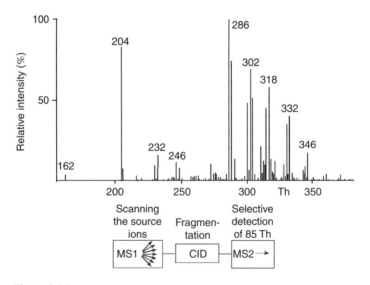

Figure 3.14
Spectrum of precursors of the 85 Th fragment as in Figure 3.13, but from a sample containing longer chain carnitines: 288 Th, octanoyl-carnitine, etc

the beginning of this chapter, MS/MS also allows the signal-to-noise ratio to be improved substantially in the detection of selected compounds or compound classes.

The possibility of obtaining both high selectivity and high sensitivity by MS/MS is largely used in the pharmaceutical industry to monitor and quantitate a selected compound in pharmacokinetics studies, which is the field that uses the largest number of tandem mass spectrometers.

3.5.3 Ion–molecule reaction

If a reactant gas is introduced to the collision cell, ion–molecule collisions can lead to the observation of gas-phase reactions. Tandem in-time instruments facilitate the observation of ion–molecule reactions. Reaction times can be extended over appropriate time periods, typically as long as several seconds. It is possible also to vary easily the reactant ion energy. Evolution of the reaction can be followed as a function of time, and equilibrium can be observed. This allows the determination of kinetic and thermodynamic parameters and has allowed, for example, the determination of basicity and acidity scales in the gas phase. In tandem in-space instruments, the time allowed for reaction will be short and can be varied over only a limited range. Moreover, it is difficult to achieve the very low collision energies that promote exothermic ion–molecule reactions. Nevertheless, product ion spectra arising from

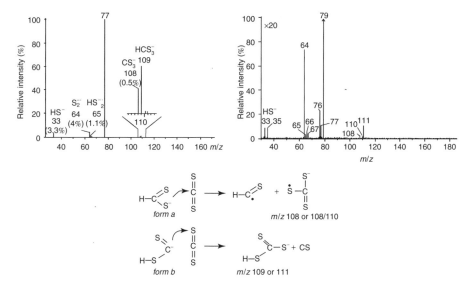

Figure 3.15
(*Top left*) Spectrum obtained when HCS_2^- ions formed in the source by chemical ionization of CS_2 are selected with the first quadrupole and allowed to collide in the collision cell filled with neutral CS_2. (*Top right*) Spectrum obtained when $HC^{32}S^{34}S^-$ at 79 Th is selected instead of the m/z 77 $HC^{32}S_2^-$ as for the first spectrum. The observed ions and their isotope shifts can be explained by the mechanism displayed. (Reproduced (modified) from Ref. 35 with permission)

ion–molecule reactions can be recorded. These spectra can be an alternative to CID for characterizing ions.

For a simple example of an ion–molecule reaction, Figure 3.15 displays the spectrum obtained when HCS_2^- ions are selected with the first quadrupole of a triple quadrupole instrument, QqQ, and allowed to collide with CS_2 molecules in the central quadrupole collision cell q.[35]

The mechanism displayed in Figure 3.15 explains the observed ions produced as well as the isotope pattern observed when the 79 Th $HC^{32}S^{34}S^-$ is selected with the first quadrupole rather than the 77 Th $HC^{32}S_2^-$. Indeed, the first reaction produces $HCS_3^{\bullet-}$, which contains one sulfur atom originating from the $HC^{32}S_2^-$ reactant. When $HC^{32}S^{34}S^-$ is selected, only one of the two sulfur atoms will be included in this $HCS_3^{\bullet-}$ product ion, which is dedoubled to 108 and 110 Th in equal abundances. The second reaction produces the ion at m/z 109, which contains both sulfur atoms from the reacting ion from the source. It is thus shifted entirely to m/z 111 when $HC^{32}S^{34}S^-$ is selected as the reactant ion.

3.6 References

1. Cooks R.G., Beynon J.H., Caprioli R.M., *et al.*, *Metastable Ions*, Elsevier, Amsterdam, 1973.
2. McLafferty F.W., *Tandem Mass Spectrometry*, Wiley, New York, 1983.
3. Busch K.L., Glish G.L. and McLuckey S.A., *Mass Spectrometry/Mass Spectrometry: Techniques and Applications of Tandem Mass Spectrometry*, VCH, New York, 1988.
4. Price P., *J. Am. Soc. Mass Spectrom.*, **2**, 336 (1991).
5. Glish G., *J. Am. Soc. Mass Spectrom.*, **2**, 349 (1991).
6. Schwartz J.C., Wade A.P., Enke C.G., *et al.*, *Anal. Chem.*, **62**, 1809 (1990).
7. Libert R., Hermans D., Draye J.P., *et al.*, *Clin. Chem.*, **37**, 2102 (1991).
8. Cooks R.G., *J. Mass Spectrom.*, **30**, 1215 (1995).
9. Hayes R.N. and Gross M.L., *Methods Enzymol.*, **193**, 237 (1990).
10. Shukla A.K. and Futrell J.H., *Mass Spectrom. Rev.*, **12**, 211 (1993).
11. de Hoffmann E., *J. Mass Spectrom.*, **31**, 129 (1996).
12. Harrison A.G. and Lin M.S., *Int. J. Mass Spectrom. Ion Phys.*, **51**, 353 (1983).
13. Nacson S. and Harrison A.G., *Int. J. Mass Spectrom Ion Processes*, **63**, 325 (1985).
14. Busch K.L., Glish G.L. and McLuckey S.A., *Mass Spectrometry/Mass Spectrometry: Techniques and Applications of Tandem Mass Spectrometry*, VCH, New York, 1988, p. 50.
15. Yamaoka H., Dong P. and Durup J., *J. Chem. Phys.*, **51**, 3465 (1969).
16. Beynon J.H., Boyd R.K. and Brenton A.G., *Adv. Mass Spectrom*, **10A**, 437 (1986).
17. Neumann G.M. and Derrick P.J., *Org. Mass Spectrom.*, **19**, 165 (1984).
18. Wysocki V.H., Kenttamaa H.I. and Cooks R.G., *Int. J. Mass Spectrom. Ion Processes*, **75**, 181 (1985).
19. Schwartz R.N., Slawsky Z.I. and Herzfeld K.F., *J. Chem. Phys.*, **20**, 1591 (1952).
20. Mabud M.A., DeKrey M.J. and Cooks R.G., *Int. J. Mass Spectrom. Ion Processes*, **75**, 285 (1985).
21. Weiske Th., Wong Th., Krätschmer W., *et al.*, *Angew. Chem. Int. Ed. Engl.* **31**(2), 183 (1992).
22. Gran W.H. and Duffenbach O.S., *Phys. Rev.*, **51**, 804 (1937).
23. Mathur B.P., Burger E.M., Boswick D.E., *et al.*, *Org. Mass Spectrom.*, **16**, 92 (1981).
24. McLafferty F.W., Todd P.J., McGilvery D.C., *et al.*, *J. Am. Chem. Soc.*, **102**, 3360 (1980).

25. Asr T., Beynon J.H. and Cooks R.G., *J. Am. Chem. Soc.*, **94**, 6611 (1972).
26. Douglas D.J. and Shushan B., *Org. Mass Spectrom.*, **17**, 198 (1982).
27. McLafferty F.W., *Int. J. Mass Spectrom. Ion Processes*, **118/119**, 221 (1992).
28. Goldberg N. and Schwarz H., *Acc. Chem. Res.*, **27**, 347 (1994).
29. Zagorevskii D.V. and Holmes J.L., *Mass Spectrom. Rev.*, **18**(2), 87 (1999).
30. Flammang R., Laurent S., Flammang-Barbieux M., *et al.*, *Rapid Commun. Mass Spectrom.*, **6**, 667 (1992).
31. Mergaert P., D'Haeze W., Geelen D. *et al.*, *J. Biol. Chem.*, **49**, 29217 (1995).
32. Wolucka B., McNeil M.R., de Hoffmann E., *et al.*, *J. Biol. Chem.*, **269**, 23328 (1994).
33. Wolucka B. and de Hoffmann E., *J. Biol. Chem.*, **270**, 20151 (1995).
34. de Hoffmann E. and Auriel M., *Org. Mass Spectrom.*, **24**, 748 (1989).
35. Gimbert Y., Arnaud R., Tabet J.C., *et al.*, *J. Phys. Chem.* **102**, 3732 (1998).

4

Mass Spectrometry/ Chromatography Coupling

In order to analyze a complex mixture, e.g. natural products, a separation technique — gas chromatography(GC), liquid chromatography(LC) or capillary electrophoresis (CE) — is coupled with mass spectrometry(MS). The separated products must be introduced one after the other into the spectrometer, either in the gaseous state for GC/MS or in solution for LC/MS and CE/MS. This can occur in two ways: the eluting compound is collected and analyzed off-line; or the chromatograph is connected directly to the mass spectrometer and the mass spectra are acquired while the compounds of the mixture are eluted. The latter method operates on-line.

The most obvious advantage drawn from coupling a separation technique with mass spectrometry consists of obtaining a spectrum used for identifying the isolated product. However, that is not the only goal that can be reached. The ideal detector should:

1. Have no alteration of the chromatographic resolution, which means not producing within the detector a mixture of the products separated before the detection.

2. Have the highest possible sensitivity.

3. Be universal, which means detecting all of the eluted products.

4. Furnish the maximum structural information possible, at best to allow positive identification of all the eluted compounds.

5. Be selective, i.e. to allow the identification of target products in a mixture. This characteristic is automatically present if item 4 is verified.

6. Give out a signal proportional to the concentration.

7. Have a constant, or at least predictable, response factor.

8. Have as low as possible cost/performance ratio.

9. Not be harmful to the product.

10. Not produce artifacts.

11. Allow the deconvolution of chromatographic peaks, i.e. the decomposition of unresolved peaks into constituents.

The importance of the last characteristic comes out of the statistical analysis of the probability of having two products within one chromatographic peak. Consider a mixture of N constituents. We will define the resolution, R, or the peak capacity in chromatography as being the maximum number of distinct compounds practically distinguishable. So, if a chromatogram is registered on a 1 m long sheet of paper and if we can consider that two compounds are detected separately when the distance between the peak centroids is at least 1 mm, we will be able to detect a maximum of 1000 compounds. This example supposes a constant peak width, which is very rare. According to our definition, we will say that the resolution $R = 1000$. If p is the probability of not having more than one compound per peak, it can be shown that these values are linked by the equation:[1,2]

$$N = (2R \ln 1/p)^{1/2}$$

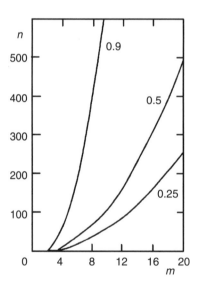

Figure 4.1
Graphical representation of the probability that there be more than one compound per peak. The peak capacity is along the n-axis. The m-axis is the number of acceptable constituents that have the probability indicated next to each curve of having only one compound per peak. The statistical analysis used to produce this graph is slightly different from that leading to the equation in the text. (Redrawn from ref. 3 with permission)

Let us calculate the maximum number N of compounds acceptable if we require $p = 90\%$, i.e. nine chances out of ten to have only one compound per peak, in the case of a resolution of 1000: we find $N = 14$. Out of these 14 peaks, statistically one out of ten will contain two unseparated compounds. This example shows that the statistical probability of having two compounds per peak is very high. Figure 4.1 shows these results graphically from a more elaborate statistical analysis.

This statistical result might surprise the users of 'classical' chromatographic methods, but it does correspond to reality for those who use GC/MS, LC/MS or CE/MS.

4.1 Elution Chromatography Coupling Techniques

The separation methods furnish a given flow of eluate (liquid or gaseous), generally under atmospheric pressure. One way or another, the eluted substances must find their way into the mass spectrometer source, where a high vacuum is necessary. We saw in the Introduction that the mean free path could be evaluated according to the following equation, where the pressure p is expressed in pascal and the distance L is expressed in centimeters:

$$L = \frac{0.66}{p}$$

A pressure of about 10^{-4} Torr (0.01 Pa or 100 nbar or 10^{-7} atm) ensures a mean free path of about 50 cm, and constitutes the upper limit of tolerable pressure for the source. Under this pressure, 1 cm^3 of gas under atmospheric pressure becomes 10^7 cm^3. A 1 cm^3 min^{-1} flow thus requires a pumping flow of 10^7 cm^3 min^{-1} under this pressure, i.e. 166 l s^{-1}; we thus obtain an approximation of the pumping capacity required by this chromatographic flow.

If the eluent is liquid, 1 mol will yield about 24 l of gas under atmospheric pressure. To gain an idea, suppose that the eluent is water and its flow rate is 0.1 cm^3 min^{-1}. A gas flow of $(0.1/18) \times 25\,000 = 139$ cm^3 min^{-1} under atmospheric pressure will be observed, which would require a pump flow at the source of $139 \times 166 = 23\,000$ l s^{-1}. This flow is exceedingly high: the eluate from a liquid chromatograph cannot be evaporated entirely in the source. We will examine the existing solutions to this problem.

In fact, the actual capacities of pumps in mass spectrometry range from 50 to 1000 l s^{-1}, so that the maximum gas flow hovers around 5 cm^3 min^{-1} under 1 atm pressure.

4.1.1 Gas chromatography/mass spectrometry (GC/MS)

Packed columns yield a carrier gas flow of about 20 cm^3 min^{-1}, whereas capillary columns achieve only about 1 cm^3 min^{-1}.

In mass spectrometry, filled columns yield a higher flow than is tolerated in the source, whereas the capillary columns fall within the acceptable values, which simplifies the gas flow problem.

The sample concentration in the carrier gas also is an important variable. If Q is the quantity (in µg) of a compound with molecular mass M injected into the column, d is the gas flow rate at the exit of the column (in $cm^3\ min^{-1}$) and l is the peak mid-height width (in s), a good approximation for the average concentration is the ratio between the injected quantity (converted into a corresponding gas volume) and the volume of the carrier gas that passes in a length of time l. As the gas volume of 1 mol under atmospheric pressure is about $25\,000\ cm^3$ at ordinary temperatures, the volume V_e $(Q/M, µmol)$ of injected sample is

$$V_e = \left(\frac{Q}{M}\right) \times 25\,000 \times 10^{-6}$$

Hence, for 10 µg injected and a molecular weight of 100 Da, we have $0.0025\ cm^3$. If the chromatographic peak width at mid-height is 10 s, the volume V_p of carrier gas for a flow rate of $20\ cm^3\ min^{-1}$ will be

$$V_p = \left(\frac{20}{60}\right) \times 10 = 3\ cm^3$$

We see that, under these conditions, the eluted sample represents less than one-thousandth of the total volume. This volume ratio is independent of the pressure and temperature. Note that an increase in the chromatographic resolution means a smaller width for the peak, and thus an improvement in this ratio. It is clearly more favorable with a capillary column. Obviously, improving this ratio before introduction into the mass spectrometer is very interesting. Ionization of the carrier gas also must be avoided, because it induces focusing difficulties and, in sources with high potential differences, discharges. Consequently, helium generally is chosen.

Two types of coupling interface between the gas-phase chromatograph and the mass spectrometer will be examined: open coupling and direct coupling.[4,5]

Interfaces that allow enrichment of the carrier gas with the eluted substance also exist. They are important if packed columns are used. However, they are now used rarely and will not be described here.

4.1.1.1 Open coupling

An open-split coupling is shown in Figure 4.2. The chromatographic column leads to a T-shaped tube that contains a smaller diameter tube. A platinum or deactivated

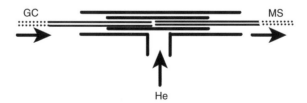

Figure 4.2
Open-split coupling. The helium current protects the eluent from air oxidation

fused-silica capillary also leads to this tube and goes into the mass spectrometer source. The capillary is kept in a vacuum-sealed device and is heated to avoid condensation. The T-shaped tube is closed at both extremities but is not sealed, so that the pressure remains equal to the atmospheric pressure. A helium current passes through it to avoid any contact with atmospheric oxygen, which could oxidize the eluted compounds. The tube that enters the mass spectrometer has a diameter and length that are chosen so as to furnish into the source a flow close to the acceptable maximum with respect to the gaseous conductance of this source and the pumping capacity. Thus, a 0.15 mm diameter and 50 cm long capillary heated to 250°C carries 2.5 ml min^{-1} of eluted gas into the source. This is sufficient in practice to pump everything that comes out of a capillary column.

This coupling does not cause any enrichment of the eluted compounds. It allows one to work under the usual chromatographic conditions, one end of the column being under atmospheric pressure. It does not require any special setting and changing the column is very easy. In principle, this system can be used with any type of column as long as no enrichment is necessary.

4.1.1.2 *Direct coupling*

This coupling consists of having the capillary column directly entering the spectrometer source by a set of vacuum-sealed joints. Its yield is always 100%. The pumping is not problematic because the capillary is necessarily very long. A length of at least 15 m is necessary for a column with an inside diameter of 0.25 mm.

Compared with open coupling, it has the inconvenience of not allowing elimination of the solvent and making the change of the column complicated. This latter inconvenience can be limited by connecting one end of the chromatographic column to a glass capillary that enters the source through a Teflon pad pierced with an opening so that the junction is sealed.

The chromatography is carried out between atmospheric pressure at the injector and a vacuum at the other end of the column, provided that the column is long enough. We saw earlier that no major inconvenience results from this, except for the comparison of chromatographic traces.

4.1.2 High-performance liquid chromatography/mass spectrometry (HPLC/MS)

The coupling of liquid chromatography is more delicate because gas-phase ions must be produced for mass spectrometry. Liquid chromatography normally is used for compounds that are not volatile and are not suitable for GC.

The problem is made even more difficult by the need to eliminate the elution solvent. Suppose, for example, that this solvent is water and that the column being used has a very small diameter that allows the flow rate to be limited to 0.1 ml min^{-1}. This corresponds to 0.1 g min^{-1} of water or 5.6 mmol min^{-1}, which yields a gas flow rate of 135 cm^3 min^{-1} at atmospheric pressure, which is still too much to be injected in a source under vacuum.

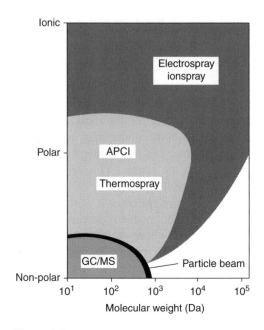

Figure 4.3
Application range of different coupling methods

Different methods are used to tackle these problems.[6-8] Some of these coupling methods, such as moving belt coupling or the particle beam (PB) interface, are based on the selective vaporization of the elution solvent before it enters the spectrometer source. Other methods, such as direct liquid introduction (DLI)[9] or continuous-flow fast atom bombardment (CF-FAB), rely on reducing the flow of liquid that is introduced into the interface in order to obtain a flow that can be pumped directly into the source. In order to achieve this it must be reduced to one-twentieth of the value calculated above, i.e. 5 μl min^{-1}. These flows are obtained from HPLC capillary columns or from a flow split at the outlet of classical HPLC columns. Finally, a series of HPLC/MS coupling methods such as thermospray (TSP), ionspray (ISP) or atmospheric pressure chemical ionization (APCI) can tolerate flow rates of about 1 ml min^{-1} without requiring a flow split. Introducing the eluent entirely into the interface increases the detection sensitivity of these methods. Electrospray ionization (ESI) can accept flow rates from 10 nl min^{-1} levels to 0.2 ml min^{-1}. Because it is more dependent on concentration than on the total quantity injected, there is little advantage of using it at high flow rates. Figure 4.3 shows the application range of different mass spectrometry chromatography coupling methods (except DLI and CF-FAB) as a function of the mass and nature of the eluted compound. Figure 4.4 shows an example of an application of HPLC/APCI coupling.[10]

4.1.2.1 Particle beam (PB) interface coupling

The particle beam interface[11,12] is a device that allows one to separate the solvent from the eluted molecules issuing from a liquid chromatographic column rapidly

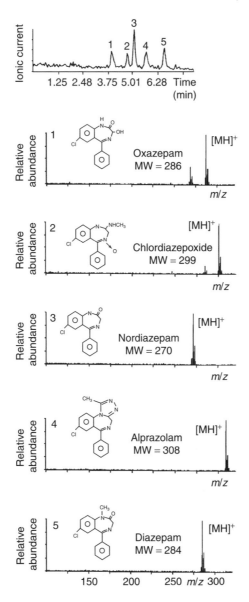

Figure 4.4
The HPLC/APCI separation of a mixture of five benzodi-
azepines. A 1.2 ml min^{-1} CH$_3$CN−CH$_3$OH−H$_2$O (40 :
25 : 35, v/v) flow is used for the analysis. (Reproduced
(modified) from Ref. 10 with permission)

and efficiently. The chromatographic eluate is pumped through the capillary up to a
glass concentric nebulizer (Figure 4.5). The eluate then is transformed into a cloud
of droplets that are scattered by a helium concentric flow. The spray passes through
a desolvation chamber, the walls of which are heated and the pressure of which

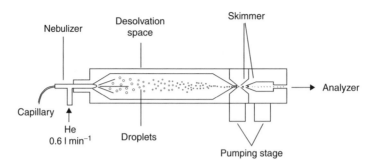

Figure 4.5
Diagram of a particle beam interface

is slightly lower than atmospheric pressure. During this travel the droplets undergo partial desolvation and yield slightly solvated eluted compound droplets.

The desolvation chamber is linked with a double-stage pumping molecular beam separator. When it leaves this evaporation chamber the mixture of helium, solvent vapor and particles undergoes a supersonic expansion into the first pumping stage (about 10 Torr). A high-speed gas beam containing particles of the eluted molecules results from this. Because the mass difference between the gas molecules and the particles is important, the particles diffuse less rapidly from the center of the beam towards the periphery. The separation is achieved by skimming the peripheral layers of the beam with a skimmer.

The skimmed solvent vapor and the helium are mechanically pumped out of the apparatus. This process is repeated in the second pumping stage (about 500 mTorr). Finally, a narrow particle beam enriched in particles with diameters smaller than 100 nm is sent into the mass spectrometer source without disturbing the vacuum. In electron ionization (EI) or chemical ionization (CI) mode, the particles injected into the source are rapidly vaporized before ionization. In FAB mode, the particle beam is directed onto a FAB nozzle covered by a matrix. Thus, the particles collide with the matrix surface and are trapped in it.

4.1.2.2 Continuous-flow fast atom bombardment (CF-FAB) coupling

The CF-FAB coupling system invented by Caprioli et al.[13,14] consists of linking the end of a capillary chromatographic column with the end of a FAB nozzle by a capillary passing through the introduction nozzle.

From 1 to 5% glycerol is added to the chromatographic solvent. The flow rate varies from 1 to 5 μl min^{-1}, which is pumped easily. The solvent evaporates but the glycerol stays at the nozzle surface with the eluted compounds and serves as a FAB matrix. Figure 4.6 is a diagram of such a coupling system and Figure 4.7 gives an application example.[15]

4.1.2.3 Coupling using atmospheric pressure ionization sources (ESI and APCI)

Electrospray ionization (ESI) and atmospheric pressure chemical ionization (APCI) sources have been described in Chapter 1. The ESI source can tolerate flows up to

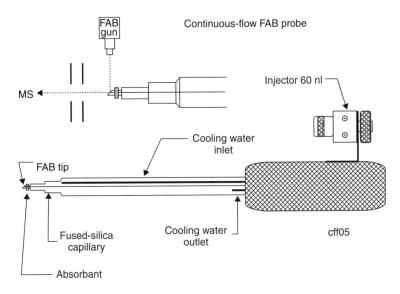

Figure 4.6
Diagram of a CF-FAB coupling system. The HPLC capillary column made
of fused silica enters through the nozzle of an introduction FAB probe.
The solvent evaporates and the glycerol remains as a FAB matrix

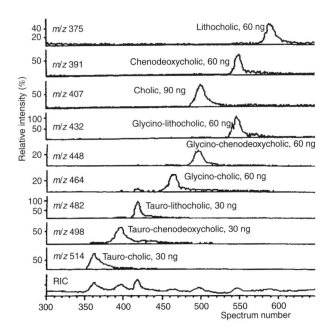

Figure 4.7
Example of an application of continuous-flow FAB and capillary column
liquid chromatography. A 1 µl aliquot containing the indicated quantities
of bile salts was injected. The total chromatography time is 11 min. The
methanol gradient in water was 50–100%, plus 1% glycerol

0.2 ml min^{-1}, allowing an easy coupling to a classical HPLC system. However, it should not be forgotten that ESI does not tolerate high salt concentrations or non-polar solvents. Acetonitrile and methanol are best suited. The tolerated flow can go down to about 100 nl min^{-1} with the nano-electrospray version, allowing coupling with HPLC using small-bore or capillary columns. Capillary electrophoresis can be coupled too, as will be discussed in the next section.

Electrospray ionization coupling can be applied as well to organic as to inorganic compounds. As an example, cisplatin (see diagram below) is a drug used in cancer therapy. By hydrolysis, chlorine atoms can be replaced by hydroxyl groups, yielding toxic metabolites. Figure 4.8 describes the HPLC/MS and HPLC/MS/MS analyses of cisplatin and its mono- and dihydroxy analogs.[16] The platinum has a complicated isotope pattern. The spectrum corresponding to peak A in the HPLC trace can be interpreted easily as corresponding to cisplatin. Indeed, its molecular weight calculated from the first isotopes of both platinum and chlorine is $194 + (2 \times 35) + (2 \times 17) = 298$ u. Adding 23 u for a sodium ion adduct leads to the observed m/z 321 peak, accompanied by the expected isotopes.

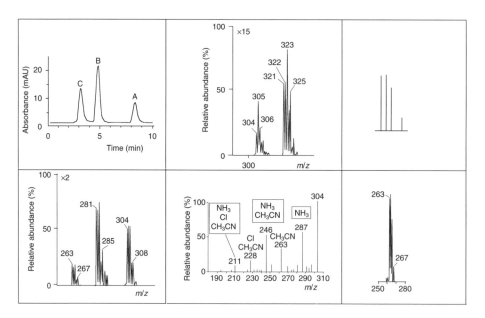

Figure 4.8
The HPLC electrospray MS and MS/MS analyses of cisplatin and its hydroxy metabolites. (*Top left*) The HPLC trace using a porous graphite column. (*Top center*) The ESI spectrum from HPLC peak A compound, the ion at m/z 321 corresponds to the sodium adduct of cisplatin. (*Top right*) Relative abundances of the platinum isotopes, without the contribution of chlorine atoms. (*Bottom left*) Spectrum corresponding to HPLC peak B; the m/z values observed are difficult to interpret. (*Bottom center*) Spectrum from peak B displaying the MS/MS fragments from m/z 304, clearly showing that the hydroxy group of monohydroxo-cisplatin has been replaced by an acetonitrile molecule. (*Bottom right*) Molecular mass region of the spectrum of peak C; the ion at m/z 263 corresponds to protonated dihydroxo-cisplatin. (Redrawn (modified) from ref. 16 with permission)

The spectrum from the compound of HPLC peak B (Figure 4.8) is not straightforward to interpret. However, MS/MS fragmentation of m/z 304 shows that acetonitrile (41 u) is incorporated and has replaced the hydroxyl group of monohydroxo-cisplatin. Indeed, the calculated mass $194 + 35 + (2 \times 17) + 41 = 304$ corresponds to the observed mass. The spectrum of HPLC peak C is easy to interpret. It corresponds to protonated dihydroxo-cisplatin.

$$
\begin{array}{ccc}
H_3N\diagdown\quad\diagup Cl & H_3N\diagdown\quad\diagup OH & H_3N\diagdown\quad\diagup OH \\
Pt & Pt & Pt \\
H_3N\diagup\quad\diagdown Cl & H_3N\diagup\quad\diagdown Cl & H_3N\diagup\quad\diagdown OH \\
\text{A: Cisplatin} & \text{B: Monohydroxo-} & \text{C: Dihydroxo-} \\
 & \text{cisplatin} & \text{cisplatin}
\end{array}
$$

4.1.3 Capillary electrophoresis/mass spectrometry (CE/MS)

Capillary electrophoresis (CE) is a family of related techniques that employ narrow-bore (20–200 µm internal diameter) capillaries to perform the separation of both large and small analytes. These separations are obtained by the use of high voltages that generate, respectively, electroosmotic and electrophoretic flow of buffer solutions and ionic species. The coupling of CE can be performed using ESI. The nanospray version yields the best results. Indeed, the flow produced by CE is typically in the range of some hundreds of nanoliters per minute. The main advantages of CE — speed, high resolution and small volumes of samples — are saved by the use of nano-ESI/MS. This separation technique allows analyses in aqueous solutions that are complementary to classical techniques such HPLC/MS. The scheme of such a coupling is represented in Figure 4.9.

The interfacing of capillary electrophoresis (CE) with electrospray mass spectrometry has progressed substantially in recent years. On-line CE/MS has been widely used for both qualitative and quantitative analysis of many chemically diverse

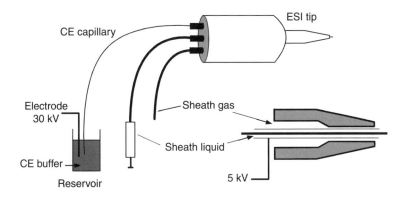

Figure 4.9
Scheme of a capillary zone electrophoresis coupling device to a mass spectrometer

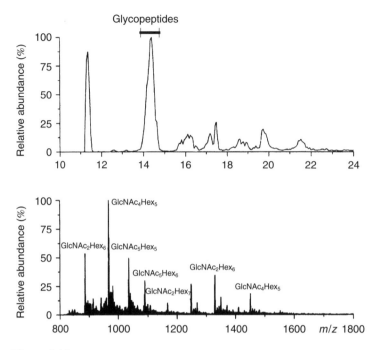

Figure 4.10
Chromatographic trace obtained by capillary electrophoresis and spectrum of the mixture of doubly or triply charged glycopeptides that elected together in the eluted peak. (Reproduced (modified) from ref. 22 with permission)

molecules. This method becomes a useful and sensitive analytical tool for the separation, quantitation and identification of biological, therapeutic, environmental and other important classes of chemical analytes. The developments of (CE/MS) have been reviewed elsewhere.[17-21]

Figure 4.10 displays the analysis of a mixture of peptides from ovalbumin obtained by CE/MS. The spectrum shown is a mean of the spectra acquired during elution of the indicated broad peak. It corresponds to a mixture of doubly or triply charged ions of several glycopeptides.[22] The displayed structures were deduced from MS/MS fragmentation spectra of these multiply charged ions.

4.2 Chromatography Data Acquisition Modes

No matter which ionization technique or chromatographic method is used, three acquisition modes exist: scanning, selected-ion monitoring (SIM; not to be mistaken with SIMS, which means secondary ion mass spectrometry) and selected-reaction monitoring (SRM).

In the scan mode, complete spectra are repeatedly measured between two extreme masses. Suppose that the time width of the chromatographic peak is 10 s at mid-height. As is shown schematically below, at least one spectrum must be measured every 5 s in order to ascertain that one of them lies entirely within the chromatographic peak. Suppose that the scan covers masses 50–550 Da at low resolution. A short calculation shows that 0.01 s per mass (500 in 5 s) can be used. All of the ions reaching the detector during that time span are counted. Increasing this time span increases the sensitivity, because the total number of counted ions increases. To increase the sensitivity, either the range of masses scanned is decreased or the scan time is increased. In the former case, analytical information may be lost, and in the latter case, a good mass spectrum may not be obtained or two eluted compounds in the same peak may not be deconvoluted (for deconvolution, see below).

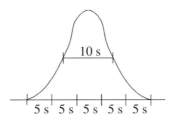

If the analysis aims at detecting target compounds of known spectral characteristics with maximum sensitivity, selected-ion monitoring (SIM) is useful. Thus, if we choose to detect a given compound by monitoring three characteristic fragments, we switch the analyzer rapidly from one mass to another. The sensitivity gain can be enormous. In fact, integrating over three masses during 5 s means that 1.66 instead of 0.01 s per mass is available.

The SRM acquisition mode allows one to obtain a sensitivity and selectivity gain with respect to SIM. The detection of selected reactions, based on the decomposition reactions of ions that are characteristic of the compounds to be analyzed, requires the use of a tandem mass spectrometric instrument. In order to carry out this type of analysis, the spectrometer is set to let through only those ions produced by a decomposition reaction in the chosen reaction region: for example, the first spectrometer selects the precursor ion with an m_p^+/z ratio that is characteristic of the compound to be detected, whereas the second spectrometer selects the fragment ion with an m_f^+/z ratio resulting from the characteristic decomposition reaction of the compound to be analyzed ($m_p^+ \rightarrow m_f^+ + m_n$) that occurs between the two analyzers.

The selectivity gain results from the fact that the fragmentation reaction implies two different characteristics of the compound under study. In fact, ions must meet two conditions in order to reach the detector: the m/z ratios of the product ions and the precursor ions must satisfy the selected values. The sensitivity gain results from the greater signal-to-noise ratio that is characteristic of MS/MS; the gain is high in spite of the weak ion transmission through the instrument.

Figure 4.11 gives an example of SRM. The problem is to detect a contaminant in a fuel sample. In the total ion reconstructed chromatogram, which corresponds to a standard GC trace, the compound cannot be seen at its retention time. If, however, selected fragmentations are monitored through detection of MS/MS product ions,

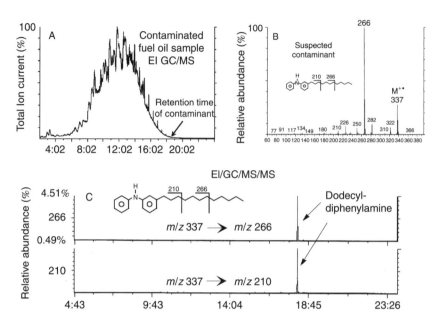

Figure 4.11
(A) Total ion chromatographic trace of a fuel oil. (B) Spectrum of the sought contaminant. (C) The SRM spectrum of the two main reactions: the sought compound is the only one detected. (Reproduced (modified) from Finnigan MAT documentation, with permission)

the only peak observed is that of the impurity, at the expected retention time. The instrument used here is a GC quadrupole ion trap (Finnigan MAT).

4.3 Data Recording and Treatment

One can no longer imagine performing GC/MS or HPLC/MS without an on-line data system. Such a system includes an acquisition processor, a magnetic recording device and a computer with its accessories.

4.3.1 Data recording

The spectrometer provides two series of data as a function of time: the number of detected ions and, simultaneously, a physical value indicating the mass of these ions. This can be, for example, the value of the magnetic field measured by a Hall probe in the case of a magnetic instrument. The ions of every mass appear with a certain distribution over a time period, as shown in the accompanying sketch. The surface under the curve is proportional to the number of detected ions, whereas the value of the magnetic field at the centroid of the peak is an indicator of the ion mass. This centroid and mass therefore must be determined. In order to achieve this, the acquisition processor accumulates the signal related to the number of ions during a short time with respect to the peak width. Usually, it samples eight times within the peak width.

As an illustration, consider this example. The spectrometer covers a range of 500 mass units within 1 s, i.e. one mass within 2 ms. During that time, the processor must carry out about eight measurements of the number of ions; 0.25 ms is allotted for every sample. In other words, it must measure 4000 samples per second. Its sampling frequency is said to be 4 kHz.

In order to achieve this, the current from the ion detector goes 4000 times a second through a resistance, and the acquisition processor reads the potential difference at the resistance ends, which is proportional to the number of detected ions, and digitizes it. This value goes to the y-axis of the mass spectrum. This signal is then associated with the simultaneous reading of the mass indicator, which yields the x-axis value. This gives a profile spectrum. To convert it into a bar graph, a suitable algorithm allows the processor to determine the limits of the peak and its centroid. The sum of the values read within these limits during the condenser discharges is proportional to the number of ions, whereas the interpolation of the indicator value at the centroid yields the ion mass. A bar graph spectrum is obtained and this is illustrated in Figure 4.12.

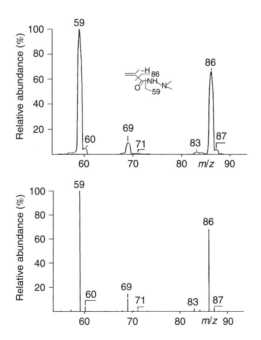

Figure 4.12
The same spectrum displayed on top as a profile of ion abundances and below as a bar graph of the centroids of the peaks

If a larger mass range is scanned per unit time, or if a higher resolution is used, the sampling speed must be increased because the number of data points per unit time increases. Good processors (1995) sample at a maximum rate of 250 kHz.

Another important characteristic of the acquisition processor is its dynamic range, which is linked in part to the signal digitization possibilities. An example illustrates this. An ion detector can typically detect between one and one million ions reaching the detector simultaneously. Its own dynamic range, i.e. the ratio of the largest to the smallest detectable signal, is thus equal to 10^6. If the analog-to-digital converter (ADC) uses 16 bits (a usual value) it can yield numerical values between 1 and $2^{16}-1$, i.e. $65\,535$. Its dynamic range is thus much lower than that of the detector itself. This problem can be overcome by alternately reading different value ranges.

4.3.2 Instrument control and treatment of results

With commercially available programs, the computer can carry out the following operations.

4.3.2.1 Data acquisition management

The acquisition program allows one to choose the scan mode or the SIM mode, the range of masses that are scanned, either low or high resolution, etc. The acquisition processor is set according to the data given by the operator concerning the analysis to be carried out. Most recent instruments are controlled entirely electronically from parameters provided by the operator. Instruments with automatic injectors thus can carry out several successive chromatographic analyses without any intervention by the operator.

Expert systems are available on recent instruments (from about 1990). They allow one to program the instrument tuning modifications or changes in the type of measurement to be carried out, based on a computer investigation of the data. For example, 'if an ion with a given m/z value is detected at a set retention time, then measure the spectrum in a negative ion mode and then come back to starting mode'. These possibilities are now revolutionizing mass spectrometry.

4.3.2.2 Interpretation of results

An interactive program, which lets the operator intervene and modify the parameters at any time, allows the investigation of the chromatographic results. Usually, it allows the following operations:

1. To show the reconstructed ion chromatogram (RIC) out of the sum of the intensities of all the detected ions. Some ions present in the background noise often can be excluded. The spectrum numbers corresponding to the chromatographic peaks are given on the x-axis. Figure 4.13 shows an example. Retention times may be indicated.

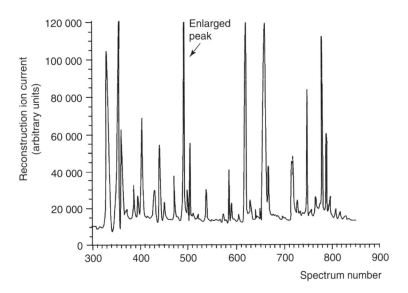

Figure 4.13
Chromatographic trace reconstructed by summing the intensities of all of
the detected masses in every spectrum, and used as arbitrary units along the
y-axis. The x-axis is the spectrum number. This is an analysis of volatile
acids as trimethylsilylated derivatives from the urine of a patient suffering
a metabolic disorder. If asked, the computer can indicate the retention time
of every peak. The peak that is pointed out is blown up in Figure 4.14

2. The operator can use magnifying effects on this chromatographic trace: he or she
can select part of the chromatographic trace and enlarge it over the whole screen, or
amplify the chromatographic trace vertically to emphasize the low-intensity peaks.
Figure 4.14 shows a blown-up version of the peak pointed out in the chromatogram
in Figure 4.13 and every spectrum recorded during the elution of that peak. Only
spectra 493 and 494 are of good quality.

3. To select one or several spectra to appear on the screen. Two spectra thus can
be compared: one measured at the beginning of elution of a chromatographic peak
and the other measured at the end. If they are identical, the probability that the peak
contains only one compound is high. However, if they are different, the peak contains
more than one compound. Figure 4.15 shows an example of this. The spectra indicate
the presence of two compounds with m/z 82 and 84. A spectrum also can be selected
within the chromatographic peak and another, representing the background noise, at
the peak edge. The second spectrum then is subtracted from the first.

4. To draw mass chromatograms. This consists of drawing a graph of the abundance
of ions of a selected m/z value, or of a chosen sum of ion abundances, as a function
of time. This process allows one to look first through a chromatograph for the
possible presence of known compounds by choosing the masses of typical ions.
Target compounds thus can be pointed out within a chromatographic envelope where

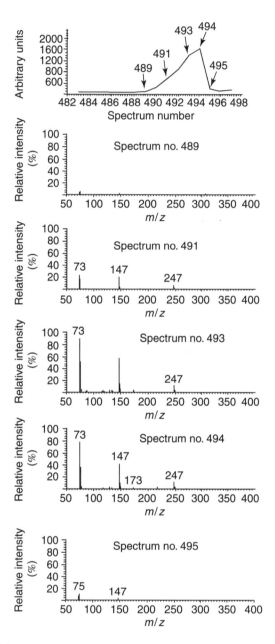

Figure 4.14
Enlarged picture of the chromatographic peak pointed out in Figure 4.13
and a few spectra measured during the elution time of that chromatographic
peak. The trimethylsilylated derivative of succinic acid can be recognized

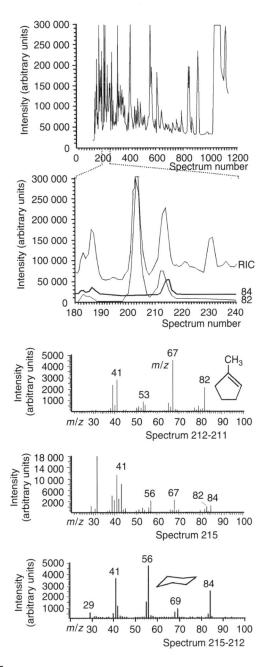

Figure 4.15
Deconvolution of a chromatographic peak containing two compounds.
Spectrum 212, where the background noise was subtracted, is different
from spectrum 215. The two spectra are obtained by the appropriate
subtraction

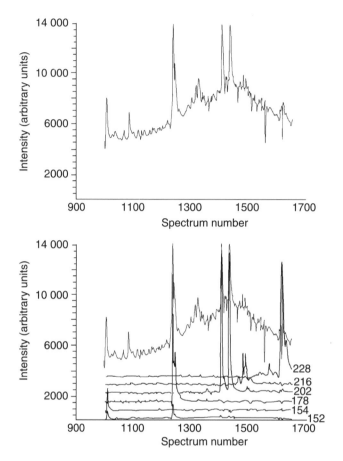

Figure 4.16

(*Top*) Chromatogram of compounds sampled over a discharge. The complexity of the mixture entails an envelope of the baseline. (*Bottom*) Selective search for polycyclic aromatics, which are important pollution agents. The following are detected: 228, benzanthracenes; 216, benzofluorenes; 202, fluoranthene and pyrene; 178, anthracene and phenanthrene. The next two are only weakly detected: acenaphthene and acenaphthylene, 154 and 152. These identifications must be confirmed by the retention times and by complete mass spectra

they cannot be identified solely from their retention times. Figure 4.16 shows an example.

The definitive identification will be the result of a combination of retention time, ion chromatogram and, as shown in Figure 4.17, the complete spectra.

Figure 4.18 shows the analysis of a hydrocarbon mixture obtained from irradiation-induced dimerization of a mixture of n-$C_{10}H_{22}$ and n-$C_{10}D_{22}$[23] as an example. The mixture is produced from the dehydrodimerization of hydrocarbons $C_{20}H_{42}$, $C_{20}H_{21}D_{21}$ and $C_{20}D_{42}$. The dimerization produces 15 eicosane isomers

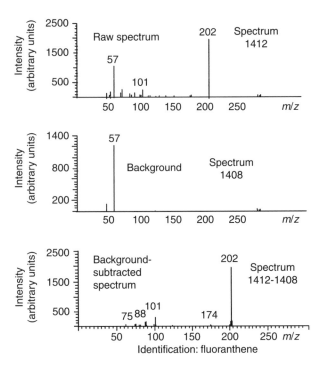

Figure 4.17
Same analysis as in Figure 4.16. Confirmation of fluoranthene identification by the full spectrum after background
subtraction

for every isotopic composition, i.e. 45 isomers in total, some of which also show
diastereoisomers. These isomers are observed through chemical ionization with
methane, which induces abstraction of a hydride or of D^-. Hydrocarbons $C_{20}H_{42}$
are observed at 281 Th as $(M - H)^+$, $C_{20}H_{21}D_{21}$ at 302 or 301 Th corresponding to
$(M - H)^+$ or $(M - D)^+$ and $C_{20}D_{42}$ at 322 Th as $(M - D)^+$. All of the 45 isomers
are observed in these ion chromatograms, whereas the flame ionization detection
(FID) trace, even though obtained at a very high resolution, shows only 20 peaks.

5. To detect target compounds that are not observable in the chromatographic trace.
Figure 4.19 illustrates this. According to the retention time, TMS-phenol (TMS =
trimethylsilyl) should appear in spectrum 342, but it is not visible on the trace
covering this region. The bottom trace in Figure 4.19 shows the intensity relative to
two ions typical of TMS-phenol: 151 (the most intense) and 166, the molecular ion.
Thus the phenol shows up! Its identification is confirmed by the spectra shown in
Figure 4.20.

Such uses of mass chromatograms end up increasing the dynamic range of chromatography. They can be used also to increase the effective resolution, as is shown
in the example in Figure 4.18.

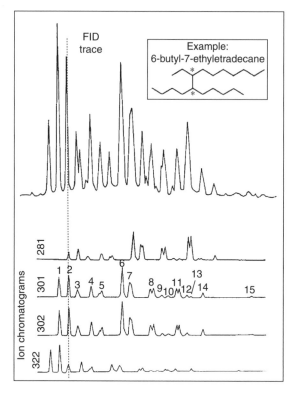

Figure 4.18
(*Top*) Gas chromatrophy trace with a flame ionization
detector. One of the isomers is shown as an example, it
has two asymmetric carbon atoms. (*Bottom*) Ion chromatograms corresponding to $(C_{20}H_{42}-H)^+$ at 281 Th,
$(C_{20}H_{21}D_{21} - D)^+$ at 301 Th or–H at 302 Th and
$(C_{20}D_{42} - D)^+$ at 322 Th. (Reproduced (modified)
from Ref. 23 with permission)

4.3.2.3 Other programs

The other programs that are usually available are described below.

Individual programs allow one to draw spectra with various formats, to carry out
comparisons of spectra or to draw complete or mass chromatograms in two or three
dimensions with various formats.

A spectrum subtraction program allows the subtraction of one spectrum from
another in order to eliminate the background noise or simply to emphasize differences
between two spectra. An example is given in Figure 4.20.

Library search programs allow one to carry out direct searches to determine
whether the spectrum that is obtained is present or not in the collection, or to carry
out inverse searches to determine if the spectrum of a given compound in a collection
occurs among those recorded during a chromatographic analysis, for example. An
example of a library search is given at the end of Chapter 2 (section 2.7.2.6).

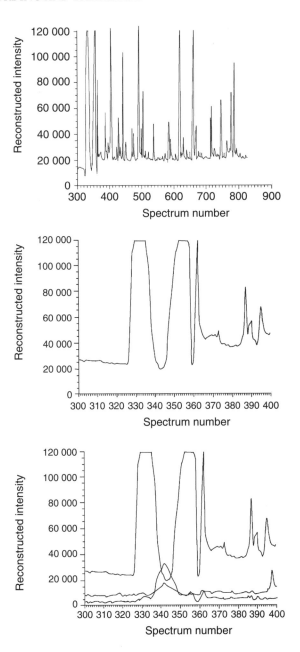

Figure 4.19
(*Top*) Complete chromatographic trace of volatile acids from the urine of a patient suffering from Wilson's disease. (*Middle*) Trace showing part of the chromatograph ranging from spectra 300 to 400. (*Bottom*) Traces showing the chromatograms of m/z 151 and 166 that are typical of TMS-phenol and thus indicate its presence

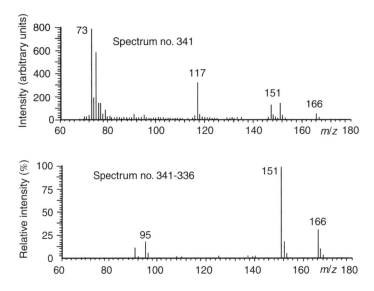

Figure 4.20
Proof of the presence of TMS-phenol that is invisible in the valley
separating the two major peaks. The second spectrum results from
subtraction of the background noise

The elemental composition for a given mass can be calculated by limiting oneself
to chemically acceptable formulae. Thus, for example, C_2H_5O and CHO_2 are acceptable for mass 45 Da but C_3H_9 is not. The calculation can be carried out with low or
high resolution. In the first case, the number of possible formulae rapidly becomes
too high, and is of interest only if limits can be ascribed to the number of atoms of
each species.

The isotopic abundances for a given formula can be calculated and compared with
experimental values.

Finally, utility programs allow one to extract some spectra from an analysis, to
erase others, etc.

4.4 References

1. Davis J.M. and Giddings J.C., *Anal. Chem.*, **55**, 418 (1983).
2. Hirschferd T., *Anal. Chem.*, **52**, 297A (1980).
3. Martin M., Herman D.P. and Guiochon G., *Anal. Chem.*, **58**, 2200 (1986).
4. Kitson F.G., Larsen B.S. and McEwen C.N., *Gas Chromatography and Mass Spectrom-
    etry: a Practical Guide*, Academic Press, New York, 1996.
5. McMaster M.C. and McMaster C., *GC/MS: a Practical User's Guide*, Wiley, New York,
    1998.
6. Ardrey B., *Liquid Chromatography/Mass Spectrometry*, VCH, New York, 1993.
7. Niessen W.M.A., *Liquid Chromatography–Mass Spectrometry* (Chromatographic Science),
    2nd edn, Marcel Dekker, New York, 1998.
8. Ardrey R.E., *LC–MS: an Introduction*, VCH, New York, 1999.

9. Arpino P.J., Baldwin M.A. and McLafferty F.W., *Biomed.Mass Spectrom.*, **1**, 80 (1974).

10. Huang E.C., Wachs T., Conboy J.J., *et al.*, *Anal. Chem.*, **62**, 713A (1990).

11. Greene F.T., in *Proce. 23rd ASMS Conference*, Houston, TX, 1975, p. 695.

12. Wiloughby R.C. and Browner R.F., *Anal. Chem.*, **56**, 2626 (1984).

13. Caprioli R.M., Fan T. and Cottrell J.S., *Anal. Chem.*, **58**, 2949 (1986).

14. Caprioli R.M., *Continuous Flow Fast Atom Bombardment Mass Spectrometry*, Wiley, New York, 1990.

15. Van Hoof F., Libert R., Hermans D., *et al.*, in *Adrenoleukodystrophy and Other Peroxysomal Disorders*, ed. by G. Uziel, R. Wanders and M. Cappa, Excerpta Medica, Amsterdam, 1990, p. 52.

16. Ehrsson H.C., Wallin I.B., Andersson A. S, *et al.*, *Anal. Chem.*, **67**, 3608 (1995).

17. Niessen W.M.A., Tjaden U.R. and VanDergreef, *J. Chromatogr.*, **636**(1), 3 (1993).

18. Issaq H.J., Janini G.M., Chan K.C., *et al.*, *Adv. Chromatogr.*, **35**, 101 (1995).

19. Tomlinson A.J., Guzman N.A. and Naylor S., *J. Capill. Electrophores.*, **2**(6), 247 (1995).

20. Figeys D. and Aebersold R., *Electrophoresis*, **19**(6), 885 (1998).

21. Banks J.F., *Electrophoresis*, **18**(12/13), 2255 (1997).

22. Kelly J.F., Ramaley L. and Thibault P., *Anal. Chem.*, **69**(1), 51 (1997).

23. de Hoffmann E., Baudson Th. and Tilquin B., *J. High Resolut. Chromatogr. Chromatogr. Commun.*, **10**, 153 (1987).

5

Analytical Information

In order to analyze the spectrum of an unknown compound, we must proceed by steps, rather like in the quiz game: "man or woman?", "young or old?", etc., asking increasingly precise questions, until the structure is deduced or until we give up.

First the origin of the sample, its history, is taken into account. This often allows the elimination of some hypotheses or narrowing of the research field. For example, a side product does not contain nitrogen if none of the reactants and none of the solvents contained it.

5.1 Mass Spectrometry Spectral Collections

Before the spectrum is examined and any interpretation efforts are started, it is strongly recommended to perform a computer or a manual library search to check whether the spectrum belongs to an existing collection. Identification of an unknown compound in this way depends directly on the quality and comprehensiveness of the collection used. However, only libraries of electron ionization spectra are efficient. Other ionization techniques yield spectra that are too much dependent on the instruments and experimental conditions.

If the spectrum is found, the research is over. However, we must take into account the fact that close isomers often have identical mass spectra, and that sometimes very similar mass spectra belong to different compounds. We cannot blindly trust identification issued from a comparison.

Three main spectra collections now exist. The first is the NIST/EPA/NIH mass spectral library, which contains 130 000 spectra of 107 000 compounds.[1] This original collection of spectra and related information is produced by the National Institute of Standards and Technology (NIST) with the assistance of expert advisors from the Environmental Protection Agency (EPA) and National Institutes of Health (NIH). It is published in the shape of complete 'bar graph' spectra and classified first according to the molecular mass and then according to the molecular formula.[2] This library also is available on CD-ROM, readable by personal computers with integrated tools for gas chromatography/mass spectrometry (GC/MS) deconvolution, mass spectrometry interpretation and chemical substructure identification. This American government publication is very cheap and of very high quality. Mass spectra for over 10 000 compounds is accessible on-line.[3]

The *Eight Peak Index of Mass Spectra* published by the Mass Spectrometry Data Center of the Royal Society of Chemistry is a popular printed index of mass spectral data that now contains some 81 000 spectra of over 65 000 different compounds.[4] These spectra are published in the shape of lists of the eight main peaks. The complete data are sorted in three different ways to allow easy identification of unknown compounds: by molecular weight subindexed on molecular formula; by molecular weight subindexed on m/z value; and by m/z value of the two most intense ions.

McLafferty and Stauffer published, in the *Wiley Registry of Mass Spectral Data*, a collection of about 230 000 spectra of over 200 000 compounds.[5] It is the largest and most comprehensive library of reference spectra. This library is available on CD-ROM, with software allowing users to create an integrated MS data acquisition and search system. Furthermore, it is accompanied by an interpretation help program called 'probability-based matching' (PBM). A very large compilation of mass spectra derived from the *Wiley Registry of Mass Spectral Data* is also available.[6] In this latter printed library, mass spectral peaks are listed according to abundance and statistical importance. It includes peaks that are the most diagnostic for each substance.

Smaller specialized libraries of mass spectrometry also exist: main polluting agents,[7] environmental contaminants,[8,9] drugs and metabolites,[10] pharmaceutical products,[11] etc.

If the spectrum is not described in a library, the information contained in the spectrum is used. The mass spectrum furnishes various types of analytical information. The first is the molecular mass. The ionization techniques that are accessible now have made this information available for almost all compounds.

The molecular mass measured by mass spectrometry can correspond to the average mass calculated using the average atomic weight of each element of the molecule, or to the monoisotopic mass calculated using the exact mass of the most abundant isotope for each element. The type of mass measured by mass spectrometry depends largely on the resolution of the analyzer. Indeed, if the instrument is unable to resolve the isotopes, the various peaks in the isotopic cluster combine and form a single peak that spreads over several masses. Thus, the mass determined by the instrument corresponds to the average mass. Sometimes it may be difficult to determine even the average mass with reasonable accuracy. On the other hand, if the resolution is high enough to distinguish the different peaks in the isotopic clusters, the mass determined by the instrument corresponds to the calculated monoisotopic mass. Furthermore, high-resolution mass spectrometry leads to very narrow peaks, allowing greater accuracy. Figure 5.1 shows an example of how resolution affects the observed mass.

The second type of analytical information is the elemental composition. Indeed, measurement of the mass with sufficient accuracy provides an unequivocal determination of the total elemental composition. This method often is referred to as 'high resolution'.

However, this information is not always available. Other elements of the spectrum provide more information, even if it is only fragmentary, on the elemental composition: isotopic abundance, mass of small fragments or of lost neutrals.

Once the crude formula has been established, or is partially known, we try to define the structure from the fragmentation. This is the subject of the following sections.

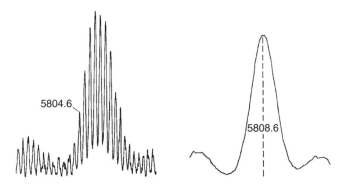

Figure 5.1
A fast atom bombardment spectrum of human insulin isotopic
cluster ($C_{257}H_{383}N_{65}O_{77}S_6$) at 6000 resolution (*left*) and at 500
resolution (*right*)

5.2 High Resolution

High-resolution instruments, including double-focusing and fourier transform ion
cydotron resonance (FTICR) mass spectrometers, are capable in principle of
measuring the mass of an ion with sufficient accuracy to allow the determination
of its elemental composition. This is possible because each element has a slightly
different characteristic mass defect, so the accurate mass measurement that shows the
total mass defect provides an identification of its elemental composition. However,
the accuracy requirements in measuring mass and thus the resolution requirements
increase very rapidly as the mass increases. Indeed, as the number of atoms increases,
the number of possible combinations also increases, with mass differences that
become increasingly small. For example, $C_{24}H_{19}N$ and $C_{21}H_{23}NS$, with masses of
321.1517 and 321.1551 u, respectively, require an accuracy of 11 ppm for their
distinction and a resolution $m/\delta m$ of 94 400 for their separation. Only four or
five models of commercial electromagnetic instruments allow such resolutions to
be reached. The FTICR instruments allow much higher resolutions to be obtained.
However, with measurement conditions that require rapid scans, such as with the
coupling of a gas or liquid chromatograph, a resolution of 10 000 is hardly achieved
by the best instruments.

Another example is illustrated in Figure 5.2. All the possible compositions encom-
passing C_{6-15}, H_{0-24}, N_{0-4} and O_{0-4} are considered at the mass 180. There are no
compositions with exact masses closer together than 0.0012 u. Thus, for a measurement at m/z 180, an accuracy of 6 ppm would be required to eliminate all the
possibilities and thus to define a particular elemental composition within the fixed
limits for the elements.

When the accurate mass measurement is used for formula confirmation or for
elemental composition analysis, all the candidates fitting the experimentally determined value and its reportable uncertainty must be considered. Setting fixed acceptable error limits for mass measurement is not correct.

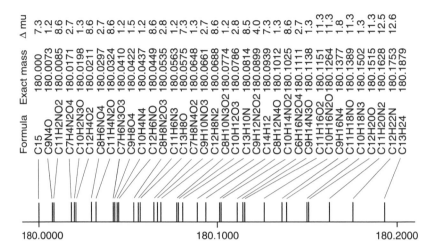

Figure 5.2
Exact masses and corresponding formulae for various possible ions of m/z 180
containing only carbon, hydrogen, nitrogen and oxygen atoms

The elemental compositions of fragments with masses lower than 100 u are easy to obtain and much information concerning the composition of the molecular ion can be deduced. Moreover, knowing the elemental composition of the fragments is of valuable help in elucidating the structure and understanding the fragmentation mechanism.

High resolution also allows an increase in the selectivity of detection in analyses aiming at screening known target compounds.

5.3 Isotopic Abundance

Most of the elements appear in nature as isotope mixtures. Thus, natural carbon is a mixture of 98.90% isotope ^{12}C and 1.10% ^{13}C isotope. Table 5.1 lists the abundances of a few elements that are important in organic chemistry. In Appendix 4, the masses and the isotopic abundances are given for all the isotopes except the rare earths.

These isotopes are responsible for the peaks in the mass spectrum appearing as isotopic clusters that are characteristic of the elemental composition. They provide important analytical data. Indeed, even without exact mass measurement, the possibilities for elemental composition determination often can be restricted by using isotopic abundance data. For example, the fragments $C_{10}H_{20}$ and $C_8H_{12}O_2$, both with a nominal mass of 140 u, produce peaks at mass 141 with 11% and 8.8%, respectively, of the abundance at mass 140 u. This is the result of a different statistical probability of having ^{13}C isotopes. These two elemental compositions thus can be distinguished in a mass analysis.

However, the actual possibilities are limited. Measurement artifacts can modify such percentages. In the case of fragments of heavier molecules, the observed intensity can result from several fragments with different compositions all present in the isotopic cluster, such as $C_{10}H_{21}$ at mass 141 in the example above.

Table 5.1   Isotopic abundance

Isotope	Relative abundance (%)	Mass (u)	Atomic mass	
			Calculated	Measured
^{1}H	99.985	1.007825	1.007976	1.00794
^{2}H	0.015	2.0140		
^{12}C	98.90	12.000000	12.011036	12.0111
^{13}C	1.10	13.003355		
^{14}N	99.63	14.003074	14.006762	14.00674
^{15}N	0.37	15.000108		
^{16}O	99.76	15.994915	15.999324	15.99943
^{17}O	0.04	16.999131		
^{18}O	0.20	17.999160		
^{19}F	100	18.998403	18.998403	18.9984
^{23}Na	100	22.989767	22.989767	22.98976
^{31}P	100	30.973762	30.973762	30.97376
^{32}S	95.02	31.972070	32.064385	32.0666
^{33}S	0.75	32.971456		
^{34}S	4.21	33.967866		
^{36}S	0.02	35.967080		
^{35}Cl	75.77	34.968852	35.452737	35.45279
^{37}Cl	24.23	36.965903		
^{39}K	93.2581	38.963707	39.098299	39.09831
^{40}K	0.0117	39.963999		
^{41}K	6.7302	40.961825		
^{79}Br	50.69	78.918336	79.903526	79.9041
^{81}Br	49.31	80.916289		

In principle, this risk does not exist for the molecular peak. In fact, a contribution of a peak with composition $(M + H)^+$ associated with a molecular $M^{\bullet+}$ peak should never be excluded. This artifact can have high abundance in the case of compounds such as amides.

However, some atoms have isotopic compositions that are more obvious. This is the case for chlorine, bromine, selenium and sulfur, for example (Figure 5.3).

The relative abundances of isotopes in a molecule or in a fragment result from their statistical distribution. Let us consider an example in order to explain the calculation of isotopic clusters. What are the relative intensities of the isotopic peaks accompanying the molecular peak of CS_2? Sulfur shows up as a mixture of three main isotopes with nominal masses 32, 33 and 34. These three isotopes occupy two possible positions in the CS_2 molecule. The total number of possible combinations is 3^2, i.e. 9:

^{32}S ^{32}S	Total mass: 64, one combination
^{32}S ^{33}S or ^{33}S ^{32}S	Total mass: 65, two combinations
^{32}S ^{34}S or ^{34}S ^{32}S	Total mass: 66, two combinations
^{33}S ^{33}S	Total mass: 66, one combination
^{33}S ^{34}S or ^{34}S ^{33}S	Total mass: 67, two combinations
^{34}S ^{34}S	Total mass: 68, one combination
	Total: nine combinations

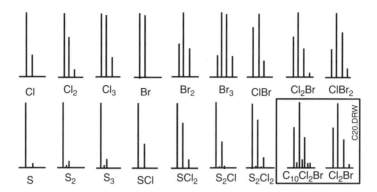

Figure 5.3
Useful isotope combinations in mass spectrometry. Isotopes of other
atoms that are possibly associated must always be taken into account,
as is shown in the framed section

Consider now the probability of observing each mass. Mass 64 results from the single
possible combination of two isotope-32 atoms. This isotope amounts to 95.02% in
nature, so the probability of having two such atoms simultaneously is $(0.9503)^2$,
i.e. 90.31%.

Mass 65 results from two possible combinations of isotopes 32 and 33, that
amount to 95.03 and 0.75%, respectively. Each combination has a probability of
occurring that is equal to $(0.9503) \times (0.0075) = 0.72\%$, which must be multiplied
by the number of combinations: $0.72 \times 2 = 1.42\%$.

Mass 66 can occur in either of two ways: either two isotope-33 atoms or one
isotope 32 and one isotope 34. The $^{33}S_2$ fragment has a probability of $(0.0075)^2 =$
0.00562% in one combination, whereas $^{32}S^{34}S$ has a probability of $(0.9503) \times$
$(0.0422) = 4.01\%$, which must be multiplied by two for the number of possible
combinations, i.e. 8.02%. In total, the probability of observing mass 66 is equal to
$8.02 + 0.00562 = 8.026\%$.

Mass 67 has a probability equal to $(2 \times 0.0075) \times (0.0422) = 0.063\%$, and mass
68 to $(0.0422)^2 = 0.1781\%$. See Table 5.2 for a summary.

We now have the possible combinations of two sulfur atoms, and the relative abundances that are calculated are those that would be observed in the case of a fragment
with composition S_2. We must now combine each of these with the carbon isotopes.
Because there is only one carbon atom in CS_2, we have 98.90% isotope 12 and 1.10%

Table 5.2

Mass	%	% of the predominant peak
64	90.31	100
65	1.42	1.5724
66	8.026	8.889
67	0.063	0.06975
68	0.1781	0.1972

isotope 13. We must now look for all the possible combinations of both of the following isotope series, with their probability, while observing that the total mass is the sum of the masses and the probability is the product of the probabilities, (see Table 5.3). Finally, we sum up the probabilities of all the ions with equal masses, (see Table 5.4):

Gathering the identical masses, we obtain the abundances of the CS_2 peaks, (see Table 5.5):

If the molecule contains another element, we should combine anew its masses and abundances.

Note that this calculation does not take sulfur isotope of mass 36 into account, it is 0.017% abundant and affects the result for the carbon sulfide of mass 80.

Table 5.3

S_2		C	
Mass	%	Mass	%
64	90.31	12	98.90
65	1.42	13	1.10
66	8.026		
67	0.063		
68	0.1781		

Table 5.4

Mass S_2	Mass C	Total mass	Probability S_2	Probability C	Total probability $((S_2 \times C)/100)$
64	12	76	90.31	98.90	89.317
64	13	77	90.31	1.10	0.99341
65	12	77	1.42	98.90	1.404
65	13	78	1.42	1.10	0.0156
66	12	78	8.026	98.90	7.938
66	13	79	8.026	1.10	0.088
67	12	79	0.063	98.90	0.062
67	13	80	0.063	1.10	0.00069
68	12	80	0.1781	98.90	0.176
68	13	81	0.1781	1.10	0.002
				Total probability (%)	99.9966

Table 5.5

Mass	Total %	% of the predominant peak
76	89.31	100
77	2.3974	2.684
78	7.9536	8.9056
79	0.152	0.17
80	0.177	0.198
81	0.0019	0.002

Mathematical methods for the calculation of theoretical relative abundances within the isotopic cluster, for comparison with experiment, usually rely on expansion of the polynomial expression based on an extension of the binomial probability distribution.[12,13] Indeed, for an element with x isotopes and relative abundances $I_1, I_2, \ldots, I_x$, and for n atoms of this element in the molecule, the problem consists of calculating, for every element, the terms of the following polynomial expression:

$$(I_1 + I_2 + \cdots + I_x)^n$$

The coefficient of the jth term then corresponds to the number of possible combinations that match a given distribution; the term itself corresponds to the probability of this combination. When considering several different types of atoms, we calculate the products for each term in these polynomials. We then match the right mass to each of these combinations, and we gather the identical masses.

Figure 5.4 shows an example of a calculation based on a method developed by Hsu and carried out on a personal computer. Some isotope distribution calculations are also available on-line.[14,15]

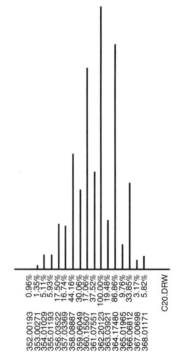

Figure 5.4
Isotopic cluster of $C_9H_{16}Se_3$ molecular ion, calculated by Hsu's program. The peaks below 1% are omitted. The numbers indicate the exact mass and the relative intensity (%)

Another method has been developed for calculating isotope distributions starting from a molecular formula and elemental isotopic abundances.[16,17] This method uses Fourier transforms to do the multiple convolutions required to determine molecular isotope distributions and calculates ultrahigh-resolution distributions over a limited mass range. Because discrete Fourier transforms can be calculated very efficiently, this new way of looking at the problem has significant practical implications. Specifically, this method for calculating isotope distributions is very fast, accurate and economical in its use of computer memory and thus can be applied to extremely large molecules.

Even more information from isotopic abundances can be obtained by tandem mass spectrometry (MS/MS), as is shown in the following example: the molecule shown in Figure 5.5 contains two sulfur atoms and its molecular peak at 340 Th is thus accompanied by an isotopic peak at 342 Th. Its relative intensity is $2 \times 4.21 = 8.42\%$ if the participation of other element isotopes is neglected. The spectra of the ions derived from fragmentation of the molecular ion and the isotopic ion (M + 2) are shown. The (M + 2) ion contains one ^{34}S and one ^{32}S sulfur atom. If both sulfur atoms are kept in a fragment, then this fragment also contains one ^{34}S and one

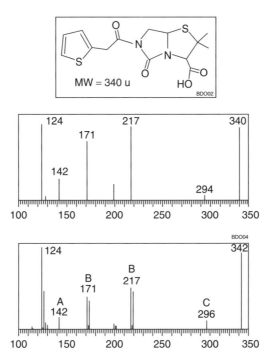

Figure 5.5
Product ion spectra obtained from fragmentation of the molecular ion and the corresponding (M + 2) isotopic precursor shown above: (A) unshifted ion, thus sulfurless; (B) dedoubled ions, containing one of the two sulfur atoms; (C) ions shifted by two mass units, containing both sulfur atoms

^{32}S sulfur atom. It is shifted by two mass units. If it contains only one of the two sulfur atoms, the probability that the atom being retained is ^{34}S is the same as the probability that it is ^{32}S; the peak is therefore dedoubled. Finally, if the fragment does not contain sulfur atoms, it shows up at the same mass in both spectra. Such an approach thus provides information on the elemental composition.

5.4 Low-mass Fragments and Neutrals

The molecular ion fragments and produces ions and neutrals that are not observed in the spectrum. The mass of the neutral product can be deduced from the difference between the mass of the parent ion and that of the observed ionic fragment.

One can deduce much information concerning the elemental composition from the masses of the neutrals or of the low-mass fragments.

In the case of fragments containing only carbon and hydrogen, masses range from 36 (C_3) to 43 (C_3H_7) u for three carbon atoms. For two atoms the limits are 24 and 29 u, for four atoms the limits are 48 and 57 u and for five the minimum is 60 u, which of course is very unlikely! We thus observe that ions or neutrals ranging from 30 to 35 u, for example, must necessarily contain atoms other than carbon and hydrogen.

Some masses can have only one reasonable formula, such as 20 u, which corresponds only to HF.

In the case of lost neutral fragments, it is important to take into account the fact that a fragment can result from several successive steps starting from the molecular ion. For example, the spectra of bile salts often contain an ion with mass $(M - 20)$ u, although they contain no fluorine. This fragment is actually the result of successive losses: $(M - H_2O - H_2)^+$.

This information can be combined with that furnished by the isotopic abundances. Mass 47 u may correspond to the crude formula CH_3O_2 or CH_3S. In the latter case, however, this ion must be accompanied by another with mass 49 u, amounting to 4.2% of its intensity and caused by ^{34}S.

All mass spectrometric data systems allow the calculation of possible compositions of fragments with given masses while limiting the possibilities to the elements contained in the molecular formula.

5.5 Number of Rings or Unsaturations

When the elemental analysis of a molecule or of a fragment is known, the number of rings or unsaturations can be calculated. Even though this does not hold only for mass spectrometry, let us recall this principle.

An aliphatic hydrocarbon has a formula C_nH_{2n+2}. Each ring or unsaturation that is present in a hydrocarbon decreases the number of hydrogen atoms by two units. Let x be the number of hydrogen atoms observed; then, if N_i is the number of rings and unsaturations, we have

$$x = 2n + 2 - 2N_i, \text{ thus } N_i = \frac{2n + 2 - x}{2}$$

For example, benzene has the formula C_6H_6 so

$$N_i = \frac{[(2 \times 6 + 2) - 6]}{2} = 4$$

Benzene indeed contains one ring and three unsaturations.

The presence of an oxygen or a sulfur atom in the molecule does not modify these numbers, as we can see when comparing CH_4 with CH_3OH and CH_3SH, or CH_3CH_3 with CH_3OCH_3 and CH_3SCH_3.

Each halogen atom 'replaces' a hydrogen atom and decreases the number of hydrogen atoms by an equal number, as is shown by CH_4 and CH_2Cl_2.

Each nitrogen or phosphorus atom increases the number of hydrogen atoms of a compound by one unit: compare CH_4 with CH_3NH_2 and CH_3PH_2 as an example.

Let n be the number of carbon atoms and nX and nN be those of halogens and nitrogen (or phosphorus), respectively. The following rule can be deduced, where N_i is the number of rings or unsaturations and x is the number of hydrogen atoms that are found experimentally:

$$N_i = \frac{(2n + 2 + nN - nX) - x}{2}$$

In mass spectrometry, this equation can be applied strictly only to the molecular ion. Protonation, deprotonation, fragmentation and other modifications change the rule. For instance, a fragment ion often results from a bond cleavage, so that the theoretical number of hydrogen atoms in a saturated fragment is decreased by one unit. Thus, for example, the saturated ion CH_3^+ contains a hydrogen atom less than the molecule CH_4. This is easily understood because the equation above leads to a half-whole value for N_i. We then only need to subtract half of that value in order to obtain the correct number.

5.6 Mass and Electron Parities, Closed-shell Ions and Open-shell Ions

5.6.1 Electron parity

Common molecules have an even number of electrons. Stable radicals are rare exceptions, such as NO. In classical chemistry, we most often meet active species that are ions with an even number of electrons, or radicals (uncharged species with an odd number of electrons). In mass spectrometry, we observe ions with an even number of electrons, but we also often meet radical ions, a species uncommon in solution chemistry and having specific characteristics.

5.6.2 Mass parity

The atomic masses used in common chemical calculations are based on averages resulting from mixtures of isotopes. In mass spectrometry, the calculation is based on the mass of the predominant isotope of each element. Because the isotopes are separated in the spectrometer, we always face several peaks with different masses and with intensity ratios defined as described earlier. Thus, for example, dichloromethane has a classical molecular mass equal to $12.01 + 2 \times 1.00 + 2 \times 35.45 = 84.91$ Da. The molecular mass in mass spectrometry (if mass defects are neglected) is $12 + 2 + 2 \times 35 = 84$ u. Several isotopic peaks are observed in the spectrum, the second most important being observed at m/z 86 with an intensity equal to 64.8% of that of the m/z 84 peak.

Organic molecules are normally made up of atoms of C, H, N, O, S, P and halogens, and we limit the following discussion to these elements. Molecular masses that are considered here are calculated by using the value of the atomic mass of the predominant isotope of each element, as is usual in mass spectrometry.

The nitrogen rule requires that the molecular mass is always even when the number of nitrogen atoms is even or zero. This results from the fact that nitrogen has a different mass parity and valence electron parity: mass 14 u, five peripheral electrons. Both of these parities are identical in the case of any other atom. It should be noted that this holds only if we consider the mass of the predominant isotope. Thus, the 'chemical' mass of bromine is 80 u, which is an even number, but its predominant isotope is that of mass 79 u, which is an odd mass. In the same way, isotopically labeled compounds do not always obey this rule.

Another approach to the problem consists of saying that adding one nitrogen atom in a molecular formula entails adding also one hydrogen to the 'saturated' formula. However, nitrogen has an even mass, and the number of new atoms with a unit mass is one. Thus, for example, CH_3CH_3 has a molecular mass of 30 u and CH_3NHCH_3 has a mass of 45 u, which is an odd mass.

5.6.3 Relationship between mass and electron parity

First, consider a molecule that contains no nitrogen and is ionized through electron ionization. The ionization process consists of expelling one electron in order to produce a radical cation:

$$M + e^- \longrightarrow M^{\bullet +} + 2e^-$$

At the beginning the molecule M has an even mass because, hypothetically, it contains no nitrogen and also has an even number of electrons. After ionization, the mass has not changed but the number of electrons has decreased by one unit, and has become odd. We have obtained a radical cation, which is represented by a dot combined with a plus sign: $M^{\bullet +}$.

If this molecule M is ionized through chemical ionization, we obtain a pseudo-molecular ion, $(M + H)^+$. The *pseudomolecular* ion has an odd mass: the mass, which is even, is increased by one because of the new proton. The number of electrons in the ion is the same as that of the neutral molecule, i.e. even. It is a normal cation, not a radical cation. The same deductions apply to negative ions.

An odd number of nitrogen atoms brings about an odd molecular mass (in Da) such as is defined in mass spectrometry: NH_3, 17; CH_3NH_2, 31; etc. Thus, in the case of an odd number of nitrogens, the earlier rule must be inverted: for the ion, the mass parity is the same as the electron parity.

From an analytical point of view, we see that an odd molecular mass, based on the predominant isotopes, indicates the presence of an odd number of nitrogen atoms.

Figure 5.6 shows the spectrum of butanol brought down to its five most intense ions.

The following rule is confirmed:

> When there is no or an even number of nitrogen atoms, any ion with an even mass has an odd number of electrons and is a radical cation or radical anion. Any ion with an odd mass, in turn, has an even number of electrons and is a cation or an anion. The reverse holds true for an odd number of nitrogen atoms. There is no exception within the C, H, N, O, S, P, alkali metal and halogen elements.

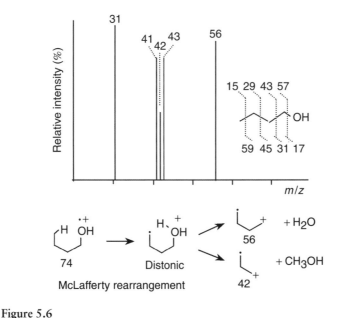

Figure 5.6
Mass spectrum of butanol limited to its five most abundant fragments. As is shown in the scheme, the ions resulting from the direct cleavage of a single bond in the molecular ion lead to odd mass fragments. These ions have a vacant valence and are thus cations with an even number of electrons, such as $CH_3{}^+$, $CH_3CH_2{}^+$, and the lost neutral is a radical: there is always a vacant valence. However, a rearrangement followed by the cleavage of a bond leads to a neutral molecule and to a radical cation, and thus to an odd number of electrons, but an even mass. Ions resulting from the rearrangement have a charge site separated from the radical site and are termed 'distonic ions'

5.7 Quantitative Data

The goal of quantitative analysis in mass spectrometry consists of correlating the intensity of the signals with the quantity of the compound present in the sample. Several quantitative analysis methods using mass spectrometry have been developed and many applications using these methods have been described.[18]

5.7.1 Specificity

The degree of specificity of quantitative analysis using mass spectrometry depends on how the spectrometer is used and even more so on the signal that is used during the correlation. For example, we can use the ion current as a signal to determine the concentration of the compound that is studied as long as there is no interference with other substances (in other words, as long as the substance being analyzed is pure or separated beforehand using chromatography).

Suppose that we know exactly the structure and the mass spectrum of the substance whose concentration we want to determine, we can use as a signal one or several ions specific to this substance. It is important to choose these ions among those with high masses (higher than 200 Th if possible). In fact, the significance of a signal depends on its ionic abundance and on its mass. It is logical to expect intense peaks corresponding to high masses to be more typical of the molecule than any low-mass ion.

An in-depth study of mass spectra of over 29 000 different compounds containing only the elements C, H, N, O, F, Si, P, S, Cl, Br and I showed that, on average, the probability of the presence of a peak in a spectrum halves every 100 Th.[19] Thus, the probability of having in a mixture compounds with ions typical of the same m/z decreases as the mass increases.

Another specificity criterion lies in the link between the intensity ratio of these characteristic ions and that of their respective abundance in the mass spectrum of the pure substance. In fact, the probability of finding in a mixture two compounds with several identical characteristic masses in the portion of the spectrum above 100 Th becomes almost zero (unless these two compounds are isomers). Several characteristic ions are necessarily used when the quantitative analysis of complex biological mixtures containing many molecules from the same family is undertaken.

Many methods exist that improve the specificity of quantitative analysis using mass spectrometry. These methods can be classified into either of two categories: those that act upon the sample and those that act on the spectrometer.

The first type of method that increases the specificity is based on a simple preliminary purification — an extraction into a solvent that is less polar, an acid–base separation, etc. — in order to eliminate every possible interference. Another possibility is to use different characteristic ions with higher masses after having formed a higher mass derivative from the compound. A further approach consists of making the spectrometer more selective, e.g. by increasing the resolution, by resorting to a different ionization technique or by applying another data acquisition mode such as selected-reaction monitoring (SRM), which was discussed in Chapters 3 and 4.

All of these techniques can, of course, be used simultaneously with chromato-graphic separation techniques. Methods that allow an improvement in the mass spec-trometer specificity thus can be worthwhile alternatives to long procedures for sample preparation. However, quantitative analysis carried out directly on neat samples may undergo a matrix effect characterized by a variation in the sample response because of the effect of the matrix on the abundance of ions within the source.

5.7.2 Sensitivity and detection limit

Sensitivity in mass spectrometry is defined as the ratio of the ionic current change to the sample change in the source. The recommended unit is Cb μg^{-1}. It is impor-tant always that the relevant experimental conditions corresponding to sensitivity measurement should be stated.

The detection limit should be differentiated from sensitivity. It is the smallest sample quantity that yields a signal that can be distinguished from the background noise (generally a signal equal to ten times the background noise). It should be noted that this minimum quantity is not enough to obtain an interpretable mass spectrum. Figure 5.7 shows an example detection limit obtained using mass spectrometry when measuring dihydrocholesterol. In these experimental conditions, the detection limit is lower than 1 pg but 1 ng of the compound is necessary to obtain an interpretable mass spectrum.

The limit of detection depends considerably on the abundance of the ionic species that is measured. The more abundant the measured ionic species with respect to all of the ions derived from the analyzed molecule, the higher is the limit of detection, as shown in Figure 5.7. The goal is thus to produce a signal that is as intense as possible. Several methods allow this goal to be reached, such as modification of the ionization conditions, reversal into the negative mode, the use of other softer ionization techniques or the derivatization of the sample, in order to increase the number of ions produced in the source or to reduce their fragmentation.

Another factor influencing the sensitivity corresponds to the time length of the signal integration. Of course, the longer that time, the more intense is the signal. The data acquisition mode thus influences the sensitivity by this factor. Three acquisition modes exist: scan mode, selected-ion monitoring (SIM) mode and selected-reaction monitoring (SRM) mode. These three modes are cited in ascending order of their effect on the sensitivity.

As a reminder, the scan consists of measuring complete spectra between two limit masses several times. The detection of selected ions consists of tuning the analyzer to focus onto the detector only ions of a specific m/z ratio. If several ions with different directly m/z ratios have to be detected, the analyzer goes rapidly from one mass to another. This method is clearly more sensitive than the scan mode because the time length of the signal integration is greater and the increase in the resulting signal-to-noise ratio allows a sensitivity gain of up to a 1000-fold.

The detection of selected reactions is a technique that requires MS/MS, which makes SRM even more sensitive and more selective than SIM. In order to carry out this type of acquisition, the instrument is tuned to transmit only those ions

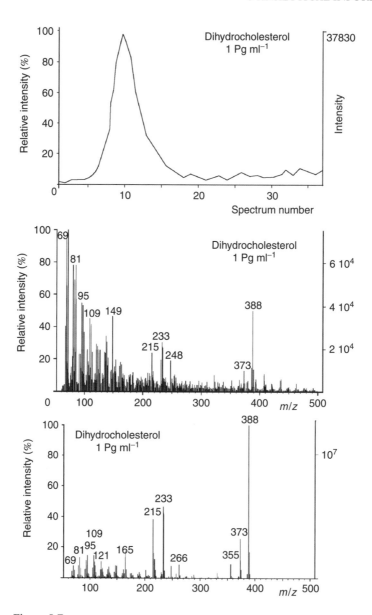

Figure 5.7
Detection limit obtained in mass spectrometry in the case of dihydrocholesterol

derived from a fragmentation reaction in the chosen reaction region. The sensitivity
gain obtained by using this method compared with SIM is due to the increase in the
signal-to-noise ratio that is characteristic of MS/MS. Figure 5.8 shows the sensitivity
and selectivity gain of SRM compared with SIM.[20]

The increased sensitivity of the SIM and SRM data acquisition results in lower
flexibility. The full-scan mode furnishes complete data so that new structural

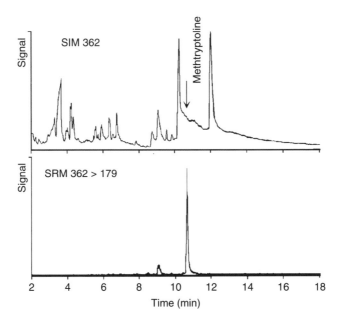

Figure 5.8
Detection of the heptafluorobutyryl derivative of methtryp-
toline using gas chromatography (*Top*) SIM of the 362 Th
negative ion. (*Bottom*) SRM of the 362 → 179 Th fragmentation. (Reproduced (modified) from Ref. 20 with permission)

information can be extracted in order to answer questions that may arise after the
initial experiment. This is not possible for data acquired in the SIM or SRM modes.

5.7.3 External standard method

The role of a standard is to determine the mathematical relationship, in the concentration range to be measured, between the selected signal intensities and the mixture
composition. This external standard method consists of preparing a synthetic sample
containing a known quantity of the molecule to be measured (M_{ste}) and then introducing a precise volume of this solution into the spectrometer and recording the
intensity of the response signal (I_{ste}).

Then, without any modification of the analytical conditions, an equal volume of
the solution containing the molecule to be quantified (M_x) is introduced into the
spectrometer and the intensity of its response signal (I_x) is measured. Because the
volumes that are introduced are equal, there is a proportionality between the response
intensities and the quantities as long as the response signal intensity remains linear
with respect to the concentration and as long as the signal intensity is zero at zero
concentration:

$$M_x = I_x \times \frac{M_{ste}}{I_{ste}}$$

or $M_x = I_x \times RF_x$ if the response factor is defined as

$$RF_x = \frac{M_{ste}}{I_{ste}}$$

In electron ionization, the response is normally linear with respect to the concentration over a wide range, often six orders of magnitude. This is not true for the other ionization techniques because of the influence that the sample quantity can have on the number of ions produced and on the fragmentation, and thus on the production yield of the various ionic species. For instance, in chemical ionization mode the formation of adducts appearing at higher sample pressures changes the relative intensities in the spectrum.

Thus, verification using a calibration curve is necessary. This calibration curve allows one to calculate the quantities of a compound to be measured in the unknown samples, to confirm the method specificity and also to define its sensitivity. In order to do this, equal volumes of a series of synthetic samples containing an increasing quantity of the molecule to be measured are introduced into the mass spectrometer and the intensities of their response signals are recorded.

This allows the determination of the mathematical relationship, in the concentration range to be measured, between the selected signal intensity (I_x) and the quantity of the molecule to be measured (M_x) that is present in the mixture. Ideally this relationship should correspond to the equation of a straight line with a slope equal to unity in order to ensure maximum precision. Figure 5.9 shows an example of quantitative analysis of phenobarbital within a mixture using GC/MS.

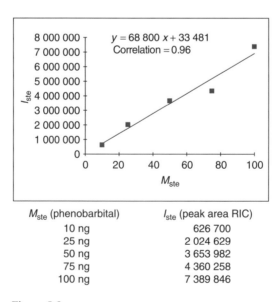

M_{ste} (phenobarbital)	I_{ste} (peak area RIC)
10 ng	626 700
25 ng	2 024 629
50 ng	3 653 982
75 ng	4 360 258
100 ng	7 389 846

Figure 5.9
Calibration curve obtained by external standardization of phenobarbital present within a mixture, in order to quantify it

5.7.4 Sources of error

Experimental procedures for quantitative mass spectrometric analysis usually involve several steps. The final error results from accumulation of the errors in each step, some steps in the procedure being higher error sources than others. A separation can be made between the errors ascribable to the spectrometer and its data treatment and the errors resulting from sample handling.

Normally, the sample handling errors are higher than those due to the mass spectrometer and thus make up the greater part of the final error. Handling errors can be numerous, such as the error in measuring a sample volume and the error in the introduction into the spectrometer.

The errors due to the mass spectrometer are also numerous, such as the variation in source conditions and instability of the mass scale. For statistical reasons, every measurement of a signal intensity carries a minimal intrinsic error. This error is inversely proportional to the square root of the number of ions detected for that signal. In order to optimize the reproducibility or the precision of the measurements, a maximum number of ions must be detected for every ionic species.

All of these error sources, other than the minimal intrinsic error, can be reduced by using the internal standard method. The absolute measurement of the signal is replaced by the measurement of the signal ratio for the molecule that is measured and for the internal standard. The same compound can play the role of the internal standard for the quantification of various compounds within the mixture.

For an evaluation of error sources in quantitative GC/MS determinations, the reader is referred to a paper by Claeys et al.[21]

5.7.5 Internal standard method

This method is based on a comparison of the intensity of the signal corresponding to the product that has to be quantified with that of a reference compound called the internal standard. This method allows the elimination of various error sources other than the minimal intrinsic error due to statistical reasons. In fact, if we choose as an internal standard a molecule with chemical and physical properties as close as possible to the properties of the molecule to be measured, the latter and the internal standard undergo the same loss in the extraction steps and in the derivatization or the same errors in the introduction of the sample into the mass spectrometer when the source conditions are varied. Because both compounds undergo the same losses and the same errors, their ratio remains unchanged during the procedure. Knowing the quantity of the internal standard that is added from the start and the relative proportion of the quantity of both compounds allows these losses and errors to be neglected. It is important to add the internal standard as early as possible in the procedure in order to obtain the maximum precision.

The method consists first of carrying out measurements on synthetic samples containing the same known quantity of the internal standard and increasing quantities of the compound to be measured. With these results a calibration curve is constructed. This allows a mathematical relationship to be obtained between the intensities of the signals corresponding to the compound to be analyzed and the internal standard (I_x/I_{sti}) and the quantity of compound present in the sample (M_x). As a reminder, maximum precision is obtained if the relationship corresponds to the equation of a straight line with a slope equal to unity.

The measurements then are carried out on the unknown samples that had a constant quantity of internal standard added to them before they were treated according to the experimental procedure. The quantity of compound in each unknown sample can be measured using the calibration curve. Figure 5.10 shows how useful an internal standard can be for measuring purposes. This figure uses once again the example of quantitative analysis of phenobarbital using an internal standard (anthracene).

The internal standard should show physical and chemical properties that are as close as possible to those of the molecule that has to be measured. It must be pure, absent from the sample and, of course, inert towards the compounds in the sample. The internal standards can be classified into three categories: structural analogs that are labeled with stable isotopes, structural homologs and compounds from the same chemical family. These various types of internal standards are classified here in descending order according to their usefulness and according to their price. In fact, the starting material for labeled compounds is fairly cheap, but most require many steps in their total synthesis and are thus very expensive. In the case of standards corresponding to structural homologs or in the case of compounds belonging to the same chemical family, the ions that are used must have masses differing from that of the compound that must be measured if direct introduction is used. However, if the introduction is carried out by chromatographic coupling and if the retention time

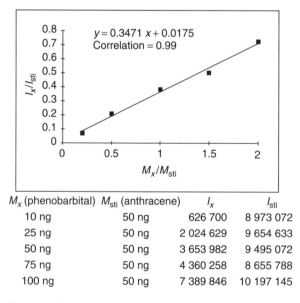

M_x (phenobarbital)	M_{sti} (anthracene)	I_x	I_{sti}
10 ng	50 ng	626 700	8 973 072
25 ng	50 ng	2 024 629	9 654 633
50 ng	50 ng	3 653 982	9 495 072
75 ng	50 ng	4 360 258	8 655 788
100 ng	50 ng	7 389 846	10 197 145

Figure 5.10
Calibration curve for the quantitative analysis of pheno-barbital using anthracene as an internal standard

of the compound is different from that of the internal standard, then the ions can have identical masses.

It is also worth mentioning that, depending upon the nature of the internal standard, weighted or non-weighted linear regression analysis to construct the calibration curve should be considered.[21]

5.7.6 Isotopic dilution method

Isotope dilution mass spectrometry can be considered as a special case of the internal standard method: the internal standard that is used is an isotopomer of the compound to be measured, e.g. a deuterated derivative. Note that an internal standard is necessary for every compound to be measured. This internal standard is as close as possible to perfection because the only property that distinguishes it from the compound to be measured is a slight mass difference, except for some phenomena that involve the labeled atoms, such as the isotopic effect. In that case, we have an absolute reference, i.e. the response coefficients of the compound and of the standard are identical. This method is often used to establish standard concentrations. The basic theory of this method rests on the analogy between the relative abundance of isotopes and their probability of occurrence.[22]

The method consists of examining the spectrum of the compound that must be measured in order to select an intense characteristic peak that is used to measure the analyte. A known, exact quantity of labeled internal standard is added to the sample with an unknown concentration. When the labeled internal standard is added, the peak corresponding to the characteristic peak is moved to a different position in the spectrum, according to the number and nature of atoms that were used in the labeling. The ratio of these two signal intensities is used to measure their relative proportion.

Suppose that there are N atoms or molecules in the mixture that yield a peak that is characteristic of the mass m.

Suppose that there are M atoms or molecules of the labeled compound in the substance that yield a characteristic peak for the mass $m + n = o$, where n corresponds to the mass displacement caused by the introduction of isotopes into the molecule.

The ratio R_{mo} of the ion intensities over the masses m and o is given by

$$R_{mo} = \frac{N P_m + M Q_m}{N P_o + M Q_o}$$

where P_m, P_o, Q_m and Q_o represent the relative isotopic abundances normalized over the isotopes for the natural product and the labeled product with masses m and o respectively. Because P_m, P_o, Q_m, Q_o and M are known, and R_{mo} can be deduced from the spectrum of the mixture containing the natural compound and the labeled compound, then the value of N can be calculated precisely (Figure 5.11).

If A represents Avogadro's number, x and y represent the quantity of natural product and the quantity of labeled product added to the sample, respectively, and

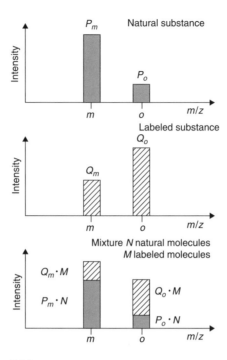

Figure 5.11
Principle of isotopic dilution illustrated by a molecule
having two isotopes of mass m and o

E and F represent the molecular masses of the natural product and of the labeled
product, respectively, then the equation is

$$R_{mo} = \frac{(Ax/E)P_m + (Ay/F)Q_m}{(Ax/E)P_o + (Ay/F)Q_o}$$

or, simply

$$R_{mo} = \frac{(x/y \times P_m/E) + Q_m/F}{(x/y \times P_o/E) + Q_o/F}$$

This mathematical equation corresponds to the equation of a curve. The straight line
represents only a special case. The nature of the calibration curve that is obtained
depends on the normalized relative isotopic abundances and thus on the nature of
the isotopes that are introduced, on the increase in the molecular weight and on how
enriched the labeled compound is.

   If P_o and Q_m are zero (there is no interference between the natural compound at
high mass o and the labeled compound at low mass m), then the equation becomes

$$R_{mo} = \frac{x}{y} \times \frac{P_m}{Q_o} \times \frac{F}{E}$$

This equation corresponds to the equation of a straight line passing through the origin
but with a slope that can be other than unity.

Hence, in theory, the calibration curve for a certain analysis can be calculated without any reference to experiments on synthetic samples containing increasing quantities of the compound to be measured. If the isotopic abundances are known with precision, the method allows one to determine the exact quantity of the substance to be measured without having to set up a calibration curve. In practice, disagreements between the theoretical and practical calibration curves are possible.

Quantitative applications using a direct inlet probe are analogous to those using a gas chromatography or liquid chromatography inlet, except for one important difference: the separation carried out before introducing the sample when using chromatographic coupling. This allows one to increase the specificity of measurement by the acquisition of an extra datum, which is the retention time. The lack of resolution of the direct inlet mode causes peaks to overlap and reduces the sensitivity. This can negatively influence the precision of the measurement.

5.8 References

1. http://www.nist.gov/srd/nist1.htm.
2. *NBS, EPA/NIH Mass Spectral Data Base*, National Standard Reference Data System 63 plus supplements 1 and 2, National Bureau of Standards, Faithers burg, MO, 1978, 1980, 1983.
3. http://webbook.nist.gov/chemistry.
4. *Eight Peak Index of Mass Spectra*, 3 vols, Royal Society of Chemistry, Cambridge, 1991.
5. McLafferty F.W. and Stauffer D.B., *The Wiley Registry of Mass Spectral Data*, 6th edition, Palissade Corporation, New York, distributed by Wiley–Interscience, New York, 1998.
6. McLafferty F.W. and Stauffer D.B., *Important Peak Index of the Registry of Mass Spectral Data*, Wiley, New York, 1991.
7. Middelditch B.S., Missler S.R and Hines H.B., *Mass Spectrometry of Priority Pollutants*, Plenum Press, New York, 1981.
8. Stemmler E.A. and Hites R.A., *Electron Capture Negative Ion Mass Spectra of Environmental Contaminants and Related Compounds*, VCH, New York, 1988.
9. Hites R.A., *Handbook of Mass Spectra of Environmental Contaminants*, Lewis, Boca Raton, FL, 1992.
10. Pfleger K., Maurer H.H. and Weber A., *Mass Spectral and GC Data of Drugs, Poisons, Pesticides, Pollutants and their Metabolites*, VCH, New York, 1992.
11. Ardrey R.E., Allen A.R., Bal T.S., *et al.*, *Pharmaceutical Mass Spectra*, Pharmaceutical Press, London, 1985.
12. Yergey J.A., *Int. J. Mass Spectrom. Ion Processes*, **52**, 337 (1982).
13. Hsu C.S., *Anal. Chem.*, **56**, 1356 (1984).
14. http://www.sisweb.com/mstools.htm.
15. http://www.shef.ac.uk/~chem/chemputer/isotopes.html.
16. Rockwood A.L. VanOrden S.L. and Smith R.D., *Anal. Chem.*, **67**(15), 2699 (1995).
17. Rockwood A.L. and VanOrden S.L., *Anal. Chem.*, **68**(13), 2027 (1996).
18. Millard B.J., *Quantitative Mass Spectrometry*, Heyden, London, 1978.
19. McLafferty F.W. and Venkataraghavan R., *Mass Spectral Correlations*, American Chemical Society, Washington, DC, 1982.
20. Yost R.A., *Adv. Mass Spectrom.*, **10B**, 1479 (1985).
21. Claeys M., Markey S.P. and Maenhaut W., *Biomed. Mass Spectrom.*, **4**, 122 (1977).
22. Pickup J.F. and McPherson K., *Anal. Chem.*, **48**, 1885 (1976).

6

Fragmentation Reactions

Most of the chemical reactions occur in the condensed phase or in the gas phase under conditions such that the number of intermolecular collisions during the reaction time is enormous. Internal energy is quickly distributed by these collisions over all the molecules according to the Maxwell–Boltzmann distribution curve.

6.1 Electron Ionization and Fragmentation Rates

Using conventional electron ionization mass spectrometry, everything occurs in a vacuum such that any collision is highly unlikely. The molecule receives energy from an electron beam and is ionized into a radical cation. The ion that is thus formed is subjected to an electric field that directs it towards the analyzer. For instance, suppose a singly charged 100 u mass ion is accelerated by a 1000 V potential difference in the source. This ion has a mass of

$$100 \times 1.66 \times 10^{-27} \text{ kg} = 1.66 \times 10^{-25} \text{ kg}$$

At the source outlet, its kinetic energy is

$$\frac{mv^2}{2} = (1000 \text{ V}) \times (1.6 \times 10^{-19} \text{ C}) = 1.6 \times 10^{-16} \text{ J}$$

The square of its speed is then

$$v^2 = \frac{(2 \times 1.6 \times 10^{-16})}{(1.66 \times 10^{-25})} = 1.93 \times 10^9 \text{ J kg}^{-1}$$

and hence $v = 4.39 \times 10^4$ m s^{-1} = 1.58×10^5 km h^{-1}. Thus the time spent in the source ranges from 10^{-6} to 10^{-7} s, and in the analyzer about one or two orders of magnitude longer.

Let us remember also that the ionization is usually carried out with a 70 eV electron beam, and that 1 eV corresponds to 96.48 kJ mol^{-1}. The ion excess energy can be several electronvolts. In the condensed phase, the cooling of an excited ion or molecule results from the collisions or from the emission of photons. Under high vacuum, only the latter possibility remains. The time delays necessary for cooling by radiation can be measured.[1] In the case of electronic excitations, the radiation occurs

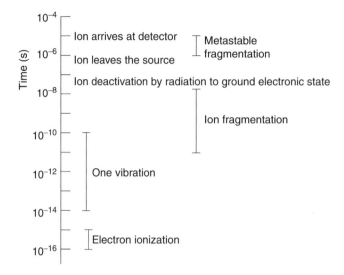

Figure 6.1
In electron ionization, the ionization occurs in a very short time. The energy
is redistributed through the vibrations in the ion. After 10^{-8} s (1/100th of a
microsecond), the ion deactivates to the ground state if it did not fragment.
This occurs well before the ion leaves the source and explains why electron
ionization is reproducible from one spectrometer to another

in the ultraviolet (UV) or in the visible range and typically occurs after 10^{-8} s. The
rotation and vibration excitations emit in the infrared range after time lengths from
1 ms to a few seconds. Figure 6.1 summarizes the time events after electron ionization.
The ionization occurs in a very short time. The energy is redistributed during the
vibrations and the ion may or may not fragment. After about 10^{-8} s, the ion will
go back to the ground electronic state by radiation and almost will fragment no
more. This occurs well before the ion leaves the source and explains why electron
ionization spectra are reproducible from one instrument to another.

Here is the problem: a great quantity of energy is transferred to isolated molecules.
The notion of temperature — the statistical distribution of energy — loses all meaning. Also, in a few microseconds, the whole process is over. The reaction has or has
not occurred and the products are detected.

The reactions that are observed are unimolecular fragmentations. Recombinations
are impossible because the collisions, under the usual conditions, are non-existent.
The rearrangements and the opening of rings are detectable only if they are followed
by a fragmentation. These reactions thus depend only on the structure and on the
energy contained within the molecule.

Both of these characteristics (isolated unimolecular reaction and very short time)
imply that the products of these reactions are the kinetic products. Hence expla-
nations based on the stability of the fragment ions may be misleading. However,
many fragmentation reactions correspond to zero- or low-energy reverse reactions,

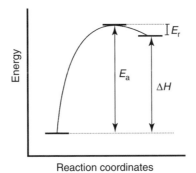

Figure 6.2
The recombination activation energy E_r of
a radical with a cation is low. As is shown,
this means that $\Delta H = E_a - E_r \approx E_a$

such as recombination of a radical with a cation. According to the principle of microreversibility, the fragmentation activation energy is equal to the endothermicity (Figure 6.2). In this case, the reasoning based on thermodynamics often is correct.

In thermal reactions, the reaction rate depends on the activation–deactivation equilibrium and on the decomposition rate of the activated complex:

$$M + N \underset{k_{-1}}{\overset{k_1}{\rightleftharpoons}} M^* + N$$

$$M^* \overset{k_2}{\longrightarrow} P_1 + P_2$$

Values of k_2 are difficult to deduce for thermal reactions in the condensed phase because the rates measured are always combinations of these three steps. In the gas phase, however, k_2 is directly measurable but the values are not directly transferable to the reactions in the liquid phase because the reacting species in the gas phase are not solvated.

6.2 Quasi-equilibrium and RRKM Theories

Two almost identical theories explaining the phenomena observed in the case of unimolecular reactions in the gas phase at high vacuum were proposed in 1952. One of them, the 'quasi-equilibrium theory' (QET), was suggested by Rosenstock *et al.*[2] and applies to mass spectrometry. The other is labeled by the initials of its authors, RRKM, standing for Rice, Rampsberger, Kassel and Marcus,[3] and deals with neutral molecules.

Both are based on assumptions and postulates. The first assumption is that the rotation, vibration, translation and electronic movements are independent of each other. The second assumption states that the movement of the nuclei can be expressed by classical mechanics. However, some quantum mechanics corrections are used.

The first postulate states that all the microscopic states are equally probable. In other words, all of the degrees of freedom participate in the energy distribution with the same probability. The second postulate states that the system can be described by movements on a multidimensional surface, and that a border surface exists that separates the reactants from the products. This surface can be crossed only in one direction: any reactant that crosses the transition state is irreversibly transformed into products.

The electron impact ionization of a molecule M to give a molecular ion in ground and excited electronic states ($M^{\bullet+}$ and $M^{\bullet+*}$ respectively) occurs over a very short time. An electron accelerated by a 10 V potential difference has a speed equal to 1.88×10^8 cm s^{-1}. It thus flies a distance of 1.88 Å, or 1.88×10^{-8} cm, in 10^{-16} s. This time is the interaction time of an ionizing electron with a molecule. The ionization must occur within that period. This is verified experimentally. This ionization is thus a vertical transition. The process is much more rapid than the time of one vibration, which is about 10^{-14} s in the case of the fastest ones. The distances between atoms thus do not change during the ionization.

This vertical ionization requires a higher ionization energy than the adiabatic process, as is shown in Figure 6.3. Note that adiabatic in this case does not have the same meaning as in thermodynamics. The total energy transferred to the ion can, of course, be higher than that of the vertical transition. Generally, the molecular ions of excited electronic states do not survive in the source because they lead to fragmentation or they return to their ground electronic state by emission of a photon. The molecular ions that leave the source to form stable or metastable ions thus generally contain a weak internal energy.

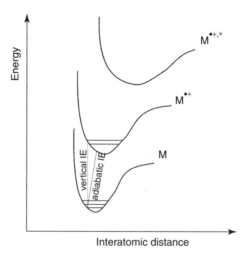

Figure 6.3
Morse curves for electron ionization. A vertical transition in ionization energy (IE) is observed if the interatomic distances have no time to adjust. The opposite is represented by the oblique line. Ionization also can lead to higher energies, including an excited electronic state

Electronic energy is unable by itself to lead to fragmentation of the molecular ion. The fragmentation requires a previous internal conversion of electronic energy to vibrational energy.

After the ionization, the energy distributes itself over the various degrees of freedom in a statistical fashion. The fast exchange of internal energy is done not only between the various degrees of freedom of the same electronic state but also between all the degrees of freedom of all the electronic states. These exchanges lead to the conversion of electronic energy acquired during ionization into vibrational and rotational energy of the ground electronic state of the molecular ion $M^{\bullet+}$. Consequently, the excited electronic states of the ionized molecule may be populated initially but relaxation to the ground electronic state occurs prior to their fragmentations. Thus, the fragmentation processes of the molecular ion are induced from this same electronic state in which vibrational and rotational energy is accumulated.

It can be shown experimentally that the statistical energy distribution is carried out within a time span corresponding to a few vibrations, i.e. less than 10^{-10} s. Note that this time span is very short with respect to the time spent in the spectrometer source, at least 10^{-7} s.

As soon as an oscillator contains more energy than a certain E_0 value that is characteristic of it, this oscillator becomes the reaction coordinate and the molecule undergoes the fragmentation reaction. Moreover, this reaction occurs faster if the internal energy distributed over this oscillator in excess over E_0 is higher.

Both theories lead to the following expression for the rate constant:[4,5]

$$k(E) = \frac{1}{h} \frac{Z^{\neq}}{Z^*} \frac{P^{\neq}(E - E_0)}{\rho_E}$$

where h is Planck's constant, Z is the partition function for adiabatic degrees of freedom, $\neq$ refers to the activated complex, $*$ refers to the active ionic species, $P^{\neq}(E - E_0)$ represents the total number of states corresponding to the activated complexes between energies zero and $E - E_0$ and ρ_E represents the density of states of the ion at energy E.

The following simplified equation also is used. It yields results that are flawed with important errors when it is applied to low or high energies.

$$k(E) = \nu \left(\frac{E - E_0}{E} \right)^{n-1}$$

where n is the number of vibrational degrees of freedom, ν is the frequency factor, E is the internal energy of the ion and E_0 is the transition energy. Structural modifications in a molecule bring about variations of ν and E_0; n also varies from one molecule to another. The ν factor is an inverse function of the activated complex steric requirements; ΔS^* is all the weaker when the activated complex is highly ordered:

$$\nu = \frac{E - E_0}{h} \cdot e^{\Delta S^*/R}$$

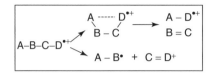

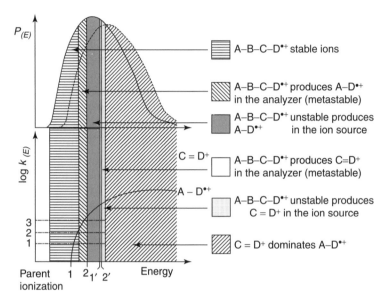

Figure 6.4
Warhaftig diagram. The x-axis is the internal energy of the ions. The
y-axis of the top diagram, P, is the proportion of ions with the energy
represented on the x-axis. Point 1 indicates the activation energy of the
first fragmentation reaction. An energy 2 is necessary, however, for the
ion to decompose fast enough for the fragment to be observable: a rate
constant higher than that indicated by the horizontal dotted line 2

Thus, referring to the spectrum in Figure 5.6, butanol can fragment, for example,
according to the two following paths:

$$\sideset{}{}{\overset{\cdot+}{C}}\!OH \longrightarrow \overset{\cdot}{\diagdown\!\!\diagup} \;+\; =\overset{+}{O}H$$

$$m/z\ 74 \qquad\qquad\qquad m/z\ 31$$

$$\overset{H\ \cdot+}{\underset{}{C}OH} \longrightarrow \quad +\ H_2O$$

$$m/z\ 74 \qquad m/z\ 56$$

The first reaction can occur owing to the transition state of any conformation,
also called a loose complex. However, the second reaction must go through a
transition state with a definite conformation called a tight complex. This brings

about a reduction in the activation entropy for the second reaction. The growth of the rate constant with respect to energy is thus greater in the case of the first reaction than in the second. However, the activation energy of the second reaction is weaker than that of the first because the bond cleavages are partially compensated for by the formation of new bonds in the transition state.

As opposed to thermal reactions, the distribution of energy among the ions is not regular. Figure 6.4 represents a Warhaftig diagram deduced from these theories. An increment in the energy of the ionizing electrons induces a displacement of the energy distribution curve towards higher energies.

On the curve of rate constants as a function of energy, three limits are indicated by dotted lines: line 1 shows a rate constant lower than 10^6 s^{-1} and corresponds to an average lifetime higher than 10^{-6} s. The corresponding ions reach the detector intact, without any fragmentation. Those corresponding to rate constants higher than 10^7 s^{-1} (at line 2 on the graph) fragment before leaving the source: only their fragments are detected. Those located between these limits fragment after leaving the source but before reaching the detector. These metastable ions are detected only in peculiar circumstances. The dotted line 3 shows the intersection between both rate curves: on the left, the reaction with rearrangement is faster; on the right, the opposite is true.

As is shown by these curves, the proportion of ions that decompose during the time of flight is small. Most of the ions are either stable within the measurement time or dissociated within the source: the game is over before leaving the source. As a result, modifications of the time of flight within the instruments through changes in the acceleration potential within the source or by switching from one instrument to another do not modify the appearance of the spectrum substantially. Thus spectra measured by magnetic instruments where the acceleration voltages hover around a few kilovolts are not very different from those obtained using quadrupole instruments where the acceleration voltages are up to a few dozen volts, or from those obtained with ion cyclotron resonance (ICR) instruments where the time spent by the ions in the instrument can be long.

This justifies the fact that spectral libraries used for comparative identifications are based on electron ionization. This method furnishes the spectra that are most easily compared.

6.3 Ionization and Appearance Energies

The ionization of a neutral molecule requires a minimum energy called the ionization energy (IE) or ionization potential. Mass spectrometers allow one to determine this ionization energy. The principle consists of increasing the energy of the ionizing electrons in the source and observing which minimum energy allows one to observe the molecular ion. In practice, analytical mass spectrometers yield an energy dispersion that is much too wide for a precise determination. Instruments dedicated to this goal yield results in agreement with those obtained by photoionization.

The appearance of a fragment ion occurs from an energy that includes the ionization energy of the neutral molecule, the activation energy and the kinetic shift. The molecular ion dissociates into fragments only if it contains an excess energy

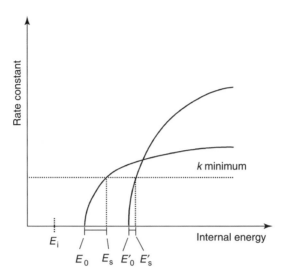

Figure 6.5
E_i: ionization energy of the precursor ion. E_0: the ion
contains enough excess energy in order to fragment but
the reaction is not fast enough to allow observation of
the fragment. E_s: the excess energy is enough for the ion
to dissociate before leaving the source. The difference
$E_0 - E_s$ is called the kinetic shift

great enough to allow decomposition, i.e. enough energy to overcome the activation
barrier. The corresponding energy is called the appearance energy (AE) or appearance
potential and corresponds to E_0. This is sufficient for observing the product ion
provided that it is observed over a long time, as is possible, for example, in Fourier
transform mass spectrometry (FTMS). In classical instruments the fragments are
observed after a very short time of about 10^{-8} s, as we have seen. The parent excess
energy must be enough to ensure a dissociation rate of that magnitude. Figure 6.5
illustrates this principle.

6.4 Fragmentation Reactions of Positive Ions

The terminology and the symbolism suggested by McLafferty[6,7] are used here because
they have become universal. Moreover, the name McLafferty is associated with a
rearrangement that we shall discuss later.

6.4.1 Fragmentation of odd-electron cations or radical cations (OE$^{\bullet+}$)

In radical cations, the charge is delocalized over the whole molecule. The most
favored radical and charge sites in the molecular ion are assumed to arise from the
loss of the electron of lowest ionization energy in the molecule. Consequently, when
we write fragmentation reactions on paper, we represent the charge as localized on

the site with the weakest ionization energy. As the following order n- $>\pi$- $>\sigma$-electrons is observed for ionization, the heteroatoms with weak ionization energies carry the charge preferentially. The symbol $^{•+}$ at the end of the molecules means an odd-electron ion without specifying the localization of either the radical or the charge site. However, use of either $^{•}$ or $^{+}$ within the molecule implies localization of the radical or the charge.

Positive charge and radical sites are electron-deficient sites. Cleavage reaction initiated by the positive charge site involves attraction of an electron pair because positive charge corresponds to the loss of an electron. The move of an electron pair induces heterolytic cleavage with migration of the charge site. Cleavage reaction initiated by the radical site arises from its strong tendency for electron pairing. An odd electron is donated to form a new bond whereas the transfer of a single electron induces homolytic cleavage with migration of the site of the unpaired electron.

Thus, the charge and the radical sites induce different cleavage reactions. The indication i and α involve all types of reactions initiated respectively at a charge or a radical site.

The fragmentation of these radical cations without any rearrangement or without any cleavage of an even number of bonds, such as occurs in rings, necessarily leads to an even-electron, or closed shell, ion and to a neutral radical. The parity rules were discussed in Chapter 5, section 5.6.

6.4.1.1 Direct dissociation (σ)

The expulsion of an electron from a σ bond can bring about the direct dissociation of the latter. We then speak of 'σ fragmentations'. One of the fragments keeps the charge and the other is a radical:

$$R-R' + e^- \longrightarrow R-R'^{•+} + 2e^-$$

$$R-R'^{•+} \left\langle \begin{array}{l} R^{•} + R'^{+} \\ R^{+} + R'^{•} \end{array} \right.$$

According to *Stevenson's rule*, if two charged fragments are in competition to produce a neutral radical by electron attachment, the radical having the highest ionization energy will be produced. The other ion, whose corresponding neutral radical has a lower ionization energy, will hold its charge and thus will be the observed fragment. Indeed, this reaction can be considered as a competition between two cations to carry away the electron:

$$R^{+} \cdots e^{-} \cdots R'^{+}$$

This rule is illustrated by the two examples in Figure 6.6. However, Stevenson's rule must be applied only to the *competitive* formation of fragment ions. Further

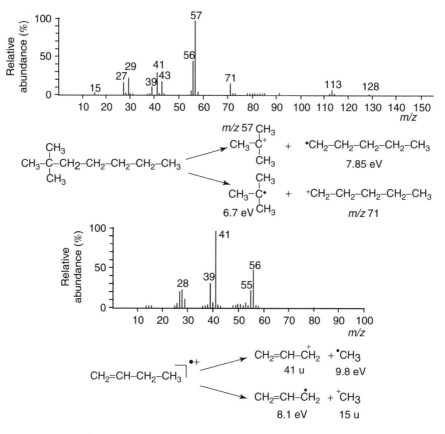

Figure 6.6
The ionization energy of the *t*-butyl radical is lower than that of the pentyl radical. Thus, the *t*-butyl ion is preferentially observed in the top spectrum. The same holds true for the comparison between allyl and methyl radicals

dissociation or additional formation by other pathways could, respectively, decrease or increase the abundance of the fragment ions and lead to erroneous interpretations.

6.4.1.2 Cleavage of a bond adjacent to a heteroatom (i)

The bond adjacent to a heteroatom can be broken by a '*charge-site-initiated reaction*', i.e. by attraction of an electron pair from this bond, and we talk about an induced cleavage (*i*):

$$R-CH_2\overset{\curvearrowleft}{\overset{\bullet+}{Y}}-R' \xrightarrow{\ i\ } R-CH_2{}^+ + \ ^\bullet Y-R'$$

This cleavage of the bond adjacent to the heteroatom can be seen as a direct dissociation assisted by inductive electron withdrawal due to the difference in electronegativity, but it occurs after the ionization.

In principle, Stevenson's rule still applies. By this rule, the cleavage of the adjacent bond could involve radical migration and charge retention if the ionization energy of $^{\bullet}$YR$'$ is less than that of RCH$_2$$^{\bullet}$. This counterpart reaction is classified as a special case of a σ bond cleavage and can be written as follows:

$$R-CH_2-\overset{\bullet+}{Y}-R' \overset{\sigma}{\longrightarrow} R-CH_2^{\bullet} + {}^{+}Y-R'$$

6.4.1.3 Cleavage of the alpha bond

The alpha bond to the radical cation site can be broken by a '*radical-site-initiated reaction*', i.e. by a transfer of the unpaired electron to form a new bond to an adjacent atom (α atom) with concomitant cleavage of another bond of this atom. The new bond compensates energetically for the cleaved bond. In this case we speak about a 'radical-site-initiated α fragmentation':

$$R-CH_2-\overset{\bullet+}{Y}-R' \overset{\alpha}{\longrightarrow} R^{\bullet} + CH_2=\overset{+}{Y}-R'$$

A half-arrow indicates the displacement of a single electron. Note that when the heteroatom corresponds to an oxygen atom, the positively charged oxygen in this last cation is isoelectronic with nitrogen and thus forms three covalent bonds.

Sometimes, several competitive cleavages are possible. Among the different possibilities, the loss of the radical of the highest ionization energy is generally observed. However, if several alkyl chains can be lost as radicals, the loss of the longest chain is favored.

By Stevenson's rule, the cleavage of the alpha bond could involve radical retention and charge migration if the ionization energy of $^{\bullet}$CH$_2$YR$'$ is more than that of R$^{\bullet}$. This counterpart reaction is classified as a special case of a σ bond cleavage and can be written as follows:

$$R-CH_2-\overset{\bullet+}{Y}-R' \overset{\sigma}{\longrightarrow} R^{+} + {}^{\bullet}CH_2-Y-R'$$

The following spectra illustrate these different mechanisms. In the mass spectrum of *t*-butyl ethyl ether presented in Figure 6.7, the fragmentation of the adjacent bond is observed because the *t*-butyl ion is very stable. The high electronegativity of the oxygen allows this fragment to carry away the electron; the ion with *m/z* 45 (CH$_3$CH$_2$O$^+$) is not observed. The other fragmentation path, α cleavage, leads to the loss of one methyl at *m/z* 87, immediately followed by a rearrangement that brings about the loss of ethylene and leads to the ion with m/z 59.

As a comparison, Figure 6.8 shows the mass spectrum of ethyl 2-butyl ether. The α fragmentation induced by the radical site has two possibilities for losing a methyl group and only one for the loss of an ethyl group, giving the 87 and 73 Th fragments, respectively. We see that the loss of the larger hydrocarbon radical is preferred, followed by the loss of an ethylene molecule via a hydrogen rearrangement, which dominates the spectrum at 45 Th. In the same way, the loss of a methyl followed

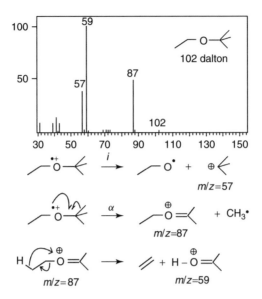

Figure 6.7
Fragmentation of *t*-butyl ethyl ether

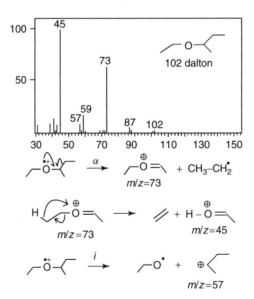

Figure 6.8
Spectrum of ethyl 2-butyl ether

by the elimination of ethylene leads to the 59 Th fragment. Cleavage of the adjacent bond, which leads to the butyl cation (57 Th), also is observed.

Figure 6.9 displays two examples where, respectively, cleavage of the α bond initiated by the radical and cleavage of the adjacent bond initiated by the charge are in competition with their counterpart reactions.

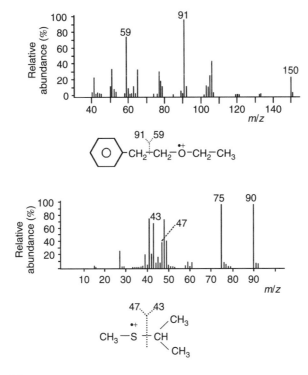

Figure 6.9
(*Top*) In this spectrum, the radical-site-initiated α cleavage is observed at
m/z 59. The α bond cleavage with charge migration is also observed at
m/z 91. (*Bottom*) In this spectrum, the adjacent bond cleavage initiated by
the charge is observed at m/z 43. The counterpart reaction corresponding
to the adjacent bond cleavage with charge retention is also observed at
m/z 47

6.4.1.4 *Competition between the cleavages of adjacent and alpha bonds*

The cleavage of the adjacent bond occurs more easily if the heteroatom is a large
atom. In the case of neighboring atoms, the more electronegative one leads more
easily to cleavage of the adjacent bond. The α cleavage becomes predominant for
electron donors. The following order is observed:

$$\text{Br, Cl} < \text{R}^\bullet, \pi \text{ bond, S, O} < \text{N}$$

Thus halogens preferentially cause the loss of the radical $\text{X}^\bullet$ through cleavage of the
adjacent bond, whereas amines preferentially lose a radical through cleavage of the
alpha bond.

As an example, compare the spectra of butylamine and butanethiol (Figure 6.10).
In the case of butylamine, the ion that is most intense corresponds to a radical-
initiated α cleavage leading to the $\text{CH}_2=\text{NH}_2{}^+$ ion at 30 Th:

$$\overset{+\bullet}{\text{NH}_2} \xrightarrow{\ \alpha\ } \text{CH}_2=\overset{\oplus}{\text{NH}_2} + \text{ }$$
$$m/z = 30$$

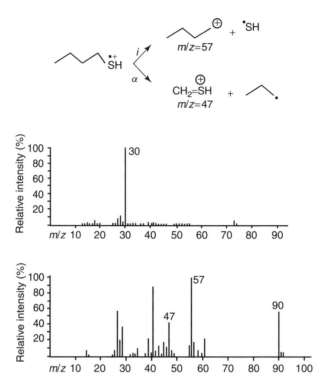

Figure 6.10
Spectra of butylamine (*top*) and butanethiol (*bottom*)

However, in the case of butanethiol, the main fragmentation occurs through a cleavage of the adjacent bond and corresponds to the loss of an HS$^\bullet$ radical, with the most intense ion being the butyl cation at 57 Th. The radical-initiated α cleavage leading to the CH$_2$=SH$^+$ ion at 47 Th also is observed but is clearly less important.

6.4.1.5 Fragmentation of radical cations with rearrangement

The rearrangements that can occur are very numerous and often make the interpretation of the spectra very difficult. However, some are frequent, very specific and well understood.

The McLafferty rearrangement consists of the transfer of a hydrogen atom to a radical cation site using a six-atom ring as an intermediate. The radical cation that results now has a radical site far away from the cation site: it is a *distonic radical cation*. This rearrangement then is followed by either a radical-site-induced or a charge-site-induced fragmentation, yielding in both cases a neutral molecule and a new radical cation. In the absence of nitrogen, these ions have an even mass and are easily detected in the spectrum.

Figure 6.11 shows the spectrum of 2-hexanone. For ketones, both paths *i* and α lead to identical products except for the charge. Once again the cation corresponding

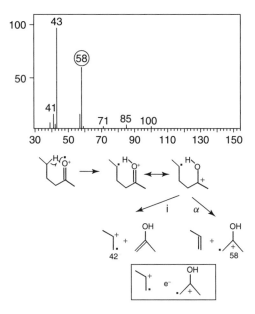

Figure 6.11
McLafferty rearrangement through a six-atom
ring intermediate. Figure 5.6 shows another
example and details the rules linking the mass
and electron parities

to the radical with the lowest ionization energy is predominant. In this case the enol
is better stabilized by the resonance with the lone pair on the oxygen. Note that the
radical cation with an odd number of electrons appears at the even mass-to-charge
ratio of 58 Th, in agreement with the parity rules.

The predominant peak at m/z 43 in Figure 6.11 corresponds to the alpha cleavage
initiated by the radical:

$$\ce{+ CH3-CO}$$
$$m/z = 43$$

6.4.2 Fragmentation of cations with an even number of electrons (EE$^+$)

Using electron ionization, the molecular radical cation is formed in the source. This
radical cation fragments into a radical and a cation with an even number of electrons
or, through rearrangements or multiple steps, into a neutral molecule and a new
odd-electron cation. The latter often are recognized easily in the spectrum because
their mass is even in the absence of a nitrogen atom.

The soft ionization techniques, such as fast atom bombardment (FAB), thermospray,
chemical ionization (CI), electrospray, matrix-assisted laser desorption/ ionization

(MALDI), atmospheric pressure chemical ionization (APCI), etc., produce molecular species with an even number of electrons, most often by the addition or abstraction of a proton. As opposed to radical cations, which are met almost exclusively in mass spectrometry, the even cations are common in 'classical' chemistry and their reactions seem much more familiar to chemists. These molecular species are generally more stable than the radical cations produced by electron ionization (EI). From an analytical point of view, the spectra are much simpler. However, compared with EI, the rearrangements are more frequent and more varied. Usually these spectra yield less information and are more difficult to interpret than the EI spectra. However, in addition to yielding the molecular mass more easily, such spectra may be more sensitive to small changes in the structure: isomers yielding identical EI spectra often give rise to different CI spectra.

For example, *cis-* and *trans-*1,4-cyclohexanediols yield essentially identical EI spectra. In CI, the *trans* isomer yields an $(M + H - H_2O)^+$ peak that is relatively more intense than that for the *cis* isomer,[8] a phenomenon that allows one to distinguish one isomer from the other. The difference is ascribed to the possibility of forming a hydrogen bond in the case of the *cis* isomer and to the assisted fragmentation and cyclization in the case of the *trans* isomer:

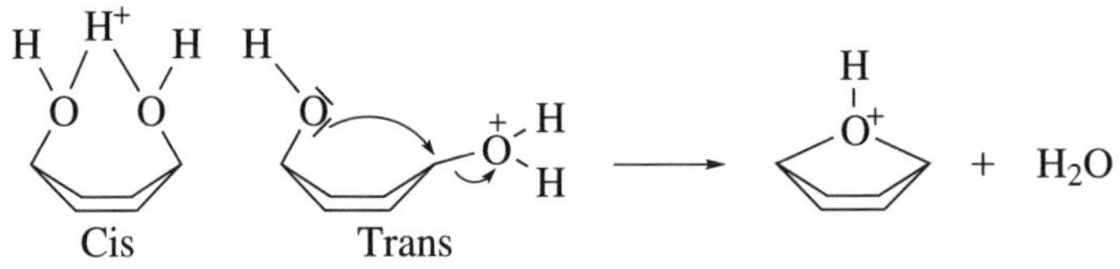

We shall label the even-electron cations using the symbol EE^+, standing for 'even electron number', and the symbol $OE^{\bullet+}$ will label the radical cations, standing for 'odd electron number'.

The production of a radical cation from an even-electron, or closed shell, ion is necessarily accompanied by that of a radical corresponding to the homolytic cleavage of a bond:

$$EE^+ \longrightarrow OE^{\bullet+} + R^{\bullet}$$

This process is usually highly endothermic and thus improbable. A statistical estimate based on many spectra shows that an even-electron ion yields even fragments in about 95% of cases.

We thus have the following situations:

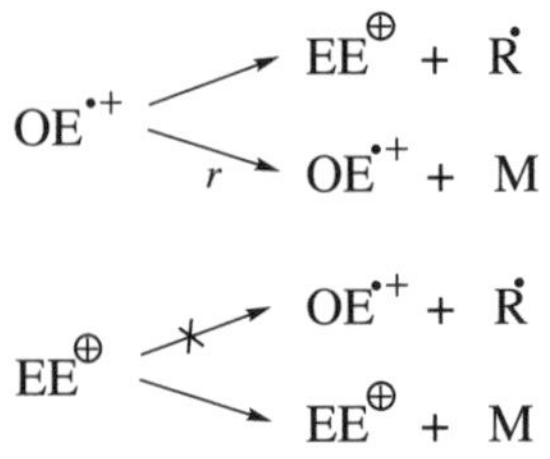

A cation with an odd number of electrons can fragment along two paths: either the production, through a cleavage i or α, of an EE^+ cation and a radical, or the production of a new radical cation ($OE^{\cdot+}$) and a molecule M after rearrangement r. A cation with an even number of electrons (EE^+) usually can produce only a new EE^+. The same rules apply to negative ions. These two types of ions thus can be recognized by recalling that, *in the absence of nitrogen*, the following hold true:

$$OE^{\cdot+} \text{ or } OE^{\cdot-}: \text{ even mass} \quad \text{and} \quad EE^+ \text{ or } EE^-: \text{ odd mass}$$

6.4.3 Fragmentations obeying the parity rule

Cleavage of the bond that is adjacent to the charged site, while observing a migration of the charge, is a common fragmentation process that occurs, especially when it allows the elimination of a small stable molecule. For example, protonated alcohols in soft ionization methods often lose water:

$$R-OH + H \longrightarrow R-OH_2^+$$
$$R-OH_2^+ \longrightarrow R^+ + H_2O$$

McLafferty[9] proposed classifying the reactions of even-electron ions that obey the parity rule (an even ion yields an even ion + neutral fragment) as follows:

(1) Cleavage of a bond with charge migration:

$$\longrightarrow CH_3-CH_2^+ + H_2O$$

(2) Cleavage of a bond with cyclization and charge migration:

$$+ \ HN=CR_2$$

(3) Cleavage of two bonds in a cyclic ion with charge retention:

$$+ \ H_2$$

(4) Cleavage of two bonds with rearrangement and charge retention:

$$\longrightarrow CH_2=CH_2 + CH_3-\overset{+}{O}\big\langle{}^H_H$$

Type 1 and 2 reactions are the most frequent in the absence of collision activation. The first type occurs especially easily when the protonated site is less basic, as is shown by the following data concerning n-butyl–XR molecules.[10] Of course, the ease with which these molecules separate from an organic cation or from a proton evolves in the same order.

−XR	NH$_2$	φ	SH	OH	I	Br	Cl
$[(M+H)-HXR)]^+/$ $(M+H)^+$	0.04	0.08	0.15	19.5	240	>250	>250
Proton affinity (kcal mol^{-1})	207	183	175	164	145	141	140

This reaction is also influenced by the stability of the cation that is formed, as is shown by comparison of the following data concerning various butylamines. A neutral molecule is lost even more easily when the cation product is stable. Thus, for example, in CI a protonated t-alkylamine loses ammonia more easily than a primary alkylamine.

	n-butyl	s-butyl	t-butyl
$[(M+H)-NH_3]^+/(M+H)^+$	0.4	0.07	2.3

The type 2 reaction is in fact the same reaction that is favored by the cyclization possibility if there is a heteroatom in a convenient position. The latter probably also provides anchimeric assistance for the expulsion of a neutral fragment. The example of cyclohexanediols provides a good illustration. As is shown by the following data, the comparison of ω-amino alcohols $H_2N(CH_2)_nOH$ with various linear chain lengths also offers an application of this principle:

n	2	3	4	5	6
$[(M+H)-H_2O]^+/(M+H)^+$	0.42	0.21	0.20	0.15	0.07

Moreover, the $(M+H-NH_3)^+$ ion is detected only for $n=4$ or 5. In the case of $n=5$:

Concerning type 3 and 4 reactions, Cooks and co-workers[11] noted that they are observed especially when the rearrangement is a four-center one. The following examples illustrate this:

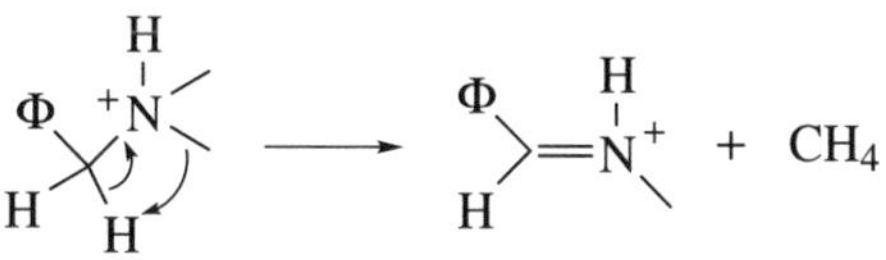

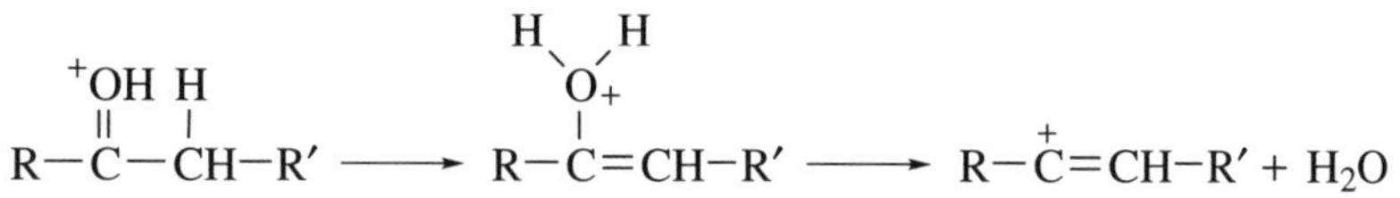

Note also that in CI a ketone can lose water, which never occurs in EI. Moreover, esters often yield the ion corresponding to the protonated acid:

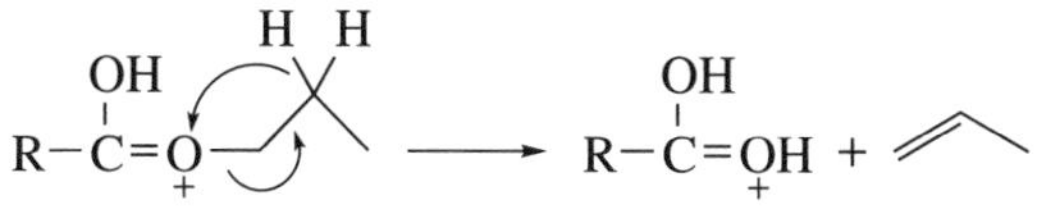

The presence of two heteroatoms complicates the mechanism. In fact, propyl acetate yields protonated acetic acid as a base peak, whereas methylbutyrate yields a $(RCOOH + H)^+$ peak that is hardly detectable: the proton is derived from the alcohol. Experiments on propyl acetate labeled with deuterium at various positions on the propyl chain yield the following percentages concerning the origin of the proton during the rearrangement. This distribution seems fairly statistical:

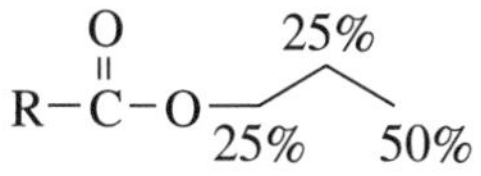

The reaction may seem complicated but its analytical usefulness is very high. For example, some triglycerides in CI yield three protonated acids and the pseudomolecular peak $(M + H)^+$.

Iminium ions with short chains fragment by forming an ion–neutral complex.[12] The slow step of the process is the 1,2 proton transfer leading to a secondary carbenium:

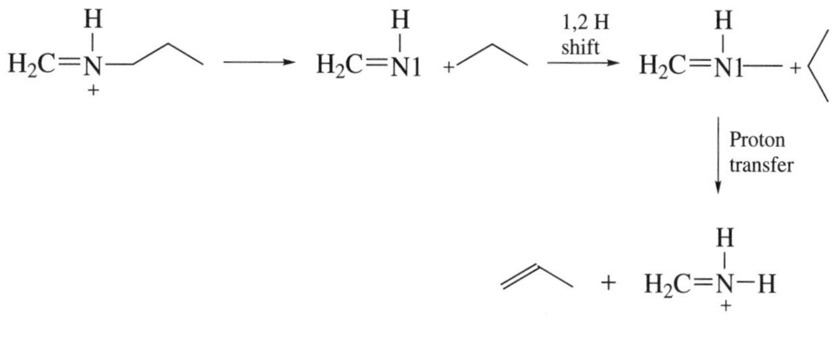

If the alkyl chain is longer, a McLafferty rearrangement, now occurring on an even-electron ion, becomes the dominant process. Labeling experiments show that the hydrogen originates from the γ-position. The rearrangement thus results exclusively from a 1,5 shift, occurring through a six-atom ring:

A similar mechanism could occur in other onium cases (oxonium, sulfonium, etc.).

6.4.4 Fragmentations not obeying the parity rule

The reactions considered up to now have been limited to those of the McLafferty classification with regard to the parity rule. The reactions of even-electron ions (EE) that do not obey the parity rule are much rarer and more difficult to predict. They are often observed in ions with extended π systems, but they often imply complex rearrangements, as is shown in the case of tropylium:

Note, however, that the main fragmentation path of the tropylium ion is the loss of acetylene.

It seems that the stability of the radical can play an important role. Thus nitroso compounds easily yield $\cdot\overline{N}{=}\overline{O}$ which is a stable radical.

6.5 Fragmentation Reactions of Negative Ions

The use of anions in mass spectrometric analysis was developed much later than the study of positive ions: the production of negative ions using the classical EI method is smaller than the production of positive ions by several orders of magnitude and commercial instruments were dedicated to the detection of positive ions. However, the situation has now changed completely, owing both to the discovery of the conversion dynode, which allows negative ions to be detected with high sensitivity, and to the development of many ionization techniques that produce negative ions efficiently.

The ions most often observed are even-electron anions $(M{-}H)^-$. They are efficiently produced in the source only from compounds containing acidic functions. This allows some selectivity for their detection in mixtures.

In general, even-electron negative ions contain less energy than the positive ions (as explained in Chapter 1) and thus produce less fragments. This explains why the most important studies and applications of negative ion fragmentations make use of collision-induced dissociation tandem mass spectrometry techniques. For some compounds, the sensitivity in the negative ion mode may be much higher than in the positive ion mode.

6.5.1 Fragmentation mechanisms of even-electron anions (EE⁻)

The fragmentation of even-electron anions has been reviewed by Bowie.[13,14] Observed reaction pathways include homolytic bond cleavage, the loss of one or several H• radicals being a common process. Alkyl radical loss also is observed:[15,16]

$$\text{H}^\bullet \text{ loss: } (\text{CH}_2\text{COCH}_3)^- \longrightarrow {}^\bullet\text{CH}_2\text{COCH}_2{}^- + \text{H}^\bullet$$

$$\text{alkyl radical loss: } \text{Ph}^-\text{CHOCH}_3 \longrightarrow \text{PhCHO}^{\bullet -} + \text{CH}_3^\bullet$$

Reactions through initial formation of an anion–neutral complex often are observed. They are followed by a displacement of the anion, a deprotonation, and an elimination process, such as[17]

$$\overset{\ominus}{}\text{CH}_2\text{COCOCH}_3 \longrightarrow [\text{CH}_3\overset{\ominus}{\text{CO}}(\text{CH}_2\text{CO})] \longrightarrow \text{CH}_3-\overset{\ominus}{\text{C}}\diagdown_{\text{O}} + \text{CH}_2\text{CO}$$

Complexes of the hydride ion with neutrals are invoked to explain some observed reactions, especially the loss of hydrogen molecules from alcoholates and enolates:

$$\text{CH}_3-\text{CH}_2-\text{O}^\ominus \longrightarrow [\text{H}^\ominus(\text{CH}_3-\text{CHO})] \longrightarrow [\text{H}_2(\text{CH}_2{=}\text{CHO}^\ominus)]$$

$$\longrightarrow \text{H}_2 + \text{CH}_2{=}\text{CHO}^\ominus$$

Various rearrangements often result from internal nucleophilic condensation, intramolecular nucleophilic substitution, formation of an ion–molecule complex, etc.

The fragmentation of ethylene glycol diacetate, shown in Figure 6.12, gives examples of many of these fragmentation mechanisms.[18]

An almost identical spectrum is observed from fragmentation of the negative pseudomolecular ion of ethylene glycol monoacetoacetate. This demonstrates the reversible rearrangement, step **a** in Figure 6.13.

From the diacetate ester (Figure 6.13), an ion–molecule complex consisting of the neutral ketene and the complementary alcoholate is formed. This complex either dissociates to yield the ethylene glycol acetate anion A, which further fragments to yield the acetate anion E, or a proton transfer from the ketene to the alcoholate occurs in the complex, which after dissociation yields the ynolate ion G.

From the ethylene glycol monoacetoacetate, the acetoacetate anion is produced. The latter loses carbon dioxide to yield the acetone enolate ion F. Alternatively, an internal proton rearrangement and fragmentation yields an ion–molecule complex

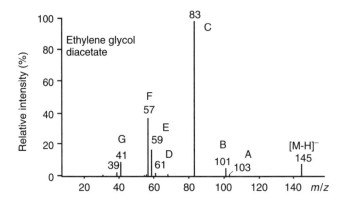

Figure 6.12
Tandem mass spectrometry product ion spectrum of the [M−H]⁻ pseudomolecular ion of ethylene glycol diacetate. Almost the same spectrum is
observed from ethylene glycol monoacetoacetate. See Figure 6.13 for an
explanation of the observed masses. (Reproduced (modified) from Ref. 18)

containing the ethylene glycol anion and acetylketene. This complex can dissociate
to yield the free ethylene glycol anion D, or the acetylketene is ionized to produce
the corresponding anion C.

6.5.2 Fragmentation mechanisms of radical anions (OE$^{\cdot-}$)

Radical anions (OE$^{\cdot-}$) undergo single cleavage reactions. Bowie[19] showed that they
occur α to the charged site or α to an atom that is conjugated with the charged site.
They also give rise to rearrangement reactions.

Radical anions are not observed often. Fullerene has no hydrogen atom in it
and thus ionization can occur only through electron capture or anion attachment.
Figure 6.14 displays its spectrum obtained under negative ion desorption chemical
ionization (DCI) conditions with a CH_4-N_2O mixture as the ionizing gas.

6.6 Charge Remote Fragmentation (CRF)

Many studies have shown[22–24] that gas-phase ion fragmentation can occur at sites
that are physically removed from the location of the charge. It is often called 'charge
remote fragmentation' (CRF) or sometimes 'remote (charge) site fragmentation'.
There is little, if any, involvement of the charge in the reaction. The charge must
be stable in its location and not be able to move to the reaction site. Generally,
the energy required to cause CRF is considerable. Change remote fragmentation has
been observed for both anions (sulfates, sulfonates, fatty acids, etc.) and cations
(long-chain amines and phosphonium). For example, palmitic acid yields, with FAB,
the anion $C_{16}H_{31}O_2^-$, which loses C_nH_{2n+2} starting from the alkyl end, as shown
in Figure 6.15. Other examples of CRF are discussed in Chapter 7. Charge remote
fragmentations have been reviewed.[25,26]

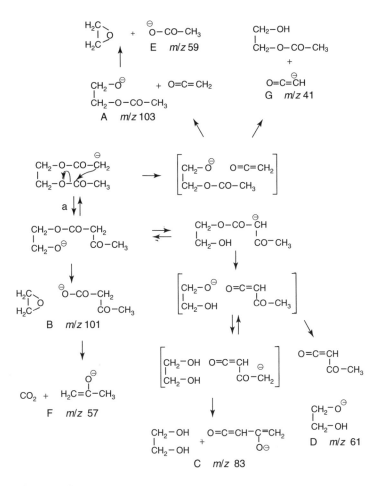

Figure 6.13
Fragmentation scheme of ethylene glycol diacetate and ethylene glycol monoacetoacetate anions. The spectrum is displayed in Figure 6.12. These two anions interconvert by the reversible pathway **a**, as demonstrated by the identity of the spectra of these two compounds. Other steps have been proved to be reversible by labeling experiments. Square brackets indicate ion–molecule complexes

6.7 Spectrum Interpretation

We have seen a series of methods allowing us to work our way from a spectrum back to the molecular structure: information on the elemental composition, the molecular mass, the number of rings and unsaturations, the relationships between the structure and the fragmentations, etc. In this section we shall study a few complementary data that are useful in the interpretation of spectra.

We saw that numerous ionization techniques exist that yield radical cations or radical anions, protonated or deprotonated molecules and various adducts. These

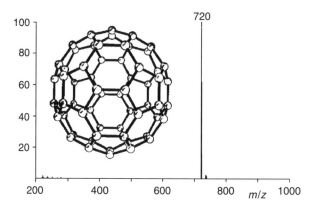

Figure 6.14
Negative ion desorption chemical ionization (DCI) spectrum of fullerene, discovered by Kroto,[20] using a mixture of CH_4 and N_2O as the ionizing gas. Small peaks are observed at 734 and 736 Th. (Reproduced (modified) from Ref. 21)

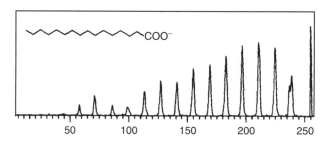

Figure 6.15
High-energy tandem mass spectrum of palmitate $(M-H)^-$ ion in the negative ion mode. (Reproduced (modified) from Ref. 22 with permission)

ions yield fragments with an even number of electrons (closed shell) or an odd number of electrons (open shell). Even though the radical cations derived from EI sources retain a privileged status in common mass spectrometry, the other ionization methods become increasingly common. Electron ionization is not possible for many categories of molecules so we shall not limit the discussion to radical cations.

Let us recall that the four main types of ions are closed-shell or open-shell, positive or negative ions.

6.7.1 Typical ions

Some ions are typical of given structures. Note first that compounds containing hydrocarbon chains give rise to a series of ions distant from each other by 14 Da $(-CH_2-)$. The mass where they appear depends on the group(s) that is(are) linked to them. Entirely saturated hydrocarbon ions appear at masses 15, 29, 43, 57, 71, 85, 99 Da, etc.

Molecules with a benzene nucleus often yield a phenylium ion at m/z 77, accompanied by a fragment corresponding to acetylene loss at m/z 51. If an alkyl chain is bonded to the benzene nucleus, ions are observed at m/z 91 that are a mixture of benzylium and tropylium structures, which produce a fragment observed at m/z 65 by losing acetylene.

Trimethylsilyl derivatives are commonly used in gas chromatography/mass spectrometry (GC/MS). When the molecule contains a hydroxyl, it fragments and yields $(CH_3)_3Si^+$ at m/z 73 and $(CH_3)_2Si^+-OH$ at m/z 75. When the molecule contains more than one $(CH_3)_3SiO-$ group, an ion of mass 147 Da is observed systematically in the spectrum, even though both groups are remote from one another. This ion has the structure shown below. It is derived from the fragmentation of a complex between the ion $(CH_3)_3Si^+$ and the neutral remainder of the molecule.

$$CH_3-\underset{\underset{CH_3}{|}}{\overset{\overset{CH_3}{|}}{Si}}-O-\overset{\oplus}{Si}\underset{CH_3}{\overset{CH_3}{<}}$$

$$m/z = 147$$

6.7.2 Presence of the molecular ion

In EI the molecular ion is observed only weakly in the case of linear saturated hydrocarbons. The presence of branching usually entails the disappearance of this peak. However, an unsaturation, especially an aromatic ring, makes the molecular ion peak more intense. The presence of electronegative saturated heteroatoms (oxygen, fluorine) normally prevents observation of the molecular ion. In fact, its intensity depends on the groups that are present. Thus an aliphatic ester yields a weak or absent molecular ion, whereas an aromatic ester usually yields an intense molecular ion peak.

Using soft ionization techniques in the positive ion mode, $(M-H)^+$ is most often observed in the case of saturated compounds and $(M + H)^+$ in the case of unsaturated compounds or heteroatom-containing compounds. However, halogen compounds or compounds containing an sp^3 oxygen often prevent observation of the molecular ion. Thus, for example, it is very difficult to form the molecular ion or the pseudomolecular ion of acetals or orthoesters.

6.7.3 Typical neutrals

The loss of a hydrogen atom is especially observed in EI starting from an aldehyde function, in the case of oxygen-containing or nitrogen-containing heterocycles or when many hydrogen atoms are bonded to carbons α to nitrogen or oxygen atoms.

The loss of a hydrogen molecule is observed starting from any ionic species (positive or negative, open- or closed-shell) every time it brings about an increased conjugation or aromaticity. It is especially common in the case of cyclic compounds.

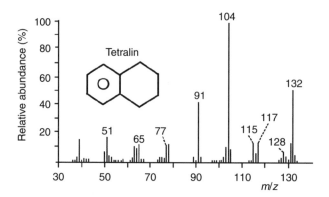

Figure 6.16
The EI spectrum of tetralin. Complete aromatization through the loss of
four hydrogen atoms is indicated by the presence of the peak at m/z 128.
The most abundant ion results from the loss of ethylene. As in the case of
most cyclanics, an intense ion is observed at (M−15), derived from the loss,
following rearrangement, of a methyl group. The typical ions benzylium
(91) and phenylium (77) lose 26 Da (acetylene) and produce fragments at
m/z 65 and 51, respectively

For example, the spectrum of tetralin (Figure 6.16) shows an intense ion corresponding to the loss of four hydrogen atoms.

The loss of 15 Da is typical for the elimination of a methyl group. Note that
saturated rings often yield an intense loss of 15 Da. For example, the EI spectrum
of cyclohexane, which has a molecular mass of 84 Da, shows a fragment with m/z
69 and a relative intensity of 25%. This ion is present also in the EI spectrum of
tetralin (Figure 6.16).

The 16 Da loss corresponds to a methane loss observed (as for the hydrogen
loss) when it produces a conjugation or aromaticity gain, i.e. especially in cyclic
compounds. Thus, steroids or bile salts commonly lose methane from an angular
methyl group. This 16 Da loss is also observed in N-oxides and sulfoxides, and then
results from an oxygen atom loss.

The fragmentation resulting from water elimination is indicated by an 18 Da
loss. It is common when a hydroxyl group is present and for various types of ions.
Molecules containing a carbonyl group do not commonly lose water except from
their conjugated acids. When studying the spectrum, remember the possible 18 Da
loss corresponding to ($H_2 + CH_4$), mostly in the cases of non-aromatic rings.

Losses of 19 or 20 Da are typical of the presence of fluorine, whose atomic mass
is 19 Da. Always beware, however, of possible consecutive losses of, for example,
water and hydrogen, which also lead to a 20 Da loss.

Some other neutral losses are typical of peculiar structures and sometimes have
great analytical usefulness. Thus, methyl esters of fatty acids lose the three carbon
atoms next to the carboxymethyl and a hydrogen atom through a specific rearrangement. This allows one to determine whether or not these carbon atoms carry

substituents.[27] Thus, for example, the methyl ester of 2,4-dimethylhexadecanoic acid, with a molecular mass of 298 Da, yields a characteristic fragment at m/z 227, derived from the loss of 70 Da corresponding to carbon atoms 2–4 and the two methyl groups they carry, and an additionnal hydrogen atom:

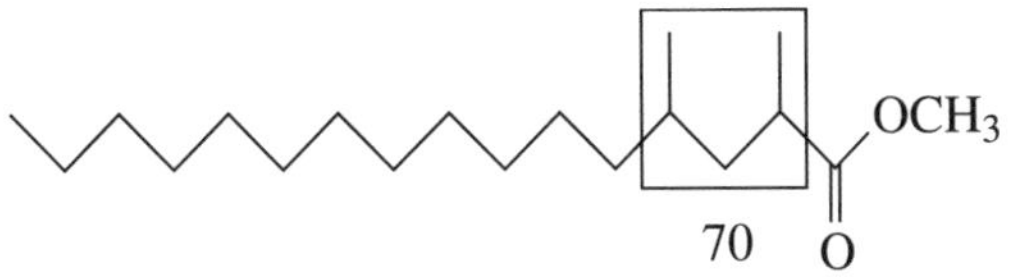

6.7.4 A few examples of the interpretation of mass spectra

Figure 6.17 shows the spectra of three substances each with a mass of 150 Da but with different structures.

The spectrum of ethyl benzoate (top) contains an ion of mass 122 Da. In the absence of nitrogen, it indicates an open-shell or odd-electron ion. This is an application of the parity rules and can be derived only from a rearrangement:

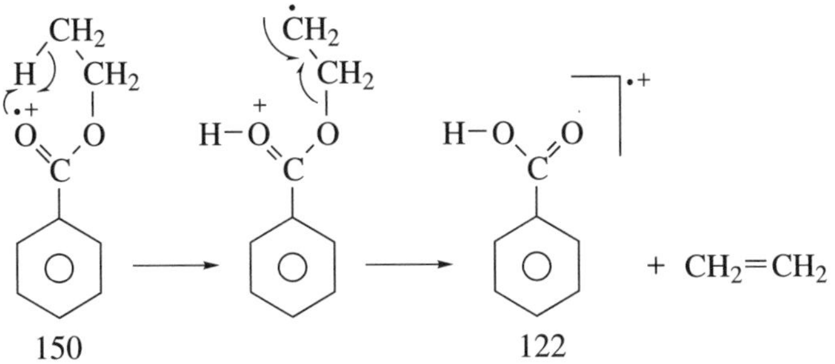

The charge-induced cleavage leads to the phenylium ion:

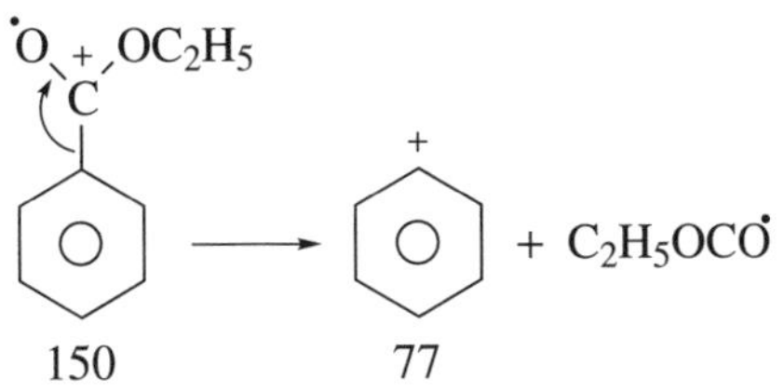

The latter loses acetylene in a retro-Diels–Alder reaction:

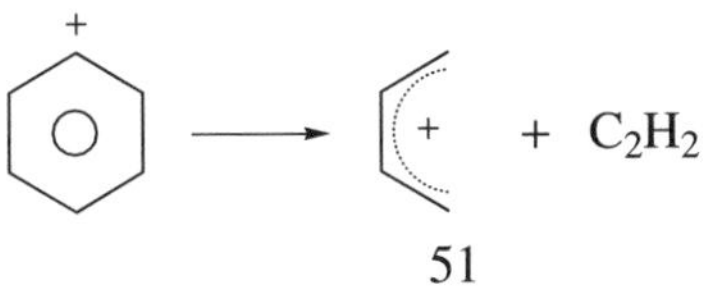

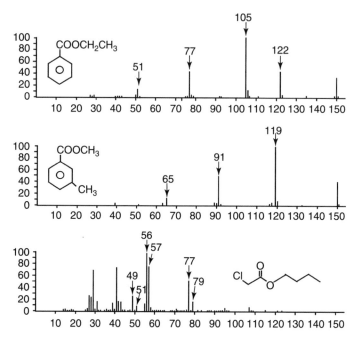

Figure 6.17
Spectra of three compounds having the same mass (150 Da) to
illustrate a few interpretative rules

Another α cleavage initiated by the radical site leads to the benzoylium ion:

Compare this spectrum with the middle one in Figure 6.17, of a methyl ester
carrying a methyl group on the benzene nucleus. Replacing the ethyl group by
a methyl group makes the McLafferty rearrangement impossible: it is no longer
possible to form an intermediate six-atom ring. The two fragmentations remain that
yield ions analogous to the previous case, but displaced by 14 Da ($105 \rightarrow 119$ and
$77 \rightarrow 91$) because of the presence of the methyl group on the benzene ring. This is
observed also after retrocyclization, the ion at m/z 51 being shifted to 65 Th.

Note that in both cases the charge is carried always by the fragments that contain
the aromatic ring: the latter in this case have the lowest ionization energy.

The bottom spectrum in Figure 6.17 was obtained from another 150 Da compound
that is very different. The molecular ion is not detected. The presence of an aromatic
group favors the observation of the molecular ion, hence this compound probably

does not contain such a group. The first intense fragment appears as a doublet at m/z 77 and 79, an isotopic cluster that is indicative of chlorine. It is formed by an α cleavage with respect to the carbonyl:

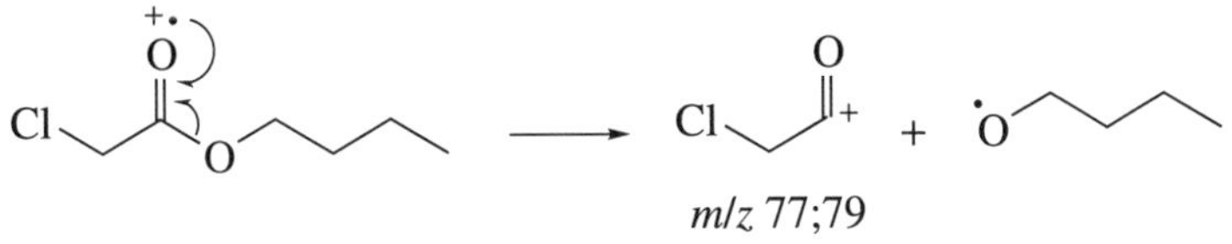

m/z 77;79

The m/z 57 ion, which does not contain a chlorine atom because the ion at m/z 59 is almost absent, is derived from cleavage of the bond adjacent to the sp^3 oxygen. The fragment at m/z 56 is an even-mass fragment: in the absence of nitrogen, this means that a radical cation is formed by a rearrangement:

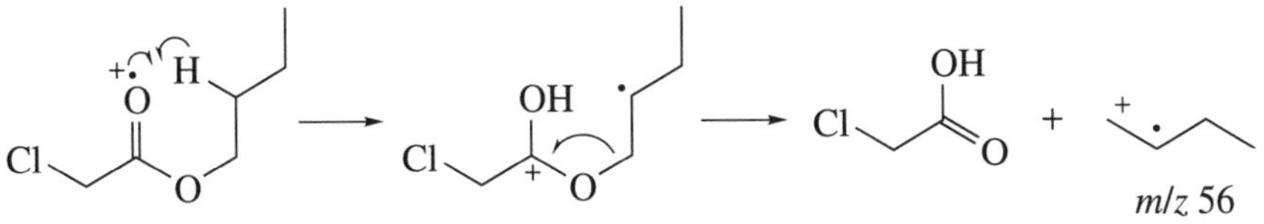

m/z 56

Once again, the possibility of having a six-atom intermediate favors this McLafferty rearrangement, which yields the most intense ion in the spectrum.

Finally, the ion at m/z 49 is accompanied by a peak at m/z 51, which indicates the presence of a chlorine atom. This means a cleavage α to the chlorine, yielding the $CH_2=Cl^+$ ion.

Other examples are shown in Figure 6.18.

t-Butylamine and n-butylamine yield only one intense fragment, the ion corresponding to the radical-site-initiated fragmentation. These fragmentations give rise to the loss of a methyl radical in the first case, which gives m/z 58, and of a propyl radical in the second case, at m/z 30:

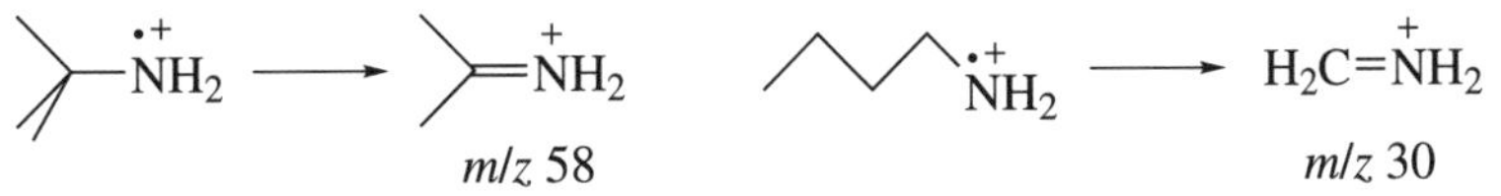

m/z 58 m/z 30

In contrast to the behavior observed with amines, the spectra of t-butanethiol and of n-butanethiol are dominated by the loss of the HS$^\bullet$ radical formed by cleavage of the adjacent bond. The ion of m/z 33 (HS$^+$) is either absent or very weak. Thus the HS group is the one that removes the electron:

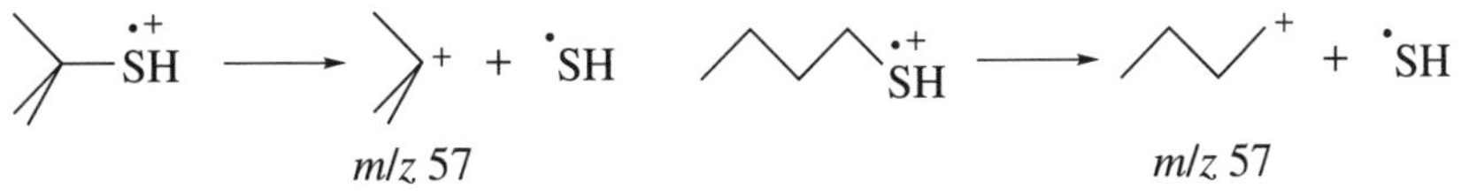

m/z 57 m/z 57

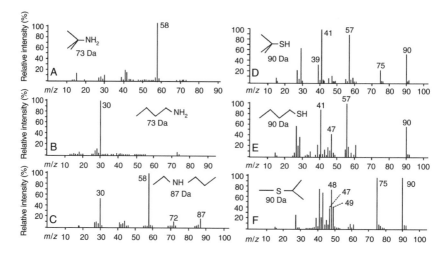

Figure 6.18
Comparison of spectra of amines, thiols and a thioether. The amine spectra display only very weak molecular ions. Thiols or thioethers have molecular ions that are relatively abundant, followed by the typical isotopic ^{34}S (4%) peak

The hydrocarbon ion of m/z 57 fragments also, in the case of both t-butyl and n-butyl, yielding successively propenium and cyclopropenium ions:

$$m/z\ 57 \quad \xrightarrow{-CH_4} \quad m/z\ 41 \quad \xrightarrow{-H_2} \quad m/z\ 39$$

These compounds yield the fragment corresponding to the α cleavage as the medium-intensity fragment, appearing at m/z 75 and 47, respectively:

$$\text{SH} \longrightarrow \text{SH} \qquad m/z\ 75 \qquad \text{SH} \longrightarrow H_2C=SH \qquad m/z\ 47$$

The secondary amine in spectrum C in Figure 6.18 shows two fragments corresponding to the α cleavage:

$$\begin{array}{l}
\xrightarrow{-\dot{C}H_3} \quad H_2C=\overset{+}{N}H \qquad m/z\ 72,\ \text{minor} \\
\text{NH} \\
87\ \text{Da} \quad \xrightarrow{-\dot{C}_2H_5} \quad \overset{+}{N}H=CH_2 \qquad m/z\ 58,\ \text{major}
\end{array}$$

Once again, the alkyl loss containing more atoms is favored. The ion at m/z 58 loses ethylene and produces the m/z 30 fragment:

$$\overset{+}{NH}=CH_2 \xrightarrow{-C_2H_4} \overset{+}{NH_2}=CH_2$$
$$m/z\ 30$$

A closed-shell ion fragments to yield another closed-shell ion through the loss of a whole molecule, in this case ethylene. In such ions with an even number of electrons, a hydrogen atom transfer no longer requires a six-atom ring.

Methyl isopropyl thioether (spectrum F, Figure 6.18) yields a 75 Th ion fragment, corresponding to a methyl loss. This fragmentation can occur through either of two paths: a methyl loss by isopropyl through an α cleavage, or loss of a methyl through cleavage of the adjacent bond. The cleavage of the other bond adjacent to the sulfur is responsible for the formation of the CH_3S^+ion at m/z 47. Ions of m/z 48 and 49 can have only the possible elemental compositions $CH_3SH^{\bullet+}$ and $CH_3SH_2^+$, respectively. The ion at m/z 48, with an even mass, results from a rearrangement and that at m/z 49 results from two successive rearrangements:

$$-S{\Big\langle}\Big]^{\bullet+} \longrightarrow -\overset{+}{SH}{\Big\langle}^{\bullet} \longrightarrow \Big[CH_3-SH\Big]^{\bullet+} + \diagup\diagdown$$
$$m/z\ 48$$

$$OR \quad -\overset{+}{SH}{\Big\langle}^{\bullet} \longrightarrow CH_3-\overset{+}{SH_2} + \diagup\diagdown^{\bullet}$$
$$\qquad\qquad H \qquad\qquad\qquad m/z\ 49$$

Note, however, that the first rearrangement, which is of the McLafferty type, occurs through a four-atom ring. In solution, such rearrangements most often imply a proton donor and a proton acceptor produced by the solvent. In the vapor phase, only intramolecular processes can be considered.

6.8 References

1. Dunbar R.C., *Mass Spectrom. Rev.*, **11**, 309 (1992).
2. Rosenstock H.M., Wallenstein M.B., Warharftig A., *et al.*, *Proc. Natl. Acad. Sci. USA*, **38**, 667 (1952).
3. Marcus R.A., *J. Chem. Phys.*, **20**, 359 (1952).
4. Longevialle P., *Principes de la Spectrométrie de Masse des Substances Organiques*, Masson, Paris, 1981.
5. Gilbert R.G. and Smith S.C., *Theory of Unimolecular and Recombination Reactions*, Blackwell Scientific Publications, Oxford, 1990.
6. McLafferty F.W., *Interpretation of Mass Spectra*, 3rd edition, University Science Books, Mill Valley, CA, 1980.

7. McLafferty F.W. and Turecek F., 4th edition, *Interpretation of Mass Spectra*, University Science Books, Mill Valley, CA, 1993.
8. Winkler F.J. and McLafferty F.W., *Tetrahedron*, **30**, 29 (1974).
9. McLafferty F.W., *Org. Mass Spectrom.*, **15**, 114 (1980).
10. Audier H.E., Milliet A., Perret C., *et al.*, *Org. Mass Spectrom.*, **13**, 315 (1978).
11. Sigsby M.L., Day R.J. and Cooks R.G., *Org. Mass Spectrom.*, **14**, 273, 556 (1979).
12. Veith H.J. and Gross J.H., *Org. Mass Spectrom.*, **26**, 1097 (1991); **28**, 867 (1993).
13. Bowie J.H., *Mass Spectrom. Rev.*, **9**, 349 (1990).
14. Bowie J.H., *Experimental Mass Spectrometry*, Plenum Press, New York, 1994, pp. 1–38.
15. Foster R.F., Tumas W. and Brauman J.I., *J. Chem. Phys.*, **79**, 4644 (1983).
16. Tumas W., Foster R.F. and Brauman J.I., *J. Am. Chem. Soc.*, **110**, 2714 (1988).
17. O'Hair R.A.J., Bowie J.H. and Currie G.J., *Aust. J. Chem.*, **41**, 57 (1988).
18. Stroobant V., Rozenberg R., el Bouabsa M., *et al.*, *J. Am. Soc. Mass Spectrom.*, **6**, 498 (1995).
19. Bowie J.H., *Mass Spectrom. Rev.*, **3**, 161 (1984).
20. Kroto H., *Angew. Chem.*, **104**, 113 (1982).
21. Richter H., Dereux A., Gilles J.M., *et al.*, *Ber. Bunsenges. Phys. Chem.*, **98**, 1329 (1994).
22. Jensen N.J., Tomer K.B. and Gross M.L., *J. Am. Chem. Soc.*, **107**, 1863 (1985).
23. Graul S.T. and Squires R.R., *J. Am. Chem. Soc.*, **110**, 607 (1988).
24. Tomer K.B., Jensen N.J. and Gross M.L., *Anal. Chem.*, **58**, 2429 (1986).
25. Adams J., *Mass Spectrom. Rev.*, **9**, 141 (1990).
26. Gross M.L., *Int. J. Mass Spectrom. Ion Processes*, **118**, 137 (1992).
27. Odham G. and Stenhagen E., *Biochemical Applications of Mass Spectrometry*, Wiley, New York, 1972, pp. 211–228.

7

Analysis of Biomolecules

7.1 Biomolecules and Mass Spectrometry

The determination of molecular weight is among the first measurements used to characterize biopolymers. Up to the end of the 1970s, the only techniques that provided this information were electrophoretic, chromatographic or ultracentrifugation methods. The results were not very precise (10–100% relative error on average) because they depended also on characteristics other than the molecular weight, such as the conformation, Stokes' radius and the hydrophobicity. Thus the only possibility of knowing the exact molecular weight of a macromolecule remained its calculation based on its chemical structure.

At that time the mass spectrometric ionization techniques of electron ionization (EI)[1] and chemical ionization (CI)[2] required the analyte molecules to be present in the gas phase and thus were suitable only for volatile compounds or for samples subjected to derivatization to make them volatile. Moreover, the field desorption (FD) ionization method,[3] which allows the ionization of non-volatile molecules with masses up to 5000 Da, was a delicate technique that required an experienced operator.[4] This limited considerably the field of application of mass spectrometry of large non-volatile biological molecules that are often thermolabile.

The development of desorption ionization methods based on the emission of pre-existing ions from a liquid or a solid surface, such as plasma desorption (PD),[5,6] fast atom bombardment (FAB)[7] or laser desorption (LD),[8] allowed a first breakthrough for mass spectrometry in the field of biomolecules analysis. Since then, the problem has no longer been the production of ions but rather that of analyzing such high-mass singly charged ions, which are technically difficult to detect with good sensitivity and difficult to analyze with good resolution.

At the beginning of the 1990s, two new ionization methods — electrospray ionization (ESI)[9] and matrix-assisted laser desorption/ionization (MALDI) — coupled to time-of-flight (TOF) analyzers[10] that avoided such inconveniences were developed and they continue to revolutionize the role of mass spectrometry in biological research. These methods allow the high-precision analysis of biomolecules of very high molecular weight.

The mass spectrometric analysis of different classes of biomolecules is reviewed in this chapter: peptides, proteins, nucleic acids, oligosaccharides and lipids. Several applications are detailed for each class.

7.2 Proteins and Peptides

Proteins and peptides are linear polymers made up of combinations of the 20 most common amino acids linked with each other by peptide bonds. Moreover, the protein produced by the ribosome may undergo covalent modifications, called post-translational modifications, after its incorporation of amino acids. Over 200 such modifications have been detected already,[11] the most important being glycosylation, the formation of disulfide bridges, phosphorylation, sulfation, hydroxylation, carboxylation and acetylation of the N-terminal acid. The most frequent are listed in Table 7.1.

Mass spectrometry not only allows the precise determination of the molecular weight of peptides and of proteins but also the determination of their sequences, especially when used with tandem techniques.[12-19]

Table 7.1   Post-translational modifications and corresponding mass variations

Post-translational modification	Mass difference (Da)
Methylation	14.03
Propylation	42.08
Sulfation	80.06
Phosphorylation	79.98
Glycosylations by:	
Deoxyhexoses (Fuc)	146.14
Hexosamines (GlcN, GalN)	161.16
Hexoses (Glc, Gal, Man)	162.14
N-Acetylhexosamines (GlcNAc, GalNAc)	203.19
Pentoses (Xyl, Ara)	132.12
Sialic acid (NeuNAc)	291.26
Reduction of a disulfide bridge	2.02
Carbamidomethylation	57.03
Carboxymethylation	58.04
Cysteinylation	119.14
Ethylpyridylation	105.12
Acetylation	42.04
Formylation	28.01
Biotinylation	226.29
Farnesylation	204.36
Myristoylation	210.36
Pyridoxal phosphate Schiff condensation	231.14
Stearoylation	266.47
Palmitoylation	238.41
Lipoylation	188.30
Carboxylation of Asp or Glu	44.01
Deamidation of Asn or Gln	0.98
Hydroxylation	16.00
Methionine oxidation	16.00
Proteolysis of a peptide bond	18.02
Deamination from Gln to pyroglutamic	−17.03

7.2.1 Fast atom bombardment (FAB), ESI and MALDI

The ionization methods that are used most often to study the peptides and proteins through mass spectrometry are FAB, ESI and MALDI. All of these techniques are characterized by the formation of stable ions (because they have only a low excess energy) and the absence of fragments.

The more recent ESI and MALDI ion sources provide a higher sensitivity and a broader mass range than the older FAB technique, and in the late 1990s have become the most important methods. Electrospray ionization produces multiply charged ions, which allow detection with a conventional mass spectrometer such as quadrupole, ion trap and magnetic instruments. The more sophisticated Fourier transform ion cyclotron resonance (FTICR) also is used with increasing success.

The MALDI source is a pulsed source that and can be used only with TOF mass spectrometers or with spectrometers that allow the storage of ions, such as ion traps and fourier transform mass spectrometers. The most frequently used is the TOF instrument. Over the last years, these MALDI/TOF spectrometers have made important progress through the introduction of delayed extraction of ions and reflector flight tubes. These instruments now have the capability of resolution equal or higher than other techniques.[20] The FTICR instruments are now using re-axialization and other improvements. With a 6 T magnet, a resolution of 2×10^6 is possible at masses in the 2000 Th range. Their complexity and cost make them only available in a very limited number of laboratories.

The development of the nanoelectrospray is the most important recent progress (1999). With injection flows in the range of 20 nl min^{-1}, long-lasting signals can be obtained from minute quantities of sample, allowing numerous tandem mass spectrometry (MS/MS) experiments to be performed for structure elucidation. Remember, indeed, that ESI sensitivity is not flow dependent up to the nl min^{-1} range.

The detection limit by ESI and MALDI depends on several factors, such as the nature of the sample and its preparation and purity, the instrument used and the quality of the operator. For peptides and proteins, the detection limit is in practice somewhere between femtomoles and picomoles, even if attomole limits have been reported.[21–23]

Because the resolution needed to separate different peaks in the isotopic cluster of a small peptide is lower than the resolution of most analyzers, the molecular weight that is measured corresponds to that calculated using the predominant isotope of each element. This is not so in the case of larger size peptides or proteins. Because the resolution required to resolve the isotopic cluster increases with the mass or with the charge carried by the ion and because the resolution of analyzers is limited, the various peaks in the isotopic cluster combine and form a single peak that spreads over several masses (15 Da at 10 000 Da and 45 Da at 100 000 Da) (Figure 7.1). Thus the molecular weight determined by these techniques corresponds to that calculated using the chemical mass of each element present in the protein. Both values are significantly different from one another. Indeed, the difference between the isotopic mass and the chemical mass of a peptide or protein is about 1 Da per 1500 Da. Hence, specifying which mass we are talking about is important. The choice is determined by the resolution of the analyzer and the mass of the ion being analyzed.

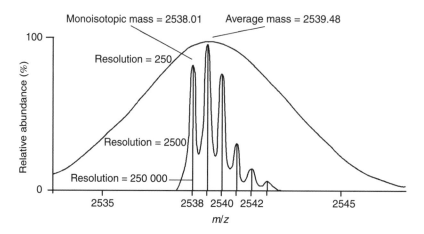

Figure 7.1

Mass spectrum of isotopic cluster of a singly protonated peptide $(C_{101}H_{145}N_{34}O_{44})$ with monoisotopic mass $= 2\,538\,015$ u and average mass $= 2\,539\,483$ Da at resolutions of 250 (resolution typically obtained for linear TOF), 2500 (quadrupole) and 250 000 (FTICR). (Reproduced (modified) from Ref. 16 permission)

The mass measurement errors also depend on a large number of factors. A typical measurement error of 0.01% can be obtained routinely. However, these errors can be higher than 0.1% in the worst case (MALDI/TOF) or lower than 0.001% in the best case (ESI/FTICR) .

The characteristic absence of fragment ions allows the analysis of complex mixtures without any previous separation. However, the analysis of mixtures of proteins with close molecular weights is limited by the resolution of the analyzer being used. For example, investigation of a mixture of a 10 000 Da protein and its oxidation product corresponding to the addition of one oxygen atom requires a resolution of at least 1000, whereas the resolution required to analyze the mixture is 10 000 if the protein has a mass of 100 000 Da. Another factor that plays a role in the possibility of analyzing this type of mixture is the relative quantity of each component. A compound whose abundance is 10% of the main component requires less resolution than the same mixture with the same compound having only a 1% relative abundance. In practice, for a 10 000 Da mass, any mixture of compounds differing by more than 10 Da can be analyzed. Otherwise, a prior separation is necessary.

Moreover, the analysis of complex mixtures is complicated by the phenomenon of competitive ionization.[24] This phenomenon, observed more in MALDI than in ESI, is characterized by the suppression of molecular ion species of some peptides when they are in the presence of other peptides in the mixture. The signal corresponding to these peptides may disappear completely and thus these peptides may not be detected even though they yielded an easily detectable signal when analyzed individually.

This phenomenon of competitive ionization has been observed much more in FAB,[25] due to the fact that FAB allows the expulsion of ions present on the matrix surface into the gas phase and that hydrophobic peptides occupy the surface layer more efficiently than the hydrophilic peptides. Because hydrophobic peptides are easier to expel into the gas phase, their molecular species peaks appear preferentially in the spectrum.

Competitive ionization may be avoided by varying the pH conditions or the matrix through chemical derivatization of the peptides contained in this mixture, or through partial fractionation of the mixture through reversed-phase liquid chromatography so that each fraction contains peptides of similar hydrophobicity.

For both ESI and MALDI, the concentration of the sample and the complexity of the contaminants play an important role in both the sensitivity and the mass accuracy. Biological samples most often are dilute solutions of peptides or proteins containing a great number of contaminants. These two problems, dilution and contaminants, are not easy to handle, especially when the total amount of sample is low, such as picomoles.

The contaminants are of different origins and include buffers, non-volatile salts, detergents and many compounds of unknown origin. Electrospray ionization can tolerate only low quantities of contaminant ions. These ions can reduce the abundance of the ions from the compound of interest and can even totally suppress them. They also very often result in the formation of adduct ions, further reducing the sensitivity by the distribution of the ion current over several species. Furthermore, they may complicate the determination of the molecular weight, or reduce the accuracy of the molecular weight if some adducts are not separated from the ions of the protonated molecule.

Generally MALDI is more tolerant than ESI to many contaminants. This can be due in part to some separation occurring during crystallization of the sample with the matrix.[26] Whatever the ionization method, the quality of the mass spectrum is higher if the contamination is reduced.

The second problem is the generally low concentration of the compound of interest in the biological samples. The volumes needed for the analysis are very low, in the microliter range for both MALDI and ESI,[27] and only part of it is actually consumed in the analysis. But the concentration has a marked influence on the observed spectra.

As a rule, a separation method should be used for both purification and concentration of the sample. The classic method for peptides and proteins is a reversed-phase high-performance liquid chromatography (HPLC) preparation of the sample, followed by a concentration step (often lyophilization) of the fraction of interest. During those steps performed on very small quantities of sample, loss of the sample can occur if care is not taken to avoid it. Lyophilization, for instance, can lead to loss of the sample by absorption on the walls of the vial. The use of microseparation methods on-line with the mass spectrometer often are preferred. Micro-HPLC[28,29] and capillary electrophoresis,[30] both coupled mainly to electrospray ionization/mass spectrometry (ESI/MS), are used more and more.

7.2.2 Structure and sequence determination using fragmentation

7.2.2.1 Fragmentation of peptides

In order to generate structural data by mass spectrometry, the molecule that is studied must undergo fragmentation of one or several bonds to match the m/z of the resulting fragments with the chemical structure. However, the various techniques that we have considered so far imply the formation of stable ions that do not yield any fragments. This property is used to facilitate the determination of the molecular weight of peptides or proteins, even when they appear in a complex mixture. However, the same property leads to a lack of information concerning the structure. This drawback is overcome by transferring at least the extra energy required by fragmentation to the stable ions produced during ionization. Although various techniques allow an energy transfer, such as photodissociation and surface-induced dissociation, the most common method remains collision-induced dissociation (CID).

The fragments that result are analyzed using tandem mass spectrometry (MS^n). This technique consists of selecting the ion to be fragmented using a first mass analyzer and sending it into a collision cell, where it collides with uncharged gas atoms. Thus the kinetic energy is transformed partly into vibrational energy and the resulting fragments are analyzed by a second spectrometer, hence the name CID tandem mass spectrometry or CID/MS/MS. If the instrument resolution is sufficient, the first analyzer can select only the isotopic peak containing only the main isotopes, such as ^{12}C and ^{16}O, which allows a fragmentation spectrum free from complex isotopic clusters (especially at high masses) to be obtained.

The tandem mass spectra may be obtained using many different instruments, such as magnetic analyzers, ion traps, quadrupoles, etc. The main difference from a practical point of view lies in the kinetic energy of the ions. In magnetic instruments the precursor ion kinetic energy is several kilo-electronvolts whereas in the other types of analyzers, such as quadrupoles or ion traps, the ion kinetic energy never exceeds 100 eV. This difference influences the fragmentation process.[31] As will be discussed in more detail below, the high-energy tandem mass spectra present a broader range of fragmentation pathways, some of which are not observed at low energy. A greater number of fragment ions often carry more information but they also increase the complexity of the spectra, thereby rendering their interpretation more difficult.

The fragmentation of peptides can be observed also by a technique named post-source decay (PSD) when reflector TOF instruments are used. In this technique, which is not only used for peptide analysis, the ions of the molecular species produced by MALDI indeed contain enough energy to fragment but this metastable fragmentation occurs during the flight between the source and the detector. With a linear TOF spectrometer the fragments reach the detector together with the precursor ions. In contrast, they will have different flight times after passing through the reflector and thus their mass can be determined. A chosen precursor can be selected by a gating system at the origin of the flight tube. The resolution for this selection is only about 100 or 200. Thus, the window at m/z 1000 will be 5 u or more. This is generally sufficient to select a peptide in a mixture but not to select one isotopic peak.

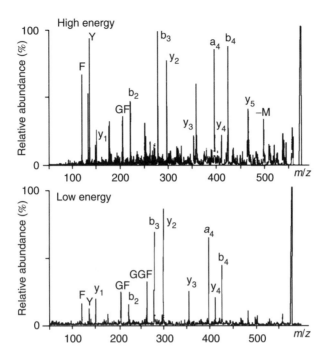

Figure 7.2
High- and low-energy fragmentation spectra of methionine-enkephalin (sequence YGGFM). The notation used is described in the text. (Reproduced (modified) from Ref. 31 with permission)

The MS/MS analysis of many peptides with known and unknown sequences allows one to identify the various existing fragmentation processes.[32] The high- and low-energy fragmentation spectra of a peptide are presented in Figure 7.2.

From a practical point of view, the fragments may be classified in either of two categories: those derived from the cleavage of one or two bonds in the peptidic chain and those that also undergo a cleavage of the amino acid lateral chain. The nomenclature suggested by Roepstorff and Fohlman[33] and later modified by Biemann[34] allows the labeling of the various fragments that are obtained.

The first fragments that were identified were produced by cleavage of a bond in the main chain. The cleavage of a bond in that peptide chain can occur in either of three types of bonds $C\alpha-C$, $C-N$ or $N-C\alpha$, which yields six types of fragments that are labeled a_n, b_n and c_n when a positive charge is kept by the N-terminal side and x_n, y_n and z_n when the positive charge is kept by the C-terminal side. The c_n and y_n fragments implicate the transfer of two extra hydrogen atoms, the first one responsible for the protonation and a second one originating from the other side of the peptide. The subscript n indicates the number of amino acids contained in the fragment. Figure 7.3 shows the various types of fragments produced through the cleavage of a bond in the peptidic chain.

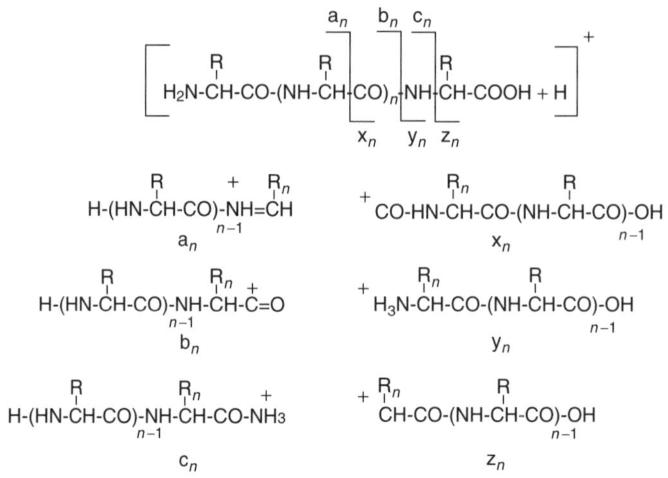

$$a_n \quad b_n \quad c_n$$

$$\left[\begin{array}{c} R \\ | \\ H_2N\text{-}CH\text{-}CO\text{-}(NH\text{-}CH\text{-}CO)_n\text{-}NH\text{-}CH\text{-}COOH + H \end{array} \right]^+$$

$$x_n \quad y_n \quad z_n$$

$$\underset{a_n}{H\text{-}(HN\text{-}CH\text{-}CO)\text{-}NH=CH} \qquad \underset{x_n}{{}^+CO\text{-}HN\text{-}CH\text{-}CO\text{-}(NH\text{-}CH\text{-}CO)\text{-}OH}$$

$$\underset{b_n}{H\text{-}(HN\text{-}CH\text{-}CO)\text{-}NH\text{-}CH\text{-}C{=}O} \qquad \underset{y_n}{{}^+H_3N\text{-}CH\text{-}CO\text{-}(NH\text{-}CH\text{-}CO)\text{-}OH}$$

$$\underset{c_n}{H\text{-}(HN\text{-}CH\text{-}CO)\text{-}NH\text{-}CH\text{-}CO\text{-}NH_3} \qquad \underset{z_n}{{}^+CH\text{-}CO\text{-}(NH\text{-}CH\text{-}CO)\text{-}OH}$$

Figure 7.3
Main fragmentation paths of peptides in CID/MS/MS

Table 7.2   Mass increments of the various amino acids

Amino acid	Code (3 letters)	Code (1 letter)	Monoisotopic mass	Chemical mass
Glycine	Gly	G	57.02147	57.052
Alanine	Ala	A	71.03712	71.079
Serine	Ser	S	87.03203	87.078
Proline	Pro	P	97.05277	97.117
Valine	Val	V	99.06842	99.133
Threonine	Thr	T	101.04768	101.105
Cysteine	Cys	C	103.00919	103.144
Isoleucine	Ile	I	113.08407	113.160
Leucine	Leu	L	113.08407	113.160
Asparagine	Asn	N	114.04293	114.104
Aspartate	Asp	D	115.02695	115.089
Glutamine	Gln	Q	128.05858	128.131
Lysine	Lys	K	128.09497	128.174
Glutamate	Glu	E	129.04260	129.116
Methionine	Met	M	131.04049	131.198
Histidine	His	H	137.05891	137.142
Phenylalanine	Phe	F	147.06842	147.177
Arginine	Arg	R	156.10112	156.188
Tyrosine	Tyr	Y	163.06333	163.17
Tryptophan	Try	W	186.07932	186.213

The mass difference between consecutive ions within a series allows one to determine the identity of the consecutive amino acids (see Table 7.2) and thus to deduce the peptide sequence. There are two exceptions: Leu-Ile, which are isomers, and Gln-Lys, which are isobars. Normally the spectra show several incomplete series of ions that produce redundant data and make the spectrum very complex and difficult to interpret.

There is a marked difference between the fragmentation observed at high and low energy. At high energy, all the fragmentations described in Figure 7.3 can be generated in principle. However, all the fragments are not observed in the spectra because the fragmentation can be influenced by the nature of the amino acids present in the sequence, as will be shown below. At low energy, the observed fragments are mostly b_n and y_n. These fragments then lose small molecules such as water or ammonia from the functional groups on the side chains of the amino acids.

Two other types of fragments found in most spectra result from cleavage of at least two internal bonds in the peptidic chain. The first type is called an internal fragment because these fragments have lost the initial N- and C-terminal sides.[35] They are represented by a series of simple letters corresponding to the fragment sequence. Fortunately, this type of ion often is only weakly abundant and, because they rarely contain more than three or four amino acid residues, they appear in the spectrum among the low masses. These peaks confirm the sequence but often are more of a nuisance than a help. Peptides containing a proline are an exception to this because the proline imino group is included in a five-atom ring and thus has a higher proton affinity than the other amide bonds in the peptide. Hence protonation and cleavage of the proline amide bond are favored to yield an internal fragment. The various fragments requiring cleavage of two bonds are shown in Figure 7.4.

The second type of fragment that results from multiple cleavage of the peptidic chain appears among the low masses in the spectrum. These are the immonium ions of the amino acids, labeled by a letter corresponding to the parent amino acid code. Even though these fragments are rarely observed for all of the peptide amino acids, those that appear yield information concerning the amino acid composition of the sample, especially when other diagnostic ions in the low-mass region are taken into account.[36] A list of immonium ions commonly found in spectra is given in Table 7.3.[37,38]

In addition to the ions described earlier, three new types of fragments that require cleavage of the peptidic chain and amino acid lateral chain were highlighted only in the high-energy spectra. These fragments are useful for distinguishing the isomers

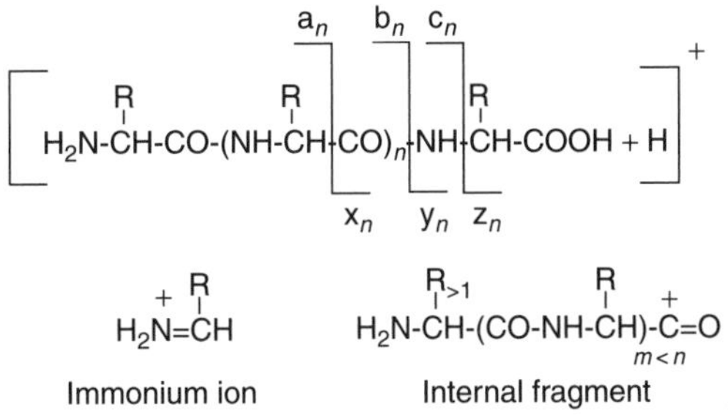

Figure 7.4
Fragments derived from a double cleavage of the peptide main chain

Table 7.3   Masses of the low-mass ions
characteristic of natural amino acids, most
often immonium ions

Amino acid	Characteristic mass
Proline (P)	70
Valine (V)	72
Leucine (L)	86
Isoleucine (I)	86
Methionine (M)	104
Histidine (H)	110
Phenylalanine (F)	120
Tyrosine (Y)	136
Tryptophan (W)	159

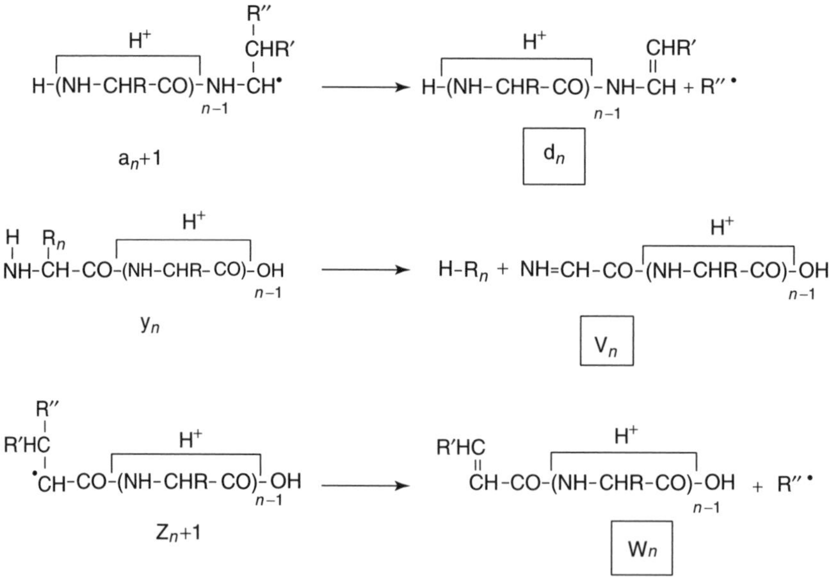

Figure 7.5
Fragmentation paths yielding the ions characteristic of amino acid lateral chains

Leu and Ile. Figure 7.5 shows the mechanisms and structures of the corresponding fragments.[39-42]

Two of these fragment types result from cleavage of the bond between the β and γ carbon atoms of the side chain of the amino acids. These fragments are symbolized as d_n or w_n, respectively, according to whether the positive charge is retained by the N-terminal or the C-terminal fragment. They are useful to distinguish the isomers Leu and Ile. Amino acids carrying an aromatic group attached to the β carbon atom (His, Phe, Tyr, Trp) either do not display these two fragments or have very low abundances.

The other type of fragment results from the complete loss of the side chain and is symbolized as v_n. This fragment containing the C-terminal moiety is intense for amino acids that do not easily yield w_n type fragments. No equivalent containing the N-terminal moiety has ever been observed.

The discussion up to now has concentrated on the CID spectra of singly charged precursor ions, but the development of ESI allows the study of the dissociation of multiply charged ions. As a rule, the fragmentation spectra of multiply proto-nated peptides, at high as well as at low energy, display fragments analogous to that observed for the monocharged species.[43,44] Multiply charged ions require lower acceleration voltages before CID because the kinetic energy is proportional to neV, where n is the number of charges, e is the electron charges and V is the acceleration voltage. At first glance they appear to fragment easier. It is best, when possible, to acquire spectra of different charge states because some fragments can be more abundant for some charge states.

The dissociation of multiply charged ions is interesting but can lead to more complicated spectra. Firstly, the presence of protons on different protonation sites induces fragmentations from a variety of starting points. Secondly, the ions obtained can have different charge states and the charge state has to be determined. Because two adjacent peaks of the isotopic cluster are separated by 1 Da, if the number of charges is n they will appear as being separated by $1/n$ Th. If the resolution of the instrument is sufficient, the charge state can be determined in this way. Tryptic peptides have two basic sites, one at each terminal, and thus easily yield doubly protonated ions: on the basic C-terminal amino acid and on the free amine group of the N-terminal. As a consequence, they mainly yield monocharged b_n and y_n ions, as shown by the example in Figure 7.6.[45,46]

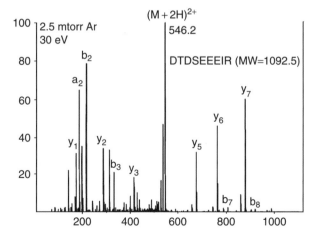

Figure 7.6
The CID ESI/MS/MS trace of a doubly charged peptide. (Reproduced from Finnigan MAT documentation, with permission)

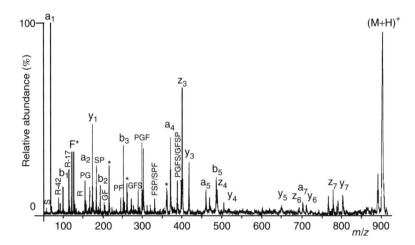

Figure 7.7
The MALDI mass spectrum of Des-Arg[1]-bradykinin, whose sequence is PPGFSPFR. The starred peaks correspond to the loss of a 28 Da fragment. (Reproduced (modified) from Ref. 49 with permission)

Fragmentation of peptides can be obtained also with a reflector TOF instrument by the PSD technique.[47,48] An example of such a spectrum is displayed in Figure 7.7. The observed fragments are close to those observed at low energy.[49]

7.2.2.2 Influence of charge position and charge delocalization

Protonation occurs mainly at the more basic sites. In peptides, the terminal amino group is basic. If more protons are added they will be located first at other basic amino acids, if present, and then on the amide groups, which will be more statistically distributed along the chain.

It has been shown that charged ions from small peptides that do not contain basic amino acids display comparable abundances for all the y_n and b_n ions. When the chain becomes longer, b_n fragments become favored.[50]

However, the presence and the position of a basic amino acid within the peptide influence the fragmentation process. The presence of basic amino acids such as Arg, Lys, His or Pro at the C-terminal amino acid induces mainly the formation of ions containing the C-terminal side (y_n, v_n and w_n), whereas the presence of these amino acids on the N-terminal side favors the formation of ions containing the N-terminal side (a_n and d_n). However, the absence of a basic site within the peptide is characterized by a distribution of ions containing one of the two sides (y_n, b_n), as shown in Figure 7.8.[51]

Similarly, increasing the localization of the positive charge on the peptide increases the 'a' and 'd' fragments, which end up outnumbering the b_n and y_n fragments if the charge is carried on the N-terminal side or outnumbering the y_n, v_n and w_n fragments if the charge is carried on the C-terminal side, as shown in Figure 7.9.[52]

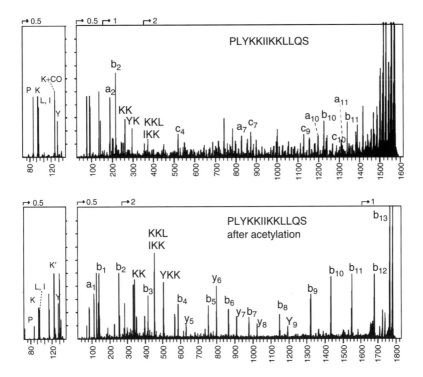

Figure 7.8
Influence of the presence and the position of the charge on the resulting fragments. (Reproduced (modified) from Ref. 51 with permission)

The rule describing the relationship between the position and the charge localization on the one hand and the fragmentation process on the other is summarized in Figure 7.10.

7.2.2.3 Strategy and examples of sequence determinations

Because mass spectrometry is entirely different from the other sequencing techniques, it offers several advantages over the latter, such as the analysis of peptides within mixtures or of peptides that are blocked on the N-terminal side or carrying post-translational modifications, etc. Mass spectrometry is thus complementary to other existing methods. Mass spectrometry also is more sensitive and more rapid that these other methods. Acquisition of the fragmentation spectrum of a peptide requires only a few minutes. On the other hand, the interpretation of this spectrum in many cases is far from simple. The complete sequence determination generally is difficult.

Interpretation of the spectra is based on the mechanisms and the fragmentation pathways described above, as shown by the following example. A CID/FAB/MS/MS trace of a peptide with sequence Gly-Ile-Pro-Thr-Leu-Leu-Leu-Phe-Lys measured at high energy is shown in Figure 7.11. This spectrum contains the complete series of type b_n ions, thus allowing one to deduce the peptide sequence from the N-terminal

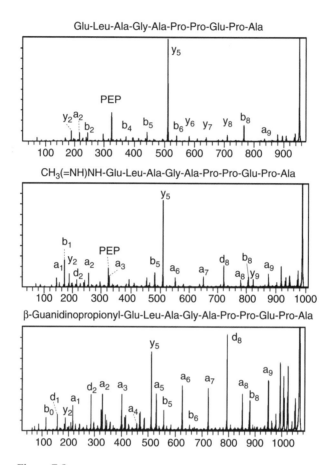

Figure 7.9
Influence of the charge delocalization on the type of fragmentation obtained. (Reproduced from Ref. 52 with permission)

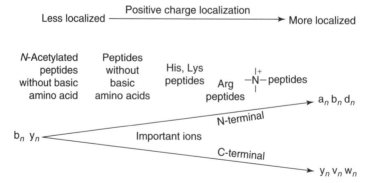

Figure 7.10
Charge and fragmentation with respect to the nature of the peptide

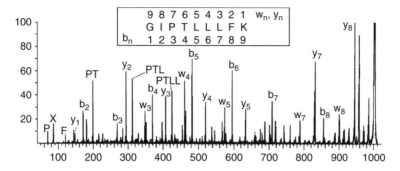

Figure 7.11
The FAB/MS/MS high-collision energy trace of a mass 1001 peptide with sequence Gly-Ile-Pro-Thr-Leu-Leu-Leu-Phe-Lys. (Reproduced (modified) from Ref. 34 with permission)

acid to the C-terminal acid, whereas the series of type y_n ions allows identification of the sequence in the reverse direction. In fact, the mass difference of 97 Da between peaks b_2 and b_3 indicates that the amino acid in position 3 corresponds to a proline (see Table 7.2). Similarly, the 147 Da difference between peaks y_1 and y_2 indicates that the amino acid in the next to last position is a phenylalanine. The m/z values of ions w_3, w_4, w_5 and w_8 imply that the amino acids in positions 3, 4 and 5 starting from the C-terminal side are leucines, whereas the amino acid in position 8 is an isoleucine. The presence of a proline induces the formation of internal fragments labeled PT, PTL and PTLL that allow one to verify the deduced sequence. The peaks labeled P, F and X represent the immonium ions and indicate the presence of proline, phenylalanine and leucine and/or isoleucine.

Interpretation of the fragmentation spectra can be difficult and time consuming. It can be simplified if the peptide is modified in such a way that one type of fragmentation is favored. This can be achieved by derivatizing the amino-terminal group with N-succinimidyl-2-(3-pyridyl) acetate, which promotes b_n fragments.[53] Modification of the peptide also can make the distinction between C- and N-terminal fragments easier.

Derivatization also allows specific residues to be detected and located by a comparison of the spectra before and after derivatization. The derivatizations most often used for this purpose are: the formation of ethyl esters from Asp, Glu and N-terminal carboxylic acids; acetylation of the N-terminal amino acid and Lys; and the Edman reaction for the N-terminal amino acid.[54]

Distinction between fragments containing the C- or the N-terminal can be achieved also by introducing an ^{18}O atom on the C-terminal of peptides obtained by tryptic digestion. Indeed, if the digestion is done using a 50/50 mixture of $H_2^{16}O$ and $H_2^{18}O$, fragments containing the carboxyl terminal are identified easily by the fact that they are dedoubled.[55] They will appear as a doublet of peaks of about the same abundance separated by $2/n$ Th, where n is the number of charges. The decision to use this method is to be taken before the digestion, whereas the other derivatization methods must be applied after digestion.

Sequence determination by mass spectrometry is based on the mass of the amino acids. However, Leu-Ile and Lys-Gln have the same masses. Differentiating Leu from Ile may be achieved using d_n and w_n fragments if they are present. To favor this type of fragment, it is possible to derivatize the peptide to localize a positive charge on the N-terminal side in order to favor the formation of d_n fragments, or to localize a positive charge on the C-terminal side in order to favor the formation of w_n fragments.[56]

The distinction between these two amino acids can be performed also at low energy by fragmenting the immonium ions from these amino acids at m/z 86. However, this method can be applied only to peptides that contain either leucines or isoleucines.[57] Differentiation between the Lys or Gln residue is performed by acetylation of the amine group using acetic anhydride.[35]

Several algorithms have been developed to interpret tandem mass spectra of peptides.[58-61] An algorithm named SEQUEST based on a different approach than the others has been developed for the interpretation of fragmentation spectra.

A correlation is searched between the fragmentation spectrum and peptide sequences contained in a database. First, the algorithm looks for all the peptides in the database that are the same mass as the precursor ions. Then, a measure of the similarity between the predicted fragments from the sequence obtained from the database and the fragments present in the spectrum of the sample allows the most probable sequence to be proposed.[62-64]

The use of ion trap or fourier transform mass spectrometry (FTMS) analyzers, which allow MS^n spectra to be obtained, presents two major advantages compared with classical MS/MS techniques. The dissociation of a fragment ion can lead to the observation of diagnostic ions not present in the fragmentation spectrum of the molecular species. Furthermore, the different fragments obtained from the molecular species are identified not only on the basis of their mass but also of their fragments.

The strategy for sequencing a protein using mass spectrometry starts with the precise determination of the molecular weight of that protein using MALDI or ESI. This result allows one to verify the sequence that is determined ultimately and also to judge the homogeneity of the sample. The protein then is subjected to reduction and alkylation of the cysteine bridges. Determination of the molecular weight allows one to determine the number of cysteines present in the protein.

In order to generate two different peptide series, the protein is digested with trypsin, which specifically cleaves on the C-terminal side of Lys and Arg, and by protease V8, which specifically cleaves on the C-terminal side of Glu and Asp (see Table 7.4).

About 0.1–2 nmol per digestion is sufficient for a CID/MS/MS analysis. After a reversed-phase HPLC fractionation, the molecular weight of each peptide is determined. Peptides with masses lower than 3000 Da that yield important signals are sequenced using CID/MS/MS. Peptides sequenced using CID/MS/MS then are assembled by matching the peptides in both series while considering the molecular weights of the large unsequenced peptides that contain several smaller sequenced peptides or the known sequence of a homologous protein.

Frequently a third enzymatic digestion is necessary to eliminate all the possible ambiguities. Using trypsin as a proteolysis enzyme allows one partially to distinguish

Table 7.4 Specific chemical or enzymatic cleavage of polypeptides (the cleavage is carboxyl side of X residue and amino side of Y residue)

Reagent	Cleavage site
Chemical cleavage	
Cyanogen bromide	Met-Y
Formic acid	Asp-Pro
Hydroxylamine	Asn-Gly
2-Nitro-5-thiocyanobenzoate	X-Cys
Phenyl isothiocyanate (Edman)	Terminal amino group of peptide
Enzymatic cleavage	
Trypsin	Arg-Y, Lys-Y
Chymotrypsin	Tyr-Y, Phe-Y, Trp-Y
Endoprotease V8	Glu-Y (Asp-Y)
Endoprotease Asp-N	X-Asp (X-Cys)

Lys from Gln, two isobars. In fact, the trypsin specifically cleaves on the C-terminal side of Lys and Arg, so all amino acids with mass 128 ending a peptide can be identified as Lys. However, the absence of a cleavage cannot be used as proof in identifying an amino acid with mass 128 as Gln, because the tryptic cleavage may be incomplete. The only way to distinguish Lys from Gln within a peptide is to acetylate the amine groups using acetic anhydride.[35] Figure 7.12 shows the strategy used in sequencing a protein using mass spectrometry.

7.2.3 Applications

The determination of a peptide or a protein molecular weight with high precision allows many problems to be solved, such as protein identification, detection of mutations within proteins, identification and localization of post-translational modifications, verification of structure and purity of synthetic peptides and proteins produced by genetic engineering, and even indirect sequencing. In opposition to direct sequencing based on gas-phase fragmentations, indirect sequencing is based on the generation of sequence-specific information from methods other than gas-phase methods, e.g. solution-phase reactions. However, in spite of these various possibilities, MS/MS is often used. Indeed, MS^n allows results based only on molecular weight determination to be confirmed and specified.

7.2.3.1 Protein identification

Classically, protein identification requires total or partial determination of their sequences by the method of Edman degradation. In some particular cases, protein identification can be based also on their recognition by specific antibodies. Edman degradation limits this method to the isolated polypeptides that are not blocked on the N-terminal side. Moreover, this technique requires a long analysis time and a large amount of sample.

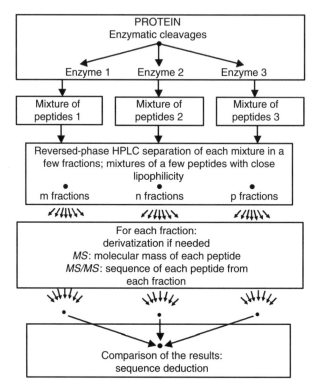

Figure 7.12
Strategy followed in sequencing proteins using mass spectrometry

Without any doubt, mass spectrometry is now the most efficient way to identify proteins.[65-67] The strategy used is depicted in Figure 7.13. The method is based on comparison of the data obtained from mass spectrometry with those predicted for all the proteins contained in a database. The efficiency of the method results from the development of mass spectrometry into a rapid and sensitive method to analyze peptides and proteins and also from the availability of larger and larger databases. In October 98, these databases contained more than 260 000 non-redundant sequences, including more than 80 million amino acids. Furthermore, the data obtained from genomic sequences after translation in the six lecture frames also can be used.

The databases based on expressed sequence tags are another usable source for search. They are composed of sequences based on cDNA fast sequencing. They are limited to short lengths of about 300 bases and contain many errors but they correspond to coding sequences. Despite their defects, they are very useful for identification of proteins by mass spectrometry.[68]

Accurate determination of the molecular weight of a protein is useful for its identification and the determination of its purity. Owing to the possibility of post-translational modifications that are not contained in the database, the correct mass often is not sufficient for identification of a protein but it does facilitate the search.

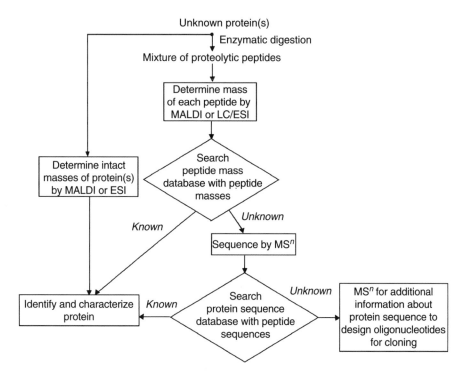

Figure 7.13
Strategy for the identification of proteins based on mass spectrometry only. This method allows peptides to be obtained from a two-dimensional gel electrophoresis with certainty, quickly and at high sensitivity

On the other hand, the peptide mixture obtained by cleavage of a protein by enzymes or specific reactants is characteristic of a protein, because the post-translational modifications only affect a small number of the peptides obtained.

The strategy thus is to cleave the protein to identify either by trypsin, V8 protease, Lys-C endoprotease or by a reactant such as BrCN. The mixture then is analyzed by mass spectrometry to obtain the molecular masses of the largest possible number of peptides. The two ionization methods MALDI and ESI are used: MALDI is the best to use if one wants to avoid chromatographic separation, because it yields very simple spectra, has a better sensitivity and is not as sensitive to the presence of contaminants; however, ESI can be coupled directly with HPLC or capillary electrophoresis (CE) if a separation is wanted. Furthermore, ESI often is used with mass spectrometers to obtain MS/MS data.

The profile of the masses of the peptides obtained by one of these methods is compared by means of a computer with a database of proteins. This is termed 'peptide mass fingerprinting'. A protein generally can be identified using the mass of four to six of its cleavage peptides having masses in the range 700–3000 Da and determined with an accuracy of 0.1–0.01%. This is improved further if the molecular weight of the protein also is provided.[69–72]

Table 7.5   Different protein identification programs available on the Internet

Name	The http address
Mass Search	http://cbrg.inf.ethz.ch/subsection3_1_3.html
Peptide Search	http://www.mann.embl-heidelberg.de/Services/PeptideSearchIntro.html
Profound	http://prowl.rockefeller.edu/cgi-bin/ProFound
MOWSE	http://www.seqnet.dl.ac.uk//mowse.html
MS-FIT	http://prospector.ucsf.edu/
Peptide Mass Search	http://www.mdc-berlin.de/~emu/peptide_mass.html

The specificity and reliability can be improved by increasing the number of peptide masses used by increasing the precision of the measured masses and by using different digestion enzymes with different specificities. The digest also can be reanalyzed after one cycle of Edman degradation, after a derivatization reaction or after deuteration to identify the N-terminal amino acid, to obtain information on the amino acid composition or to find out the number of exchangeable hydrogen atoms.

Various algorithms have been developed for the identification of proteins from the masses of the compounding peptides, and these are available through the Internet as shown in Table 7.5.

This method can be improved further using peptide sequence information obtained by MS/MS. Indeed, the sequence or even part of the sequence of a peptide is more specific than its molecular weight. This approach, termed 'peptide sequence tag', allows a protein to be identified from a partial sequence and from mass differences between this sequence and the N-terminal and the C-terminal of the peptide resulting from cleavage of the protein.[73,74]

Another approach, also based on MS/MS but that does not require interpretation in terms of the sequence of observed fragments, also has been developed. As described earlier, this method based on the SEQUEST algorithm compares the observed masses with those expected from a known sequence in the database.

These methods of course need the protein to be present in a database. If this is not so, the method of choice is to obtain some short sequences of part of the unknown protein. Oligonucleotidic probes then are synthesized and used to identify the gene coding for this protein. The gene then is cloned and sequenced.

Mass spectrometry thus allows the identification of proteins with a high degree of confidence from a very small quantity of sample. The method is fast and can be applied even to mixtures of a small number of proteins. The great number of peptides obtained after cleavage of the mixture of proteins is analyzed by liquid chromatography/tandem mass spectrometry (LC/MS/MS). The proteins of the mixture then can be identified. The possibility to be able to identify several mixed proteins is important, because the final product of many biological experiments, e.g. immunoprecipitation, is actually a mixture. Moreover, this capability can be very useful for the study of proteinic complexes and of protein–protein interactions.[75]

Polyacrylamide gel electrophoresis is certainly the method most often used for protein separation. This most powerful electrophoresis method combines a separation based on the isoelectric point in a first dimension and a size separation in a second dimension. Separation of several thousands of proteins in one analytical operation is possible. The proteins to identify then are often presented to the mass spectrometrist as a spot or a band on a gel. The methods described before remain applicable but several dedicated modifications have been proposed. They differ by the digestion step of the protein.[76-78] Either the proteins are digested directly on the gel or they are extracted first by electroelution or electroblotting.

Evolution of the gel separation technology and the databases, combined with the speed and sensitivity of the mass spectrometry analysis for identification, render possible the analysis of all the proteins from a population of cells (proteome).[66,79] This approach even allows variations in the expression of different proteins as a consequence of a stress to be studied.

7.2.3.2 Detection of mutations

Mass spectrometry allows the detection of mutations within proteins, whether they are natural or obtained through directed mutagenesis.[80-82] A point mutation within a protein leads to molecular weight variation of this protein. This variation is equal to the difference in mass between the wild-type and mutant residues. The region of protein containing the mutation can be determined by a chemical cleavage or proteolytic digestion of this protein followed by mass spectrometry of the resulting peptides. Peptides containing the mutation present the same variation of molecular weight as the intact protein. On the other hand, mutation-free peptides have measured molecular weights in agreement with those calculated for the wild-type protein.

Thus a change in a peptide molecular weight indicates the position of the mutation whereas the difference between the native peptide molecular weight and that of the mutant peptide allows one to determine the nature of the amino acid that has mutated (see Table 7.6) as long as only one possibility exists. Otherwise, MS/MS is necessary.

Hemoglobin is a protein for which a large number of mutants have been identified through mass spectrometry.[83-85] For example, a mutant of human hemoglobin (Hb) β chain, the Hb Providence mutant (β82Lys-Asp), was characterized using this strategy (Figure 7.14). After a tryptic digestion, fast atom bombardment/mass spectrometry (FAB/MS) analysis of the peptides that are obtained shows two missing peptides — T9 (sequence 67–82 with mass 1668.9 Da) and T10a (sequence 83–93 with mass 1220.6 Da) — whereas a new peak is present at m/z 2858.4. This suggests a mutation at Lys 82 that prevents any cleavage and thus prevents the formation of peptides T9 and T10a. Moreover, the new peptide mass corresponds to that of the peptide with sequence 67–93 if we allow for the mutation of Lys 82 into Asp 82.[86]

The problem is slightly different in the case of mutants obtained by directed mutagenesis. In that case, the expected modification is known and it is thus not necessary to verify the whole sequence. Comparing the molecular weight calculated from the mutant known sequence with that measured precisely using MALDI or ESI is sufficient. For example, ESI analysis of recombinant bovine somatotropin with molecular weight 21 816 Da and of five homologous somatotropins mutated by directed mutagenesis, having a Gly, Pro, Ser, Glu and Asp instead of an Asn 99, allowed it to be proved clearly that the first four mutations had been obtained by

Table 7.6 Mass differences (Da) observed between various amino acids: the residue in the row corresponds to the expected amino acid whereas the residue in the column corresponds to the mutant amino acid

	Gly	Ala	Ser	Pro	Val	Thr	Cys	Leu/Ile	Asn	Asp	Gln/Lys	Glu	Met	His	Phe	Arg	Tyr	Trp
Gly	—	14	30	40	42	44	46	56	57	58	71	72	74	80	90	99	106	129
Ala	—	—	16	26	28	30	32	42	43	44	57	58	60	66	76	85	92	115
Ser	—	—	—	10	12	14	16	26	27	28	41	42	44	50	60	69	76	99
Pro	—	—	—	—	2	4	6	16	17	18	31	32	34	40	50	59	66	89
Val	—	—	—	—	—	2	4	14	15	16	29	30	32	38	48	57	64	87
Thr	—	—	—	—	—	—	2	12	13	14	27	28	30	36	46	55	62	85
Cys	—	—	—	—	—	—	—	10	11	12	25	26	28	34	44	53	60	83
Leu/Ile	—	—	—	—	—	—	—	—	1	2	15	16	18	24	34	43	50	73
Asn	—	—	—	—	—	—	—	—	—	1	14	15	17	23	33	42	49	72
Asp	—	—	—	—	—	—	—	—	—	—	13	14	16	22	32	41	48	71
Gln/Lys	—	—	—	—	—	—	—	—	—	—	—	1	3	9	19	28	35	58
Glu	—	—	—	—	—	—	—	—	—	—	—	—	2	8	18	27	34	57
Met	—	—	—	—	—	—	—	—	—	—	—	—	—	6	16	25	32	55
His	—	—	—	—	—	—	—	—	—	—	—	—	—	—	10	19	26	49
Phe	—	—	—	—	—	—	—	—	—	—	—	—	—	—	—	9	16	39
Arg	—	—	—	—	—	—	—	—	—	—	—	—	—	—	—	—	7	30
Tyr	—	—	—	—	—	—	—	—	—	—	—	—	—	—	—	—	—	23
Trp	—	—	—	—	—	—	—	—	—	—	—	—	—	—	—	—	—	—

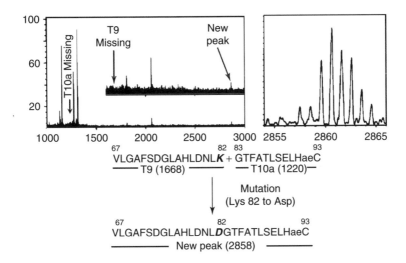

Figure 7.14
The FAB spectrum of the tryptic cleavage of the Hb Providence mutant of
human hemoglobin; aeC represents the modified cysteine as an aminoethyl
derivative. (Reproduced (modified) from Ref. 86 with permission)

comparison of the calculated molecular weights with the observed values.[87] In the
case of the last mutant, the difference between Asn and Asp (1 Da) is too weak
to be verified by this technique. In the case when the mass difference between the
substituted amino acids is smaller than the error in the measurement of the protein
molecular weight, a cleavage is necessary to obtain peptides of smaller masses. A
cleavage is necessary also when several mutations are introduced.

An interesting alternative to the strategy presented above is analysis by MALDI
using 'bioreactive probes'. The probe, used classically to receive the sample/matrix
mixture, is modified by covalent attachment of enzymes on its surface. The studied
protein is applied to this derivatized surface. After the time necessary for reaction,
a MALDI matrix is deposited. The reaction is thus blocked and after drying of
the matrix the digestion products are observed by MALDI. This approach has the
advantage of limiting the handling of the sample and eliminating sample loss.[82,88]

7.2.3.3 Identification and localization of post-translational modifications

Mass spectrometry also allows the investigation of post-translational modifications
or any unnatural chemical modifications of an amino acid leading to a mass
variation.[89–92] These modifications are difficult or even impossible to identify through
Edman degradation because they make the terminal N-amino acid inert to the
degradation reaction or the resulting derivative not easy to identify.

Considering the important role played by disulfide bridges in establishing and
preserving the three-dimensional structure of proteins, it is important to know whether
a protein contains disulfide bridges and which cysteines are linked by these bridges.

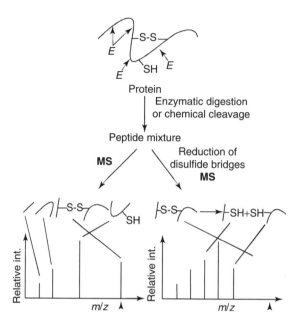

Figure 7.15
Strategy used to locate the cysteines linked by a disulfide
bridge

Mass spectrometry can locate the disulfide bridges within a protein. Analyzing the
peptides produced by enzymatic cleavage of the protein using a protease before
and after the reduction of all disulfide bridges is sufficient. Peptides containing an
intermolecular disulfide bridge disappear and yield two new peptides containing the
cysteine residues responsible for the disulfide bridge, as is shown in Figure 7.15.[93]

Both disulfide bridges of Paim I, a 73-amino-acid protease inhibitor of α-amylase,
were determined by mass spectrometry.[94] This protein was digested using *Staphy-
lococcus aureus* V8 protease and the peptides obtained were analyzed through FAB
before and after HPLC separation. In the FAB spectrum of the mixture, two ions
with masses 2523 and 3066 Da do not correspond to any peptide generated from
the protein sequence. Therefore, they may be considered to be peptides containing
a disulfide bridge. Based on their mass and on the specificity of the protease used,
these ions characterize the bridged peptides Gly 37–Glu 55 + His 66–Ser 73 and
Gly 37–Glu 55 + His 61–Ser 73.

The FAB spectrum allows this hypothesis to be confirmed. The peptides resulting
from the cleavage of this bridge appearing in the spectrum for these bonds are
easily reduced *in situ* during the FAB/MS analysis. This suggests that Cys 42 is
linked with Cys 70. The spectrum of a fraction of the mixture shows the ions
with masses 1366 and 1472 Da corresponding to peptides Ala 1–Glu 13 and Ser
14–Asp 26 and an ion with mass 2836 Da corresponding to the bridged peptide
Ala 1–Glu 13 + Ser 14–Asp 26. This suggests that the second disulfide bridge that
is possible between Cys 8 and Cys 24 is the actual one (Figure 7.16).

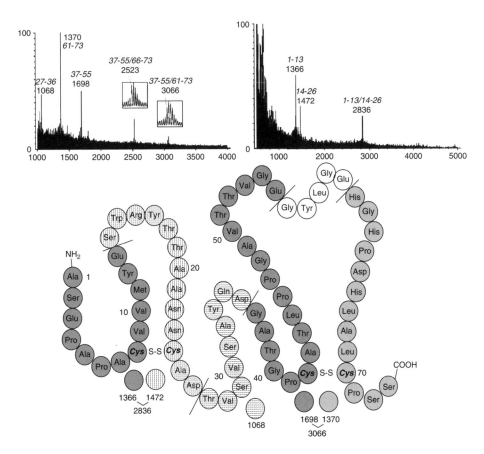

Figure 7.16
The FAB spectra of the digestion of Paim I by protease V8 and of a peptide present in this mixture but HPLC- fractionated. (Reproduced (modified) from Ref. 94 with permission)

Many other post-translational modifications have been identified and localized by mass spectrometry. The most important ones are phosphorylation[95] and glycosylation.[96] These two modifications can be detected by specific precursor ion scans with detection limits lower than picomoles. This high sensitivity makes the method particularly attractive but it has been observed that the specificity of the diagnostic ions used to detect glycosylation is not sufficient to assign a glycopeptide with certainty in a complex mixture.[97,98]

Another elegant approach to detect phosphorylated or glycosylated peptides during LC/ESI/MS makes use of 'collision excitation scanning'. When low-mass ions are scanned, the collision potential in the source is set to a high value so that low-mass ions are generated in abundance. Those ions that are diagnostic are m/z +204 for glycosylation and −79 for phosphorylation, respectively. When higher masses are scanned, the collision energy is set to a low value to observe the ions of the molecular species. A comparison of the ultraviolet (UV) trace, which locates the different peptides on a time scale, combined with a trace of the diagnostic ion used allows a retention time to be assigned to the glycosylated or phosphorylated peptides. The observation of the higher mass region of the spectra at this retention time provides the molecular masses of the modified peptides.[99,100]

It is sometimes possible to determine the nature and localization of post-translational modifications using mass spectrometry by a correct choice of the protease, yielding peptides containing only one possible site for this modification, but this is not always possible. In that case it is necessary to use MS^n because this technique also allows one to determine the nature and localization of post-translational modifications, as is shown in the example of the peptide with sequence Arg-Pro-Pro-Gly-Tyr-Ser-Pro-Phe-Arg that underwent phosphorylation.[101]

Because tyrosine 5 or serine 6 are two potential phosphorylation sites, it is necessary to use MS/MS. The tandem mass spectrum of the peptide shown in Figure 7.17 indicates that the main ions are a_n and c_n fragments for the N-terminal side and x_n

	Arg	Pro	Pro	Gly	Tyr(P)	Ser	Pro	Phe	Arg
v_n					561				
w_n		957	860			473			
y_n		998	901	806	747		417		
		918	821	724	667		417		
x_n				832	775				
a_n		226	323		623	710		954	
		226	323		543	630		874	
c_n				425	668				
d_n					694				

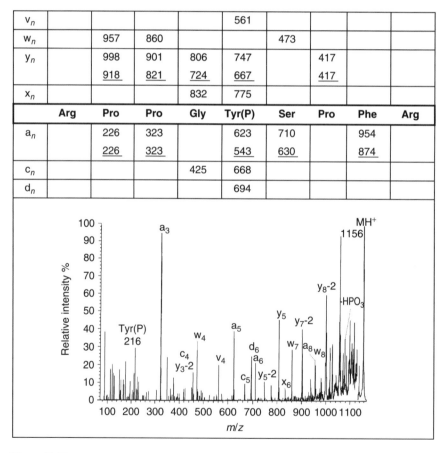

Figure 7.17
The FAB MS/MS trace of phosphorylated bradykinin [$Tyr_5(P)$]. The masses underlined in the table correspond to phosphate-free fragments. (Reproduced (modified) from Ref. 101 with permission)

and w_n fragments for the C-terminal side, allowing one to localize the modification on tyrosine 5.

In addition to the previously described ions, others at m/z 1059 and 1058, corresponding to the loss of HPO_4 and $H_2PO_4^{\bullet}$, and at m/z 216, characterizing the immonium ion of phosphorylated tyrosine, confirm this localization.

7.2.3.4 Verification of the structure and purity of peptides and proteins

Mass spectrometry also can play an important role in the verification of the structure and purity of synthetic peptides.[102,103] The development of automatic synthesizers has made the production of synthetic peptides increasingly easier. A large number of errors, however, may occur during or after the synthesis (see Table 7.7), the majority of which are detectable by mass spectrometry. The tandem mass spectrum allows one to confirm the nature of the modification but also to determine its position within the peptide.

For example, Figure 7.18 shows the MALDI spectrum of a synthetic peptide with a molecular weight of 1984.2 Da. Two compounds appear next to the sought peptide. The mass difference between these two impurities and the synthetic peptide suggests that partial oxidation of methionine and glycine addition occurred during the synthesis.

Table 7.7 Average mass differences for selected synthetic problems

Mass difference (Da)	Problem	Affected residues
−32	Desulfurization	Cys
−18	Dehydratation, diketopiperazine	Asp, Glu, C-terminus, Gln
−17	Loss of NH_3 with cyclization	Asn, Gln
−2	Disulfure bond formation	Cys
+1	Deamidation	Asn, Gln
+14	Methylation	Asp, Glu, Lys, N-terminus
+16	Oxidation	Met, Trp
+28	Formylation	Ser, Thr, Tyr, Lys, N-terminus
+32	Di-oxidation	Met, Trp
+48	Tri-oxidation	Cys, Trp
+56	t-Butyl (incomplete deprotection)	
+90	Anisylation	Asp, Glu
+99	HBTU/AA >1[a]	N-terminus
+100	t-Butoxycarbonyl (incomplete deprotection)	
+106	Thioanisylation	Asp, Glu
+222	9-fluorenylmethoxycarbonyl (incomplete deprotection)	
+242	Trityl (incomplete deprotection)	

[a]HBTU = 2-(1H-benzotriazole-1-yl)-1,1,3,3-tetramethyluronium hexaflurophosphate; AA = aminoacid.

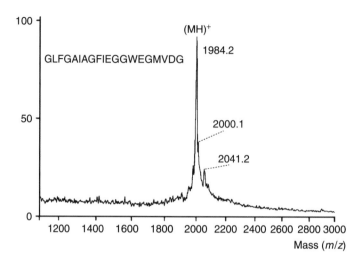

Figure 7.18
The MALDI spectrum of a synthetic peptide. (Reproduced from
Finnigan MAT documentation, with permission)

The ESI and MALDI techniques also allow rapid verification of the fidelity and
homogeneity of proteins produced by genetic engineering.[103–105] In most cases, veri-
fying the sequence is not sufficient. It is necessary to check whether all the wanted
covalent modifications are present.

Figure 7.19 outlines the general strategy to characterize recombinant proteins
by mass spectrometry. If the accurately measured molecular weight of the protein
produced is in agreement with that calculated from the DNA sequence, then the
deduced amino acid sequence is correct, assignment of the N- and C-terminals is
correct, no post-transitional modifications occurred and the amino acids are not chemically modified. In contrast, if the measured molecular weight does not fit with
the calculated one, then there is either a sequence error or a modification of the
protein.

In such a case, the protein is cleaved in peptides and the peptides are analyzed
by mass spectrometry. Peptides whose mass or sequence do not correspond to those
expected from the DNA sequence provide information on the errors or modifications
that occurred.

For example, ESI/MS allowed the analysis of various proteins derived from the
HIV virus and obtained by genetic engineering, such as the p18 protein.[106] Beyond
the main peak series corresponding to the protein with a measured molecular weight
of 14 590 Da (as opposed to a calculated molecular weight of 14 589 Da), the spectrum shown in Figure 7.20 contains two other series. The first minor series (8%)
indicated by the letter T and giving a measured mass of 12 651 Da corresponds to the
C-terminal side of p18 cleaved at position 111–112, whereas the second series (3%)
indicated by the letter D with a measured mass of 29 175 Da corresponds to the
dimer formed by the linkage of two cysteines of two different proteins.

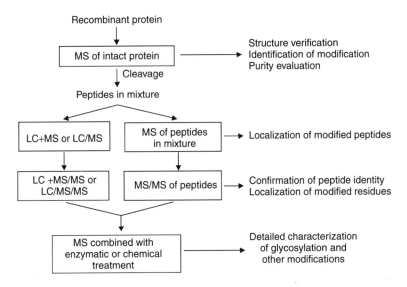

Figure 7.19
General strategy to characterize recombinant proteins

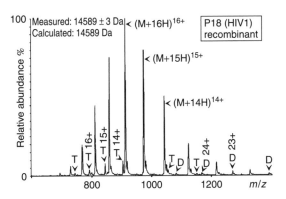

Figure 7.20
The ESI spectrum of the p18 recombinant protein obtained by genetic engineering. (Reproduced (modified) from Ref. 106 with permission)

7.2.3.5. *Protein ladder sequencing*

An approach to protein sequencing without using MS/MS has been described.[107] This two-step approach is called 'protein ladder sequencing'. First, a set of sequence-defining peptide fragments, each differing from the next by one amino acid, is generated by wet chemistry from a polypeptide chain that must be sequenced. Second, MALDI mass spectrometry is used to read out the complete fragment set in a single operation. The mass spectrum contains ions corresponding to each polypeptide species present. The mass differences between consecutive peaks correspond to amino acid residues, and their order of occurrence in the data set defines the sequence of amino acids in the original peptide chain. The sensitivity of protein ladder sequencing (picomole total amounts of peptide samples) is comparable to that of existing Edman methods, with the potential for far greater sensitivity.

Several methods were proposed to obtain this set of peptide fragments. The first method suggested is based on a rapid stepwise degradation using the Edman reagent (phenyl isothiocyanate, PITC) in the presence of a small amount of terminating agent (phenyl isocyanate, PIC). A small proportion of peptide chain blocked at the amino terminus is generated in each step. A predetermined number of cycles is performed without intermediate separation or analysis. The following scheme illustrates the PITC reaction with the terminal amino group of the peptide to liberate a cyclic derivative of the N-terminal amino acid corresponding to a phenylthiohydantoin amino acid and to leave an intact peptide shortened by one amino acid. The principle of the method and an example are presented in Figure 7.21.

R = Peptide chain

In another method, a volatile degradation reactant is used, the trifluoroethylisoth-iocyanate, to avoid any need for a complex cleaning method. Production of the complete set of fragment peptides is performed by adding, at each cycle, the same amount of the original protein. There is no need to add a terminating agent. This allows one to use, later on, a derivatization reagent of the free terminal amine to improve the detection sensitivity.[108]

The two preceding methods allow sequencing from the N-terminal side. Other methods have been developed to obtain a set of peptides characteristic of the C-terminal side. To do this, enzymatic or chemical degradation methods have been used.[109,110] If the enzymatic degradation is used, the peptide sample is divided into several parts, each treated with an increasing amount of carboxypeptidase. This allows fragment peptides covering all the sequence to be generated.

7.2.3.6 Non-covalent protein complexes and tridimensional structural information

Mass spectrometry and particularly ESI can be used for the study of non-covalent interactions of proteins. Indeed, ESI is a sufficiently soft ionization method to allow, under appropriate circumstances, the observation of non-covalent complexes of proteins formed in solution and transferred in the gas phase.[111–117]

The first interest is to obtain from the information in the gas phase results that are representative of the species present in the liquid phase. However, transposition of the results has to be done with many cautions. The forces responsible for molecular interactions in the liquid phase are not the same as those in the gas phase. Indeed, the

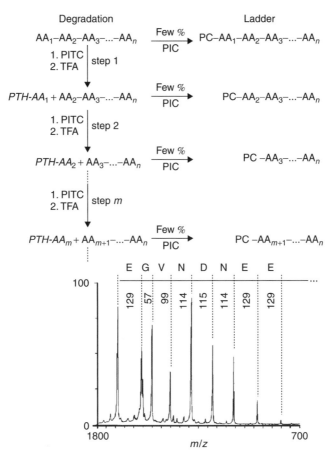

Figure 7.21
(*Top*) Protein ladder sequencing principle. Phenyl isothiocyanate (PITC) produces phenylthiohydantoin (PTH) of the terminal amino acid and a new peptide with one less amino acid. Phenyl isocyanate (PIC), in low quantity, produces N-terminal phenylcarbamate (PC) from a small fraction of each peptide. (*Bottom*) Example of sequencing of [Glu^1] fibrinopeptide B. (Reproduced (modified) from Ref. 107 with permission)

main interactions acting in solution result from Van der Waals forces, hydrophobic forces and hydrogen bonding. In the gas phase, electrostatic forces are predominant. Even if a great number of observations demonstrate that complexes in the gas phase reflect the properties in the liquid phase, no generalization can be made. Each case is unique. Several controls have to be performed in order to confirm that the gas phase observations are related to the liquid phase behavior.

Several experimental methods can be used to study non-covalent interactions. Among them, let us mention various spectroscopic methods and analytical ultracentrifugation. Each method presents advantages and inconveniences. Nuclear magnetic resonance and X-ray diffraction allow information to be obtained on the structure of proteins in solution or in the solid state. However, these two methods require large quantities of high-purity material and are slow. The advantages of mass spectrometry are rapidity, sensitivity and specificity. Mass spectrometry allows fast determination of the stoichiometry of the complex because the molecular mass is determined directly. Mass spectrometry also allows information to be obtained on the relative stability of the complex by studying the energetics of its dissociation.

Electrospray ionization has allowed the observation of a great number of non-covalent complexes: protein–protein, protein–metal ion, protein–drug and protein–nucleic acid. About one–third of the proteins exist as multimeric forms. Mass spectrometry allows the study of their quaternary structure. This has been done for alcohol dehydrogenase from horse liver and from yeast. The ESI spectra are displayed in Figure 7.22. The horse liver alcohol dehydrogenase is observed to be dimeric, whereas that of yeast is tetrameric.[118]

These observations are in agreement with structural studies based on other techniques. The observed abundances of the various forms are dependent on the experimental conditions. An increase in the energy transferred to the ions in the ESI source leads to dissociation to monomers.

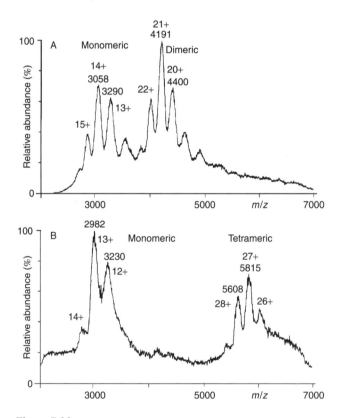

Figure 7.22
The ESI mass spectra of alcohol dehydrogenase under native form obtained from (A) horse liver and (B) yeast. (Reproduced from Ref. 118 with permission)

Some recent papers suggest that MALDI can be used to study non-covalent complexes.[119,120] However, in ESI the solution is injected directly into the ion source whereas MALDI requires crystallization of the sample with the matrix. The effect of sample handling thus could be larger. Furthermore, the energy deposited to desorb the ions is not clearly known. It is thus too early to consider MALDI as an established technique for this purpose.

From a practical aspect, the ESI conditions that allow the survival of non-covalent complexes are not the standard ones. Normally, to obtain a maximum of sensitivity and to have a stable signal, the polypeptides are analyzed in basic or acidic solution in the presence of organic solvents. However, non-covalent complexes are not stable under acidic or basic conditions or in the presence of organic solvents.

Similarly, the ESI source parameters are important to observe non-covalent complexes, and the parameters that enable them to be observed are not the common ones. During the ESI process, ions are desolvated in a first step by collisions occurring at the interface between the atmospheric pressure and the vacuum, and depend on the potential difference between the electrodes, nozzle and skimmer or capillary and skimmer. To observe fragile non-covalent complexes, this collision-activated desolvation must be reduced. It is interesting to note that most observed complexes carry a reduced number of charges. They thus appear at more elevated m/z values and analyzers with a sufficient mass range should be used.

Mass spectrometry also provides information on the tridimensional structure of proteins. Often, the information from mass spectrometry complements those obtained by other techniques such as circular dichroism, nuclear magnetic resonance or fluorescence. In some circumstances, mass spectrometry, by its speed and sensitivity, allows information to be obtained that is impossible to obtain by other techniques.

Protein analysis by ESI/MS can occur on the native (folded) or on the denaturated (unfolded) form.[121,122] Indeed, the conformation of the protein in solution has an important effect on the peak distribution observed in the ESI spectrum, as shown in Figure 7.23. The distribution of the charges on the denaturated protein appears broader and with a larger number of charges, i.e. larger z value, than for the native one. This reflects better accessibility during the ionization process to the basic amino acids in the

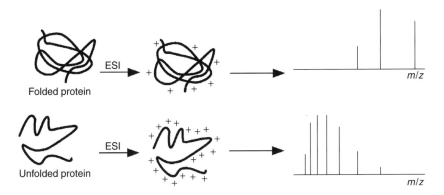

Figure 7.23
Characteristic charge distribution of native or denaturated protein

unfolded structure. This difference also allows the denaturation process to be followed over time and sometimes possible intermediates to be detected (Figure 7.24).[123]

Thus, direct information on protein structure in space can be obtained from mass spectrometry. However, this information is most often obtained in an indirect way. For example, hydrogen/deuterium exchange occurring over time in solution can be followed by mass spectrometry.[124–126] As the protein unfolds, more and more exchangeable sites are exposed to the solvent. Exchangeable hydrogens located inside the folded protein only exchange hydrogen for deuterium at a very low rate. The role of mass spectrometry is limited to the determination of the number of deuterium atoms incorporated. However these results can be interpreted in terms of global conformation and stability of the protein in solution. Information even can be obtained on selected regions of the protein.

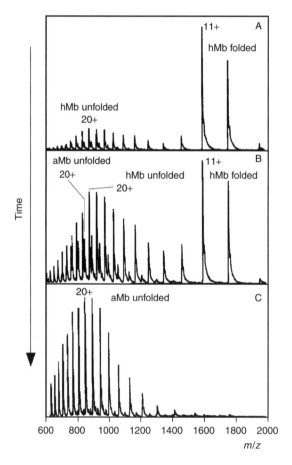

Figure 7.24
Results from the acidic denaturation of myoglobin as a function of time. Three species are observed: native hMb (heme + myoglobin), denatured hMb (heme + denatured hemoglobin) and denatured aMb (denatured apomyoglobin). (Reproduced (modified) from Ref. 123 with permission)

A similar method is based on the observation of irreversible reactions with a functional group of the side chain of an amino acid.[127,128] This method allows the accessibility of the solvent to a selected amino acid to be measured. The role of mass spectrometry is to determine the number of modified amino acids and their position in the sequence of the protein.

Another method is based on the proteolysis of the protein.[129,130] A possible cleavage site will be cleaved by an enzyme only if it is accessible. Thus, sites located inside the protein or surrounded by bulky groups will be resistant to proteolysis. In contrast, sites located in flexible areas or sites are not highly structured are easily cleaved. On the basis of the mass of the peptides obtained, mass spectrometry allows sites accessible for cleavage to be located. This method allows also the determination of the region of a protein responsible for the interaction with a ligand. Indeed, a variation in the conformation of a protein changes the accessibility towards a protease. Similarly, the interaction of a protein with a ligand results in the protection of some sites to the cleavage. The strategy in use is based on qualitative as well as quantitative comparison of the peptides obtained after proteolytic cleavage of the protein in the presence or absence of the ligand.

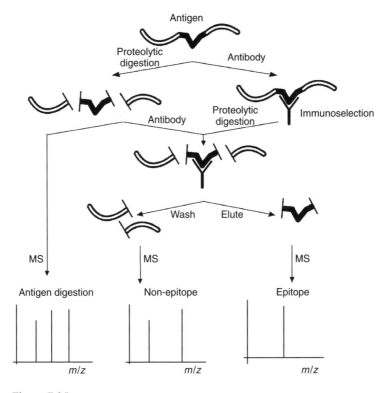

Figure 7.25
Two different methods for the identification by mass spectrometry of an epitope recognized by a given antibody. The selection by affinity for the antibody is performed either before or after the enzymatic cleavage

A specific application of this last method is the identification of epitopes recognized by a given antibody.[131,132] As illustrated in Figure 7.25, two possible approaches are used. In the first one, the protein presenting the epitope is digested by a protease and the peptides obtained that present the epitope are selected by the antibody. The mass spectrometry analysis, in general MALDI, allows peptides to be identified that are bound to the antibody and thus present the epitope.

The alternative method consists of proteolysis of the protein–antibody complex. Mass spectrometry allows those sites to be observed that are protected against the enzymatic cleavage. Such sites are implied in the protein–antibody interaction. This method allows the identification of non-linear epitopes, whereas the first method is limited to linear epitopes.

Figure 7.26 illustrates the application of the first method to an antibody directed against the recombinant protein bFGF. Twelve peptides are identified in the MALDI spectrum of the mixture obtained after enzymatic cleavage (Figure 7.26A). After immunoprecipitation with the antibody, the non-bounded peptides are washed out

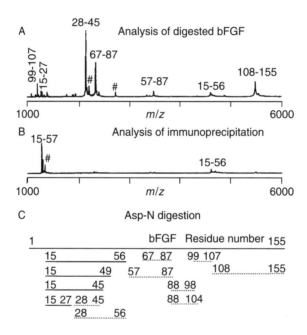

Figure 7.26
The MALDI mass spectrum obtained from the peptides resulting from enzymatic digestion (Asp-N) of the bFGF protein: (A) spectrum before and (B) spectrum after immunoprecipitation with an antibody directed against the recombinant protein. The peaks marked # correspond to either Cu adducts or doubly charged ions. (C) Schematic representation of the observed peptides. A continuous line corresponds to a peptide that links to the antibody, and a dotted line to a peptide that does not link. (Reproduced (modified) from Ref. 132 with permission)

and only four peptides are detected, as shown in Figure 7.26B. A comparison of the sequences of these peptides allows the epitope to be located somewhere in residues 15–27, as explained in Figure 7.26C.

Table 7.8 lists the mass modifications produced by the replacement of a given amino acid by another in the sequence of a peptidic chain.

Table 7.8 Amino acid substitutions and variations in corresponding masses

Substituted amino acids	Nominal mass difference	Exact mass difference	Substituted amino acids	Nominal mass difference	Exact mass difference
Leu/Ile	0	0.000	Lys/Thr	27	27.047
Lys/Gln	0	0.036	Arg/Lys	28	28.006
Glu/Lys	1	0.948	Val/Ala	28	28.031
Asn/Ile	1	0.959	Arg/Gln	28	28.043
Asp/Asn	1	0.984	Glu/Val	30	29.974
Glu/Gln	1	0.984	Trp/Arg	30	29.978
Met/Lys	3	2.946	Met/Thr	30	29.993
Thr/Pro	4	3.995	Ser/Gly	30	30.011
His/Gln	9	9.000	Thr/Ala	30	30.011
Pro/Ser	10	10.021	Gln/Pro	31	31.006
Ile/Thr	12	12.036	Met/Val	32	31.972
Asn/Thr	13	12.995	Phe/Leu	34	33.984
Lys/Asp	13	13.068	Phe/Ile	34	33.984
Ala/Gly	14	14.016	His/Pro	40	40.006
Thr/Ser	14	14.016	Val/Gly	42	42.047
Leu/Val	14	14.016	Arg/Leu	43	43.017
Ile/Val	14	14.016	Arg/Ile	43	43.017
Glu/Asp	14	14.016	Asp/Ala	44	43.990
Lys/Asn	14	14.052	Phe/Cys	44	44.059
Gln/Leu	15	14.975	Cys/Gly	46	45.988
Lys/Ile	15	15.011	Phe/Val	48	48.000
Asp/Val	16	15.959	Tyr/Asp	48	48.036
Cys/Ser	16	15.977	Tyr/Asn	49	49.020
Ser/Ala	16	15.995	Arg/Cys	53	53.092
Tyr/Phe	16	15.995	Arg/Thr	55	55.053
Leu/Pro	16	16.031	Asp/Gly	58	58.005
Met/Leu	18	17.956	Glu/Ala	58	58.005
Met/Ile	18	17.956	Arg/Pro	59	59.048
Arg/His	19	19.042	Phe/Ser	60	60.036
His/Asp	22	22.032	Tyr/Cys	60	60.054
His/Asn	23	23.016	Arg/Ser	69	69.069
His/Leu	24	23.975	Glu/Gly	72	72.021
Arg/Met	25	25.061	Trp/Leu	73	72.995
Tyr/His	26	26.004	Tyr/Ser	76	76.031
Pro/Ala	26	26.016	Trp/Cys	83	83.070
Leu/Ser	26	26.052	Trp/Ser	99	99.047
Ile/Ser	26	26.052	Arg/Gly	99	99.080
Asn/Ser	27	27.011	Trp/Gly	129	129.058

7.3 Oligonucleotides

Oligonucleotides, also named nucleic acids (DNA or RNA), are linear polymers of nucleotides. The nucleotides are composed of a heterocyclic base, a sugar and a phosphate group. There are five different bases. Cytosine (C), thymine (T) and uracil (U) are pyrimidine bases and adenine (A) and guanine are purine bases. Both DNA and RNA not only differ by their base composition (ACGT for DNA and ACGU for RNA) but also by the nature of the sugar part (2′-deoxy-D-ribose in DNA and D-ribose in RNA). The nucleosides and the deoxynucleosides are formed by binding the N-9 position of the purines or the N-1 position of the pyrimidines to the C-1′ of ribose or deoxyribose, respectively.

The structures of the main nucleosides are presented in Figure 7.27. In the polymer chain, the nucleosides are linked together by the phosphate groups attached to the 3′ hydroxyl of one nucleoside to the 5′ hydroxyl of the adjacent nucleoside. Moreover, these oligonucleotides may undergo natural covalent modifications that are most often present in tRNA and in rRNA, or unnatural ones that result from reactions with exogenous compounds. An important aspect of the research in this field consists of characterizing these structural modifications. Mass spectrometry plays an important role in highlighting these modifications and in determining their structures and positions within the oligonucleotide.

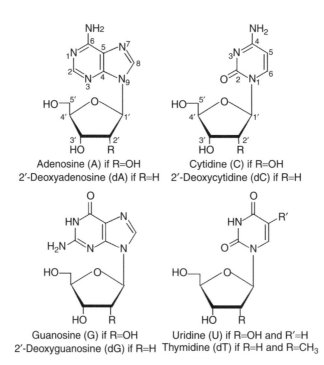

Adenosine (A) if R=OH   Cytidine (C) if R=OH
2′-Deoxyadenosine (dA) if R=H   2′-Deoxycytidine (dC) if R=H

Guanosine (G) if R=OH   Uridine (U) if R=OH and R′=H
2′-Deoxyguanosine (dG) if R=H   Thymidine (dT) if R=H and R=CH₃

Figure 7.27
Structures of the main nucleosides making up the oligonucleotides

Mass spectrometry not only allows precise determination of the molecular weight of oligonucleotides, but also in a direct or indirect way the determination of their sequences. Reviews on the mass spectrometry analysis of oligonucleotides have been published in the last few years.[133–136]

7.3.1 Mass spectra of oligonucleotides

For many years, mass spectrometry has had an important role in the analysis of nucleic acids. However, older work was limited to the analysis of nucleotides because the EI and CI methods did not allow oligonucleotide analysis. With the advent of FAB and PD, small oligonucleotides comprising up to about 10 bases have been analyzed.

Because the nucleic acids are the most polar of the biopolymers, they have benefited much from the advent of ESI and MALDI because such methods allow the analysis of very small quantities of oligonucleotides, starting at some tens of femtomoles. However, the oligonucleotides are not so easy to analyze as the peptides and proteins. One reason for this is the formation of strong bonds between the phosphodiester group and the alkali metal atoms, mostly Na and K, leading to the formation of adducts containing several of these atoms. The oligonucleotides are best detected in the negative ion mode, and the observed ions have the general formula $(M - [n + m]H + mNa/K)^{n-}$. This distribution of the ions of the molecular species over several mass values leads to a reduction of the signal-to-noise ratio. The separation of the individual ions requires high resolution. Generally, they are not separated and result in one large peak. This reduces the accuracy of the mass determination. Furthermore, mixtures containing compounds with small mass differences will not be separated. This difficulty can be overcome, however, by adding ammonium ions to the sample.[137] These will replace the alkali metal ions by binding to the phosphodiester group, but in the gas phase ammonia is released, leaving the free oligonucleotide.

A second difficulty in the analysis of oligonucleotides is their easy fragmentation. As will be discussed in more detail, the oligonucleotides tend to lose their nucleic bases by a 1,2-elimination mechanism. Fragmentation of the nucleic chain at either the 5′ or 3′ side of the sugar having lost a base then can occur. This stability problem is more important for MALDI than for ESI analysis.

Oligonucleotides can be analyzed in both positive and negative ion modes. However, negative ion mode give better sensitivity and resolution, especially in ESI.

Figure 7.28 displays the MALDI spectra of a DNA 40-mer in the negative ion mode and an RNA 195-mer in the positive mode.[138,139] The MALDI spectra are dominated by either the protonated $(M + H)^+$ or deprotonated $(M - H)^-$ ions of the molecular species. Multimers of general formula $(nM \pm H)^\pm$ are generally detected, but multiply charged ions are sometimes present, mainly for large oligonucleotides. As mentioned before, metal adducts also are often present.

The success of the analysis of oligonucleotides is strongly dependent on the choice of the matrix and the preparation of the sample. A major improvement has resulted from the introduction of new matrices specific for the oligonucleotides. An example

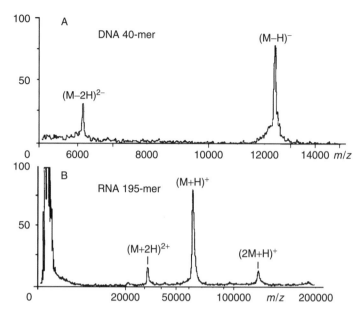

Figure 7.28
(A) The MALDI linear TOF mass spectrum of a DNA 40-mer in
negative ion mode. (B) The MALDI reflection TOF mass spectrum
of an RNA 195-mer in the positive ion mode. In both cases, the
spectra were obtained using 3-HPA as matrix and a laser at 337 nm.
(Reproduced (modified) from Ref. 138 and 139 with permission)

is 3-hydroxypicolinic acid.[140] Several strategies for sample preparation have been
adopted, with the principal goal of eliminating interference from the ubiquitous alkali
metal ions: a combination of the use of cation exchange columns to replace the alkali
metal ions by ammonium and the addition of some ammonium salt to the sample.

The response of nucleotides to MALDI depends also on their nature, their size and
their composition in nucleic bases. Oligonucleotides containing only thymine give
a much stronger signal than those composed of the other bases. Thymine is more
difficult to protonate than the other bases. It is thus also more difficult to eliminate,
which results in better stability of the polymer chain. The RNA is easier to analyze
than the DNA. This also appears to result from the fact that 1,2-elimination of the
base occurs only with deoxyribose; the presence of the $2'$-hydroxyl group in RNA
hinders the elimination reaction.

When the size of the nucleotide increases it becomes more difficult both to ionize
and to detect. Furthermore, the formation of adducts is increased and the fragmentation reaction also increases. These facts together explain why larger oligonucleotides
are detected with a lower signal-to-noise ratio and a lower resolution, or the signal
may be totally suppressed. Dependence on the size results in suppression of the
largest nucleotides when equimolar mixtures with small ones are analyzed.[141]

The error on the measured molecular mass obtained by MALDI varies between 0.01
to 0.05% for small oligonucleotides. This error increases to 0.5% or even more for large

nucleotides. Indeed, at higher masses the signals from adducts or resulting from the loss of a base are more abundant and cannot be resolved, any more, causing a broadening of the peak. The error on the centroid determination increases. One could consider that the maximum size that can be analyzed reasonably by MALDI is around 50 bases for DNA. For RNA more than 100 bases have been analyzed with success.[139]

The ESI mass spectra of a synthetic oligonucleotide CATGCCATGGCATG and tRNA of more than 24 kDa are displayed in Figure 7.29.[142,143] The ESI mass spectra display a series of peaks of different charge states resulting from the loss of several protons $(M - nH)^{n-}$ as well as metal adducts $(M - [n + m]H + mNa/K)^{n-}$ whose

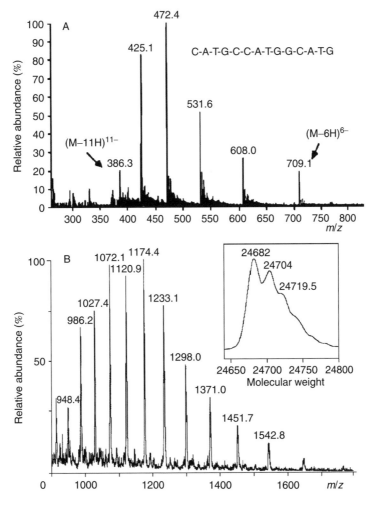

Figure 7.29
(A) Negative ion ESI mass spectrum of a 4260 Da synthetic oligonucleotide. (B) The ESI mass spectrum of tRNA^{Val1} from E. coli; the calculated mass is 24 681 Da. (Reproduced (modified) from Ref. 142 and 143 with permission)

importance depends primarily on the sample preparation. The charge state is generally around one charge per 1000 Da as a mean, depending on the analyzed nucleotides and on the experimental conditions: nature of the solution and source parameters.

As for MALDI, elimination of the metal adducts is critical to the success of the analysis. One strategy consists of precipitating the oligonucleotide with an ethanol solution containing ammonium acetate and then redisolving it in a dilute aqueous solution of ammonia. This solution also may contain divalent ion chelating agents, which allows one to observe up to a 50-mer devoid of sodium adducts; without precautions only up to one sodium per phosphate group can be observed. Another strategy is the use of a 20/80 water/acetonitrile solution containing imidazole, pyridine and acetic acid. This solution not only allows the sodium adducts to be reduced but also reduces the abundance of charge states, thereby increasing the sensitivity and simplifying the spectra.

Molecular weights determined by ESI are generally more precise that those obtained by MALDI. Errors as low as 0.01% are often obtained after appropriate sample preparation. Oligonucleotides containing more than 100 bases have been analyzed successfully by ESI.[143]

Electrospray ionization with the high resolving power of Fourier transform mass spectrometry (FTMS) makes possible the detection of adducts and subpicomole impurities that confuse lower resolution measurements, as shown in Figure 7.30. This method achieves accurate determination of molecular weights (<0.002%) and permits the verification of 50–100-mer DNA and RNA sequences.[144,145]

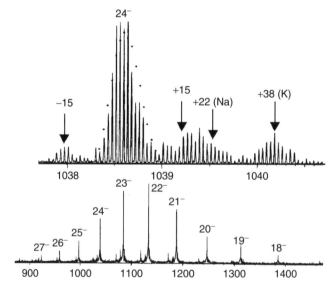

Figure 7.30
The ESI/FTMS trace of tRNA^{Phe} with a zoom of the isotopic cluster for the 24$^-$ anion. Peaks at −15 and +15 Da are due to minor impurities. (Reproduced (modified) from Ref. 145 with permission)

7.3.2 Applications of mass spectrometry to oligonucleotides

The simplest and also most common application of mass spectrometry is the accurate determination of molecular weight to verify the presence of an oligonucleotide of a given sequence that is wanted, predicted or wished. This particularly applies for synthetic oligonucleotides or oligonucleotides produced by techniques of molecular biology, such as polymerase chain reaction (PCR).

The use of oligonucleotides produced with automatic synthesizers becomes more and more frequent. However, synthetic oligonucleotides can be considerably contaminated or, worse, their sequence may be different from expected due to the depurination reaction, non-removed protecting groups, deletions, etc. The association of mass spectrometers with automatic synthesizers has greatly enhanced the degree of confidence for the oligonucleotides produced.[146,147] Indeed, the precise determination of molecular weight of the oligonucleotides produced not only allows confirmation of the expected sequence but, in cases of defects, allows identification of the modification that has occurred or the nature of the contaminants.

An important advantage of mass spectrometry over other techniques such as electrophoresis or chromatography is its high speed. For instance, the association of MALDI/TOF with automatic preparation of the samples, together with automated acquisition of the spectra, allows the analysis of more than 100 oligonucleotides in 90 min.[147]

Another important advantage of mass spectrometry is its capacity to verify the incorporation of modified nucleotides. Figure 7.31 displays the mass spectrum of a synthetic oligonucleotide whose theoretical mass is 5838 Da. In addition to the desired nucleotide, other compounds are present. The mass difference between the peaks (111–151 Da) corresponds to the masses of the nucleic bases. This suggests that depurination reactions have occurred during the synthesis.

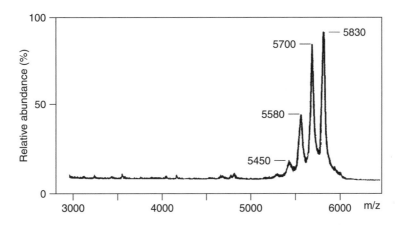

Figure 7.31
The MALDI/TOF mass spectrum of an oligonucleotide that has undergone a partial depurination reaction. (Reproduced (modified) from Ref. 147 with permission)

Determination of the size or molecular weight of nucleic acids is crucial in molecular biology. This importance is increased by the introduction of the PCR, which allows an individual to be identified (e.g. a pathogen bacteria) or the genetic links between individuals and allows the prediction or diagnosis of diseases.

Indeed, the PCR allows amplification to a detectable level of a determined region of DNA from whatever the source. Other applications in medicine, agriculture or the environmental sciences also are based on the identification of nucleic acids.

The classical methods for analyzing the nucleic acids are based on electrophoresis, either on gels or in columns. In this approach, the nucleic acids are separated according to their molecular weights and detected by markers that are colored, fluorescent or radioactive. Important drawbacks are the length of the analysis time, the low resolution achieved and the need to mark the molecules. Owing to its capacity to assign quickly and accurately the molecular weight of the nucleic acids, mass spectrometry is an interesting alternative.

Up to now, exploitation of these capabilities has been limited. Only PCR products or digest obtained with restriction enzymes have been analyzed in this way.[148,149] Some examples of detection of a mutations by mass spectrometry have been presented.

For instance, MALDI analysis allows the detection of a mutation often present in patients with cystic fibrosis (see Figure 7.32).[150] This mutation corresponds to the deletion of three base pairs, which leads to the loss of a phenylalanine residue in the protein. Using appropriate primers for the PCR reaction, the normal gene presents a 59-base fragment whereas the mutated one presents a 56-base fragment.

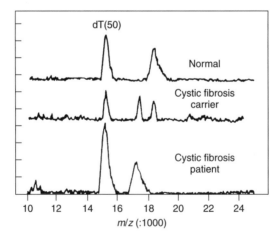

Figure 7.32
The MALDI/TOF mass spectrum of an oligonucleotide produced by PCR amplification of a part of the gene responsible for cystic fibrosis. The top spectrum is obtained from a normal individual, whereas the central and bottom spectra are obtained, respectively, from a healthy heterozygote carrier and an ill homozygote. (Reproduced (modified) from Ref. 150 with permission)

Determination of the molecular weight of an oligonucleotide allows its composition in to be assigned nucleic bases.[151]

If the oligonucleotide is small (about 5-mer) and the mass resolution at least 0.01%, the base composition can be assigned unambiguously. Improving the mass resolution combined with other information such as the number of bases, possible limitations to the composition, etc. will allow the determination of the composition in bases to be extended for larger oligonucleotides, or at least limit the number of possible candidate oligonucleotides. For instance, a measure of the molecular weight with a precision of 0.01% of a 14-mer nucleotide allows the complete composition to be assigned in bases if it is supposed that it contains only one guanine residue.

Determination of the molecular weight of oligonucleotides also will allow assignment of their sequence in an indirect way, as compared with the direct way based on MS/MS. One approach allowing the sequencing of oligonucleotides is known as 'DNA ladder sequencing'.[152,153] In a first step, a series of nucleic fragments, differing from each other by one nucleotide, is produced in solution from the unknown oligonucleotide. This is performed by enzymatic digestion by exonucleases such as SV phosphodiesterase or CP phosphodiesterase, which allow the nucleotides to be removed sequentially from 3'-OH or 5'-OH, respectively, To obtain fragments that cover the whole sequence, a small volume of the solution is set apart at different times during the digestion, or alternatively the solution is divided into parts and each one is treated with an increasing concentration of enzyme.

All these collections of fragments then are analyzed by mass spectrometry. The sequence is assigned from the mass differences between two consecutive fragments: 289 Da for dC, 304 for dT, 313 for dA, 329 for dG, 329 for A, 305 for C, 345 for G and 306 Da for U. For DNA the smallest difference is 9 Da (difference between A and T), whereas for RNA it is only 1 Da (difference between U and C). The accuracy of the mass determination is thus important. This approach allows up to about 30 bases to be sequenced, as displayed in Figure 7.33.[154]

The series of nucleotidic fragments can be obtained also by chemical digestion.[155] This allows the degradation of oligonucleotides that are non-sensitive to nucleases. However, by chemical digestion three types of fragments can be produced. Indeed, any potential site can be cleaved in either of two directions: towards the 3' end or towards the 5' end. The redundancy of these fragments always introduces possible confusion, but sometimes can be helpful to obtain the full sequence. Furthermore, fragments resulting from the cleavage at two sites are useless for sequence determination and add to the confusion.

Another sequencing approach based on mass spectrometry makes use of the sequencing products from the Sanger method, but mass spectrometry replaces the standard electrophoretic method.[141,156]

In the Sanger method, an oligonucleotide complementary to that used for sequencing is produced using a DNA polymerase. However, in the solution of the four regular nucleoside triphosphates is placed a small quantity of the dideoxy analog of one of the four nucleoside triphosphates. The result is to stop the polymerization every time a dideoxy base is included, because the 3' hydroxyl end is lacking. This

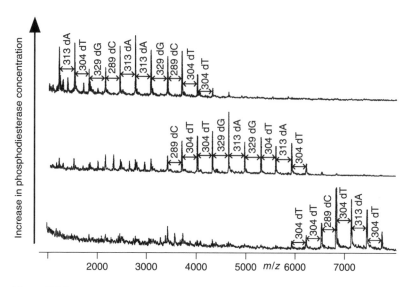

Figure 7.33
The MALDI/TOF mass spectrum of a 25-mer oligonucleotide after enzymatic digestion with a type I phosphodiesterase at three different concentrations. The different peaks correspond to successive cleavages from the 3' side to the 5' one. (Reproduced (modified) from PerSeptive Biosystems documentation, with permission)

leads to a mixture of oligonucleotides of variable length, but always ending with the same dideoxy base.

Four similar mixtures can be obtained by repeating the operation with the four possible dideoxy bases. The use of MALDI mass spectrometry to analyze the sequencing products is particularly attractive in view of the shorter analysis time that it requires: several hours are need to perform an electrophoresis, whereas obtaining a mass spectrum needs only a few minutes. The reading of the sequence from the four mass spectra corresponding to the four dideoxy bases is completely similar to that performed after electrophoresis. The mass resolution needed is not critical, as it is in ladder sequencing, because the mass differences to observe correspond to one nucleotide, or about 300 Da. This approach is limited to oligonucleotides of about 50 bases. Automatic DNA sequencers allow the routine analysis of up to 500 bases. To be able to compete, both the mass limit and the sensitivity of MALDI/TOF have to be increased by one to two orders of magnitude.

Electrospray ionization is a soft ionization technique allowing, under appropriate conditions, the transfer of non-covalent complexes from the liquid to the gas phase. The stoichiometry of the complex is easy to assign because it is its molecular weight that is actually measured. Electrospray ionization has allowed the observation of non-covalent complexes of oligonucleotides themselves (duplexes and tetraplexes), of drug–nucleotide and of nucleic acid–protein,[157–159] and thus is a useful method for the study of the hybridization of native or chemically modified oligonucleotides, for the rapid detection of possible drug–oligonucleotide interaction, etc.[160,161]

Mass spectrometry offers several advantages over classical methods for the characterization of non-covalent complexes: rapidity, simplicity and ability to work on mixtures. One of the goals in this field is to transfer gas-phase data to the liquid phase. However, this transposition has to be done with caution. Indeed, some of the complexes observed are non-specific associations formed in the gas phase or aggregates resulting from the ESI process itself. Also, some of the complexes present in the liquid phase might not be observed in the mass spectrum.

7.3.3 Fragmentation of oligonucleotides

The fragmentation of oligonucleotides in the gas phase can be a problem to be surmounted for the analysis of these molecules. However, this fragmentation can be used to determine the sequence of the oligonucleotide. The dissociation of oligonucleotides can occur in the source as the result of the energy excess that is imparted to the molecule during the desorption/ionization process. This is the case of FAB spectra of oligonucleotides. Indeed, the negative mode FAB mass spectra of oligonucleotides are characterized by the presence of two series of ions that are characteristic of the sequence: one containing the phosphate on the 5′ side and the other containing the phosphate on the 3′ side. Normally the intensity of the 5′ terminal series is greater than that of the 3′ series, and other ion series of masses 18 Da lower (water loss) or 80 Da higher (addition of HPO_3) are also present. A spectrum that is typical of a small oligonucleotide with sequence UGUU,[162] measured by negative mode FAB/MS, is shown in Figure 7.34.

Fragmentation of nucleotides can be observed also in MALDI spectra.[163,164] In principle, these fragmentations could allow the sequence to be deduced. However, coupling of MALDI to a TOF spectrometer and the variable kinetics of these fragmentations result, in the absence of specific experimental conditions, in broadening of the peak of the molecular species and a loss of resolution and sensitivity.

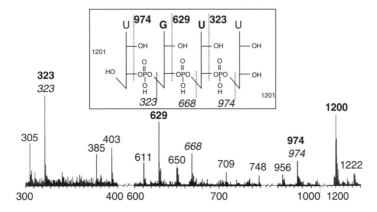

Figure 7.34
Negative mode FAB spectrum of an oligonucleotide with sequence UGUU. (Reproduced (modified) from Ref. 162 with permission)

Only fragmentations that occur as fast as the desorption can be observed in the mass spectrum at their right m/z values in a linear TOF instrument. Dissociation that occurs later on can be observed only if the extraction from the source is delayed or if a reflectron analyzer is used.

The coupling of the MALDI to a trapping analyzer, either ion trap or FTICR, allows all those fragments to be detected.[165] The formation of these fragments depends on the matrix, the power of the laser beam and the sequence of the oligonucleotide submitted to the analysis.

Electrospray ionization produces stable ions of the molecular species, except if the conditions in the source are adjusted to induce dissociation by collision in the nozzle–skimmer region.[166]

Dissociation of the nucleotides can be obtained also by adding internal energy to the stable ions. Combining this induced dissociation with MS/MS allows fragment ions to be obtained only from the selected precursor. This allows the analysis of oligonucleotides from a mixture and also suppresses the low-mass ions originating from the matrix when FAB or MALDI are used. With ESI, the spectra are simplified because only one charge state is selected. An increasing number of mass spectrometry analyses of oligonucleotides use MS/MS. Results obtained by coupling an ESI or MALDI source to an FTICR analyzer or by coupling an ESI source to a triple quadrupole instrument have been reported.[165,167] However, the large majority of reported results are obtained by coupling ESI to an ion trap mass spectrometer, which allows CID/MS^n to be performed.[168–170]

A systematic nomenclature has been developed by McLuckey to describe the fragments observed when oligonucleotides are studied by mass spectrometry. This is illustrated in Figure 7.35.[169] Cleavage of the four possible types of phosphodiester bond yields eight types of fragments, symbolized a_n, b_n, c_n and d_n if they contain the 5′-OH group and w_n, x_n, y_n and z_n for those containing the 3′-OH group. The subscript n represents the number of residues contained in the fragment, and thus indicates the position of the cleavage. The additional loss of a base is indicated by parentheses with, if possible, the identity of the base. For example, the fragment a_3-B_3(A) indicates a cleavage of the bond between the ribose carbon atom and the

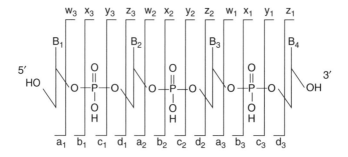

Figure 7.35
Nomenclature for the dissociation products from oligonucleotides

oxygen atom of the phosphodiester group at position 3, with the additional loss of an adenosine base at this same position.

This nomenclature is similar to that proposed for peptides. Note, however, that d_n and w_n fragments are not similar to those symbolized with the same letters for peptides. It is also interesting to mention that another nomenclature for the fragments observed in MALDI spectra has been proposed by Viarr and completed by Nordhoff.[163,171] This nomenclature, however, has the drawback that it cannot be applied systematically to all possible fragments. It characterizes three kinds of fragments, denoted X-, X*- and Y-, corresponding respectively to d_n, a_n-B_n and w_n.

As a rule, the fragmentation schemes of oligonucleotides are similar for the various types of ion sources or analyzers used. The differences observed can be explained either by the charge state or the energy content.

One of the most frequently observed fragmentations corresponds to the loss of one of the nucleic bases as a neutral or an anion. The tendency to produce the anion rather than the neutral increases for higher charge states and longer nucleotides chains.

Other often-observed fragments correspond to the cleavage of the nucleic chain to yield three series of ions of types a_n, a_n-B_n and d_n characteristic of the sequence from 5' to 3', and one series of ions of type w_n characteristic of the inverse sequence, from 3' to 5'. Of course, the abundances of these ions depend on experimental factors, such as the ionization method. The MALDI spectra display abundant d_n and w_n ions whereas ESI spectra are generally dominated by a_n-B_n and w_n ions.

Other types of fragmentations resulting, for instance, from a rearrangement or a double cleavage are also observed. The abundance of double cleavage ions increases at higher internal energy, for higher charge states and for less stable oligonucleotide ions.

Several fragmentation pathways have been proposed to explain the formation of complementary ions of types a_n-B_n and w_n.[163,172–174] The first one includes two steps with the loss of the nucleic base as neutral or anion via a 1,2-elimination, as shown in Figure 7.36. This elimination may be favored by an intramolecular base catalysis by the negatively charged oxygen atom from the 3' phosphate group. Then, this intermediate can yield the a_n-B_n and w_n fragments by cleavage of the 3'C−O bond of the phosphodiester group. The mechanism again is a 1,2-elimination, as shown in Figure 7.36. The reaction is favored by the formation of the resonance-stabilized furan cycle. It can be catalyzed by the negative charge of the 5' or 3' phosphate, removing the 4' hydrogen atom of the ribose. The type d_n ions are produced by the mechanism depicted in Figure 7.36 from type a_n-B_n ions if they contain sufficient excess of internal energy.[175]

A second fragmentation pathway has been proposed for type a_n-B_n and w_n ions. This pathway does not require the loss of a nucleic base or the proximity of the charge to the cleavage site. It thus allows observation of these ions to be explained in the absence of adjacent charge. The first step is the direct cleavage of the 3'C−O bond of the phosphodiester group to produce the a_n and w_n fragments. The a_n ions then dissociate by the loss of a nucleic base to yield ions of type a_n-B_n.

Two mechanisms have been suggested to explain this fragmentation pathway. The first is a 1,2-elimination involving the transfer of the 4' hydrogen atom to

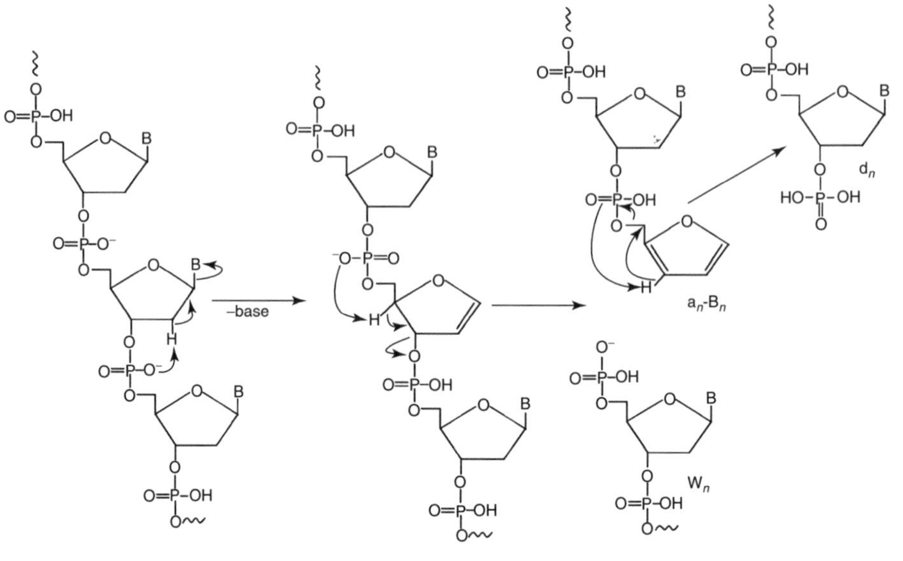

Figure 7.36
Fragmentation pathways of oligonucleotides that necessitate a charge proximate to the cleaved site

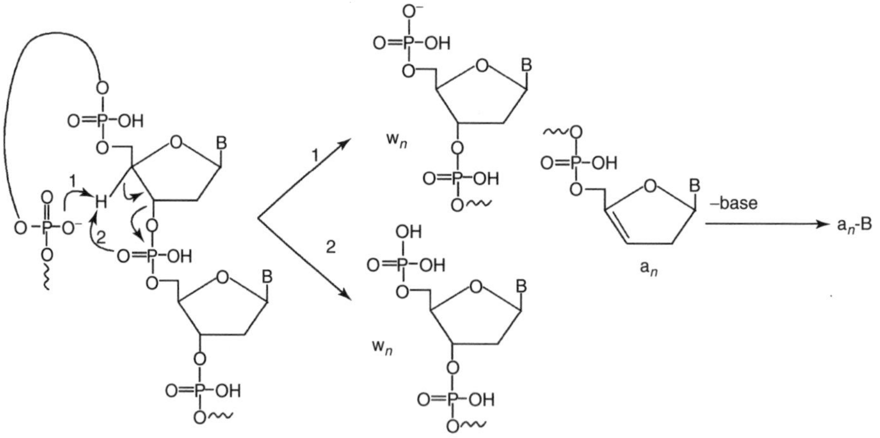

Figure 7.37
Fragmentation pathways of oligonucleotides that do not necessitate either the prior loss of a nucleic base or a charge located near the cleavage site

a remote-charged phosphodiester, as represented in Figure 7.37, reaction 1. In the second mechanism, this 4′ hydrogen atom is transferred to the oxygen atom of a non-charged 5′ phosphate group through a six-atom cyclic transition state, as shown in reaction 2 of Figure 7.37.

An algorithm has been developed for the interpretation of tandem mass spectra of oligonucleotides.[176] This greatly facilitates the interpretation of mass spectra of oligonucleotides and is based on the general fragmentation scheme of these molecules.

On the basis of the recognition of specific fragments, it first builds a sequence for both the 5′ and 3′ sides. Then, it adjusts by cross-checking the two sequences. Then, on the basis of the measured mass and the constraints of the nucleic base composition, it rejects incorrect candidate sequences.

The determination of the sequence, based on the gas-phase fragmentation of oligonucleotides and subsequent identification of the fragments, is potentially very interesting. It is the fastest way to sequence oligonucleotides. However, the fragmentation spectra are often very complicated and may be difficult to interpret. They need a sufficient resolution to separate all the fragments produced and the mass determination has to be sufficiently accurate to assign unambiguously the observed fragments.

These requirements explain why the mass spectrometry method is limited to about 20-mer to obtain structural information and to 10-mer for the complete sequencing.

7.3.4 Characterization of modified oligonucleotides

Oligonucleotides can undergo covalent modifications.[177] These can be natural ones such as those present in tRNA and rRNA but they can also result from reactions with exogenous substances and are the markers indicating possible degradations of the cells. They also can be modified chemically to create new drugs. Almost all these modifications are accompanied by variations in mass, and mass spectrometry is thus useful to identify their nature and position in the sequence.[178–180]

A general fragmentation diagram was drawn up based on the MS and MS/MS of various nucleosides measured in positive and negative modes by different ionization techniques (EI,[181] CI,[182] FAB,[183] ESI[184]), as shown in Figure 7.38. The most abundant fragmentation results from the cleavage of the bond between the nucleobase and the sugar.

Knowing this fragmentation diagram and using MS/MS, we can detect and identify the structures of the modified nucleosides directly in the nucleoside mixture obtained by enzymatic digestion or hydrolysis, without any previous separation. In fact, the analysis of the enzymatic digestion or the hydrolysis of a modified oligonucleotide contains the molecular species ion of the modified nucleoside in addition to the molecular species

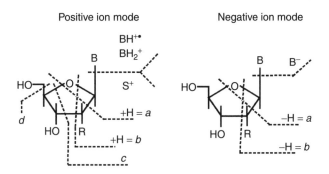

Figure 7.38
General fragmentation diagram of nucleosides

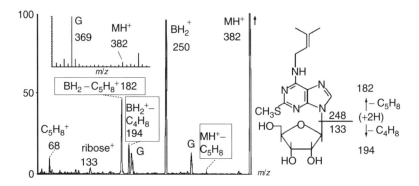

Figure 7.39
Collision-induced dissociation FAB/MS/MS trace of N^6-isopentenyladenosine
(25 ng) contained in an unpurified hydrolysate of *E.coli* tRNA^{Tyr}. (Reproduced
(modified) from Ref. 185 with permission)

ions of the natural nucleosides. The MS/MS analysis of this ion of the modified nucleoside allows one to deduce its structure from the fragments obtained.

The recognition of a molecular species ion in a spectrum may be difficult. Hence, a scan of neutral losses with mass 132 Da allows one to detect selectively the protonated molecular ions of the various nucleosides contained in the mixture, because the production of an intense BH_2^+ ion from MH^+ by the loss of a sugar molecule (132 Da for a ribose) is an important characteristic of positive ion spectra. A modified nucleoside, N^6-isopentenyladenosine, was identified in tRNA^{Tyr} of *E.coli* using this method[185] (Figure 7.39).

Other methods (described in Figure 7.40) based on mass spectrometry, such as gas chromatography/mass spectrometry (GC/MS)[186] or high-performance liquid chromatography/mass spectrometry (HPLC/MS),[187–190] also allow one to detect and identify the modified constituents of oligonucleotides. These methods are good for small sample quantities of around 1 µg, and allow the detection of a modified nucleotide within a mixture of 10^7 nucleotides.

Localization of a modified nucleotide also is done using mass spectrometry. Several strategies are proposed. A first step is to determine the nature of the modified nucleotide as described above. Then, oligonucleotides of smaller sizes (less than 10 bases) are obtained from the starting oligonucleotide using an enzyme such as RNAse T1. These different oligonucleotides then are separated and the fraction that contains the modification is analyzed by mass spectrometry. The molecular weight determination of the oligonucleotide, which presents this modification, allows its composition to be determined in nucleic bases. On the basis of this composition and sequence of gene, the modification can be localized.[191]

Another method to localize a modified nucleotide is based on sequencing of the oligonucleotide containing the modified nucleotide, using MS if it is previously purified or using MS/MS if it still lies in a mixture. The position and the molecular weight of the modified nucleotide can be determined from the fragment ions, as is shown in Figure 7.41.[192]

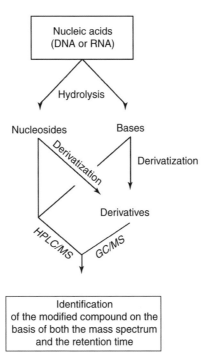

Figure 7.40
Strategy followed in order to detect and identify the modified constituents of oligonucleotides by HPLC/MS or GC/MS techniques

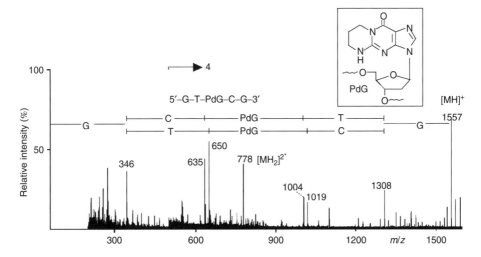

Figure 7.41
Collision-induced dissociation FAB/MS/MS spectrum of an oligonucleotide containing a modified base. (Reproduced (modified) from Ref. 192 with permission)

7.4 Oligosaccharides

Oligosaccharides are molecules formed by the association of several monosaccharides linked to each other through glycosidic bonds. The monosaccharides that are most often present in oligosaccharides are given in Table 7.9.

The determination of the complete structure of oligosaccharides is more difficult than for proteins or oligonucleotides because it requires the determination of additional specificities as a consequence of the isomeric nature of monosaccharides and their capacity to form linear or branched oligosaccharides. Knowing the structure of an oligosaccharide requires not only the determination of its monosaccharide sequence and its branching pattern, but also the isomer position and the anomeric configuration of each of its glycosidic bonds. Moreover, the monosaccharides in their cyclic forms can have pyran/furan isomers. The determination of all of this structural information can be obtained by mass spectrometry, even if generally it requires the use of various techniques. The mass spectrometry of oligosaccharides has been reviewed.[193–195]

7.4.1 Mass spectra of oligosaccharides

The use of mass spectrometry to characterize oligosaccharides is not new. Indeed, GC/MS has been used for several decades to identify monosaccharides or very small oligosaccharides. The use of GC/MS needs the prior derivatization of these molecules by methylation, acetylation or trimethylsilylation.

Today (2000), GC/MS is still the method of choice for the determination of the monosaccharide composition of oligosaccharides and is widely used. The analysis starts with the hydrolysis or alcoholysis of the oligosaccharide to monosaccharides, which after derivatization (mostly by trimethylsilylation) are analyzed by GC/MS.[196] Monosaccharide identification is based on comparison of retention time and fragmentation pattern with references.

Table 7.9 Monosaccharides present in oligosaccharides (the molecular weight of unmodified oligosaccharides may be calculated by summing the mass increments of the various residues and adding the mass of one water molecule)

Monosaccharide	Examples	Formula	Monoisotopic mass	Chemical mass
Pentose	Arabinose (Ara) Ribose (Rib) Xylose (Xyl)	$C_5H_8O_4$	132.04	132.12
Deoxyhexose	Fucose (Fuc) Rhamnose (Rha)	$C_6H_{10}O_4$	146.06	146.14
Hexose	Glucose (Glc) Galactose (Gal) Mannose (Man)	$C_6H_{10}O_5$	162.06	162.14
N-Acetyl-aminohexose	N-Acetylglucosamine (GlcNAc) N-acetylgalactosamine (GalNAc)	$C_8H_{13}NO_5$	203.08	203.19
Sialic acid	N-Acetylneuraminate (NeuAc)	$C_{11}H_{17}NO_8$	291.09	291.26

Gas chromatography/mass spectrometry also is used largely in methylation analysis, which allows the position of the glycosidic bonds to be determined on a residue in an oligosaccharide.[197] In its most used version, permethylation of the oligosaccharide is performed prior to hydrolysis. Then, the partially methylated monosaccharides produced are reduced to the corresponding alditols and peracetylated, yielding partially methylated alditol acetates (PMAA). Analysis of these molecules by GC/MS allows, on the basis of the straightforward interpretation of the fragmentation, the position of the former glycosidic bonds to be determined.

Mass spectrometry analysis of oligosaccharides without hydrolysis has started with the FAB ionization technique and has developed with ESI and MALDI. Fast atom bombardment usually generates a weak signal, whereas ESI is not as efficient for native oligosaccharides as MALDI. Indeed, native oligosaccharides do not contain either acidic or basic groups, and thus are not easily ionized in ESI. The response for native oligosaccharides is much weaker than for peptides or proteins. However, phosphorylated, sulfated or sialic-acid-containing oligosaccharides are well ionized by ESI in negative ion mode.

In general, oligosaccharides are analyzed by ESI after derivatization, either by methylation or acetylation. Reductive amination of the aldehyde or ketone group also is used.[199] The derivatized products are more lipophilic, allowing the use of more volatile organic solvents, which improves the production of ions by ESI. Reductive amination together favors the formation of positive ions by the introduction of a basic site or a permanent charge. Permethylation is preferred for derivatization because the resulting mass increase is lower and thus larger oligosaccharides can be analyzed. Derivatization not only improves the signal but also can bring more structural information, especially with MS/MS, as will be discussed below.

Whatever the ionization method used, the mass spectra of oligosaccharides analyzed by ESI, MALDI or FAB display intense ions of the molecular species resulting from protonation $(M + H)^+$ or cationization by an alkali metal ion $(M + \text{alkali metal})^+$ in the positive ion mode or from deprotonation $(M - H)^-$ in the negative ion mode. In ESI, multiply charged ions also are produced.

Alkali metal adducts often are observed, and even in negative ion mode $(M + Na - 2H)^-$ are often abundant for monocharged ions. If they can be avoided, not only does the signal increase because it is not divided any more over several species but also better MS/MS spectra can be obtained. To eliminate these adducts, glassware should not be used during sample work-up. Then, addition of acid or, better still, ammonium acetate allows their interference to be reduced further. However, often alkali metal salts are added at low concentrations to suppress the protonated species. This is easier to achieve, but fragmentation of these adducts yields less sequence information than protonation.

The FAB spectra present source fragmentation, but with ESI few fragments are observed, as shown in Figure 7.42. A MALDI source produces ions of the molecular species containing more energy, but with linear flight tubes the fragments are not observed because they arrive at the detector at the same flight time as their precursor. These fragments can be observed, however, either by delaying the extraction from

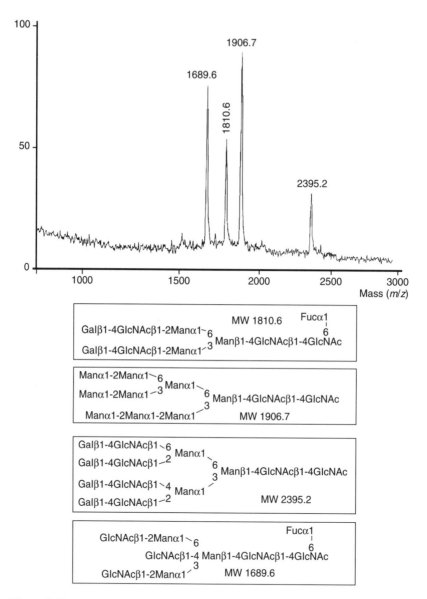

Figure 7.42
The MALDI spectrum of a mixture of four oligosaccharides derived from the
cleavage of a glycoprotein. (Reproduced from Finnigan MAT documentation,
with permission)

the source or by the use of reflector instruments. They are observed even better if
the MALDI source is coupled to a spectrometer using ion storage, such as ion trap
or FTMS instruments.

Owing to the limited number of monosaccharides present in oligosaccharides,
measurement of the molecular weight allows the composition of monosaccharides

such as pentoses, hexoses, etc. to be deduced but it is not able to distinguish between different isobaric monosaccharides. Thus hexoses, for example, cannot be differentiated on this basis into glucose, mannose, etc.

The molecular weight of the oligosaccharide will, however, allow structures to be proposed if the number of possible structures is limited. For example, oligosaccharides linked to asparagine (termed an N-linkage) from animal glycoproteins have compositions limited to a few different monosaccharides. They all have a common pentasaccharidic core made up of three mannose residues and two N-acetylglucosamines. Taking this information into account, the structures of the different oligosaccharides observed in Figure 7.42 have been elucidated. For instance, the oligosaccharide observed at m/z 1810 as a sodium adduct can have only the composition $(hexose)_5(N\text{-acetylhexose})_4(deoxyhexose)_1$ and its most probable structure is displayed in Figure 7.42. It is clear that these structures, based only on molecular weight, have to be confirmed by cross-checking with other information.

7.4.2 Fragmentation of oligosaccharides

Because the soft ionization methods used for oligosaccharides produce few fragments, collision-induced dissociation (CID) or post-source decay (PSD) must be used for structural study. These two techniques have been applied to deprotonated, protonated or alkaliated molecular ions from native or derivatized oligosaccharides using FAB,[200–205] ESI[206–209] or MALDI.[210–212]

The fragmentation of oligosaccharides is strongly influenced by a large number of factors, such as the ionization method, the analyzer, the nature of the derivatization, the nature of the molecular species, etc. However, five series of fragment ions can be observed. As shown in the spectrum presented in Figure 7.43, the first two series of ions, generally abundant, come from the cleavage of one glycosidic bond. Thus, they contain the reducing or the non-reducing end. Both of these fragments, by undergoing cleavage of a second glycosidic bond, lead to a third series of ions called internal fragments. They do not contain the two initial termini any more (cleavage at both ends). The two last series of ions, generally less abundant, are due to a double cleavage across the glycosidic ring and contain either the reducing side or the non-reducing side.

A nomenclature suggested by Domon and Costello[213] was developed to characterize the various fragments obtained by mass spectrometry whatever the method used to produce them. The fragments retaining the charge on the non-reducing side are called A, B or C, and those retaining the charge on the reducing side are called X, Y and Z, depending on whether they cut the ring or the glycosidic bond (see Figure 7.44). The subscript corresponds to the number of the ruptured glycosidic bond whereas the superscript at the left of the A and X fragments corresponds to bonds that were broken in order to observe these fragments, the bonds being numbered as is indicated in the figure. The letter α, β, etc., which may be attached to the subscript number, indicates the branch involved in the cleavage, provided, of course, that the oligosaccharide is branched.

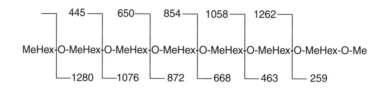

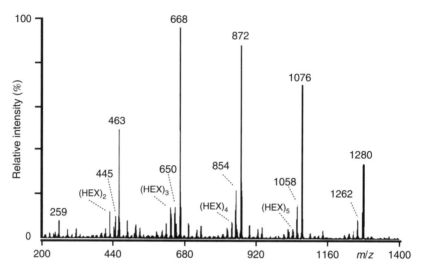

Figure 7.43
Collision-induced dissociation ESI/MS/MS trace of doubly charged sodium adduct of methylated maltoheptulose ($m/z = 760.7$). (Reproduced (modified) from Ref. 194 with permission)

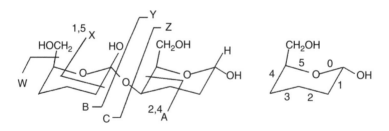

Figure 7.44
Nomenclature suggested by Domon and Costello

The fragments that are observed most often in positive mode spectra correspond to cleavage of the glycosidic bond, with oxygen atom retention on the reductive part, by the mechanism shown in Figure 7.45. This yields B and Y ions.

For the fragments derived from a double cleavage across the ring (A and X ions), their formation is favored under high-energy collisions from the sodium adducts.[214]

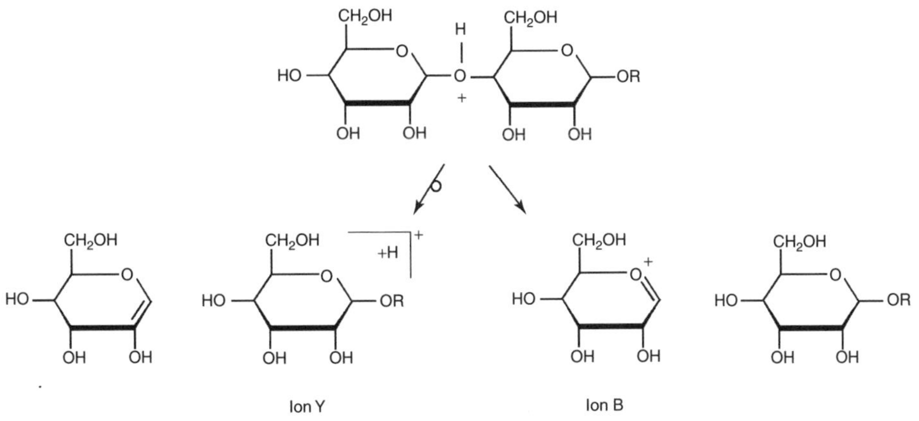

Figure 7.45
Mechanism for the formation of B and Y ions in positive mode

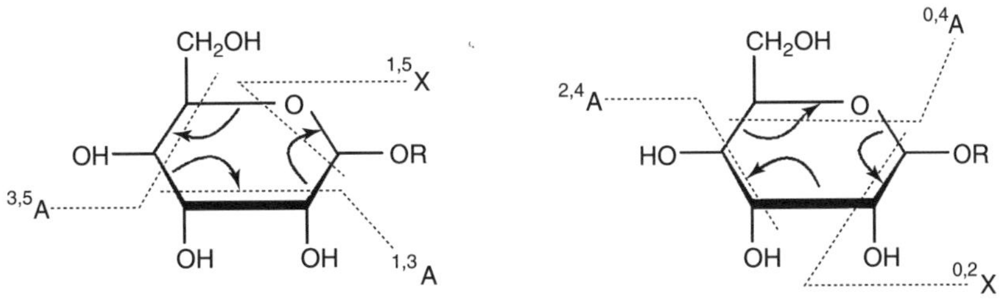

Figure 7.46
Formation mechanism of A and X ions in positive mode

These fragments are produced through a charge remote fragmentation (CRF) mechanism shown in Figure 7.46. This mechanism implies two decomposition paths yielding $^{1,5}X$, $^{1,3}A$ and $^{3,5}A$ ions in one case and $^{0,2}X$, $^{2,4}A$ and $^{0,4}A$ ions in the other case. It is also interesting to observe that the derivatized oligosaccharide carrying a preformed charge due to prior reductive amination of the oligosaccharide leads exclusively to the fragment ions that contain the reducing end.[215,216]

In general, MS/MS allows one to determine the sequence and the branching pattern of oligosaccharides. The isomer position of each of their glycosidic bonds also can be determined. On the other hand, the anomeric configuration of glycosidic bonds and the distinction of diastereoisomeric monosaccharides are seldom accessible by this technique.

The ions derived from cleavage of the glycosidic bond allow one to determine the sequence and the branching pattern of oligosaccharides. Indeed, the mass difference between fragment ions within the same series allows one to deduce the sequence of oligosaccharides. Such a determination is detailed in Figure 7.47A, which shows the spectrum of a peracetylated pentasaccharide, lacto-N-fucopentose (LNF-I).[217] This spectrum is dominated by the series of B oxonium ions. Ions with m/z 273, 561,

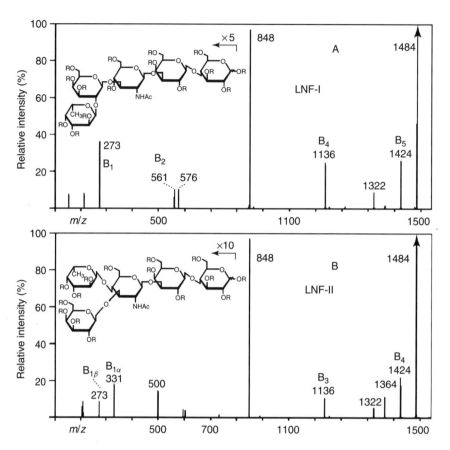

Figure 7.47
Collision-induced dissociation FAB/MS/MS traces of two peracetylated oligosaccharides: (A) LNF-I and (B) LNF-II, two branching isomers. (Reproduced (modified) from Ref. 217 with permission)

848, 1136 and 1424 correspond to fragments B_1 (Fuc-Ac 3), B_2 (Fuc-Hex-Ac 6), B_3 (Fuc-Hex-GlcNAc-Ac 8), B_4 (Fuc-Hex-GlcNAc-Hex-Ac 11) and B_5 (Fuc-Hex-GlcNAc-Hex-Hex-Ac 14). These fragments make it possible to ascribe a complete and unambiguous sequence to this oligosaccharide; however, they do not allow one to distinguish between the two diastereoisomers galactose and glucose (Hex = Gal or Glc).

The same principle may be used to determine the branching pattern. Figure 7.47B shows the FAB/MS/MS trace of a branched isomer of the pentasaccharide LNF-II. As opposed to the linear isomer, the branched isomer is characterized by two monosaccharidic B_1 ions with masses 273 Da ($B_1\beta$, Fuc-Ac 3) and 331 Da ($B_1\alpha$, Hex-Ac 4) and by the absence of disaccharidic B_2 ions at mass 561 Da. As in the linear structure, the ions with a higher mass sequence are also observed at m/z 848 (B_2, trisaccharidic: Hex-[Fuc]-GlcNac-Ac 8), 1136 (B_3, Hex-[Fuc]-GlcNac-Hex-Ac 11) and 1424 (B_4, Hex-[Fuc]-GlcNac-Hex-Hex-Ac 14) and allow one to establish the remaining sequence because the latter does not contain any further branching.

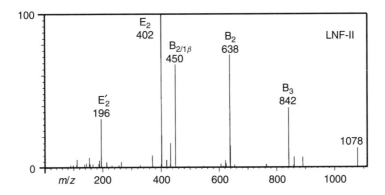

Figure 7.48
Collision-induced dissociation ESI/MS/MS spectrum of the LNF II oligosaccharide. (Reproduced (modified) from Ref. 218 with permission)

The ESI/MS/MS fragmentation spectrum of permethylated LNF-II is displayed in Figure 7.48.[218] This spectrum is somewhat more complicated than the one from the peracetylated derivative, but allows additional structural information to be obtained. As for the peracetylated derivative, this spectrum contains the ions characteristic of the sequence observed at m/z 638 (B_2, trisaccharidic: Hex-[Fuc]-GlcNAc-Me 8) and 842 (B_3, Hex-[Fuc]-GlcNAc-Hex-Me 11), which allow the branched structure and the sequence on the reducing side to be established.

However, this spectrum is different from the peracetylated one because it displays more secondary fragments. This process, due to the presence of HexNac, leads from the B_2 ion to the preferential loss of the substituent at the 3 position of the HexNac residue, in this case a tetramethylated hexose, to produce the E_2 fragment at m/z 402. From this last ion, the loss of the substituent at the 4 position, here a trimethylated deoxyhexose, yields the E'_2 fragment at m/z 196. This spectrum, as with that of the peracetylated compound, allows the sequence to be assigned but furthermore provides information that the 3-position of the HexNac bears a hexose and the 4 position a deoxyhexose.

The best results using this method are obtained with protonated molecular ions $(M + H)^+$ of derivatized or native oligosaccharides. However, the peracetylated or permethylated derivatives are superior not only in terms of response but also in their ability to identify the internal fragments derived from the double cleavage of glycosidic bonds. Compared with the fragments having the same number and the same type of residues but produced by the cleavage of only one glycosidic bond, the internal fragments present two free hydroxyls groups. Thus, these fragment ions have distinct masses and are easily differentiated. This is not the case of the native oligosaccharides and this can complicate, or even compromise, the determination of the sequence and lead to erroneous information.

The A and X ions derived from the cleavage of two bonds in the glycosidic ring and the W ions from the cleavage between carbons 5 and 6 allow a clear identification of the positional isomers of each glycosidic bond in the linear or branched oligosaccharides by their presence and/or their absence. For example, the cleavage of the ring with hydroxyls at positions 1 and 4 involved in glycosidic bonds will allow only the formation of $^{1,5}X$, $^{0,2}X$, $^{3,5}A$ and $^{2,4}A$ ions. On the other hand, as shown in Figure 7.49, the implication of hydroxyls at positions 1 and 2 will allow only the formation of $^{1,5}X$ and $^{1,3}A$ ions. The diagnostic fragments useful in determining the positional isomers of the glycosidic bonds are listed in Table 7.10.

The example shown in Figure 7.50 includes the structure of the oligosaccharide, the tandem mass spectrum of its $(M + Na)^+$ adduct and a summary table of the diagnostic peaks.

The best results that are obtained by this method use the molecular species ions $(M + \text{alkali metal})^+$ of native or derivatized oligosaccharides measured at high energy.

The tandem mass spectrometric analysis of oligosaccharides thus allows one to obtain valuable structural information. However, determination of the complete structures from such information alone is difficult. This task is still more difficult if residues contain highly labile groups such as sialic acid, N-acetylaminohexoses or fucose, because the spectra are dominated by fragments from these residues. The use of the MS^n capabilities of the ion trap or FTICR instruments is thus a great help.[219,220] There are three major advantages of MS^n over traditional MS/MS: dissociation of a fragment can lead to new fragments not observed from the ion of the

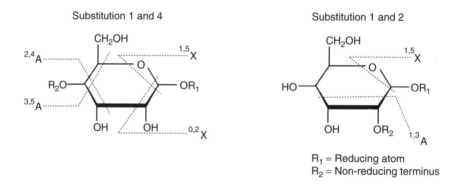

Figure 7.49
Diagnostic fragments for monosaccharides whose hydroxyl groups in either position 1 and 2 or 1 and 4 participate in glycosidic bonds

Table 7.10   Potentially diagnostic ions in determining the positional isomers of glycosidic bonds

Positional isomer	Fragments						
	W_i	$^{0,2}X_i$	$^{1,5}X_i$	$^{0,4}A_i$	$^{1,3}A_i$	$^{2,4}A_i$	$^{3,5}A_i$
1 and 2	−	+	+	−	+	−	−
1 and 3	−	+	+	−	+	+	−
1 and 4	−	+	+	−	−	+	+
1 and 6	+	+	+	+	−	−	+

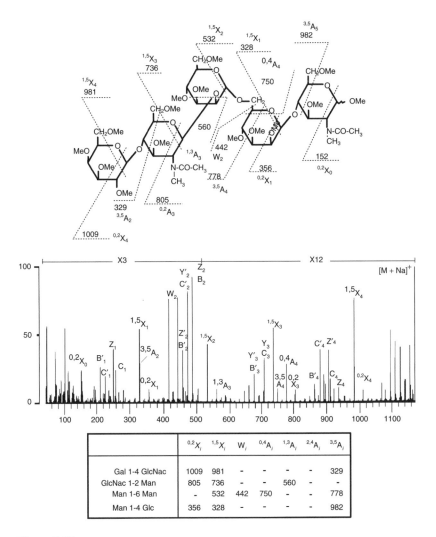

Figure 7.50
Collision-induced dissociation FAB MS/MS trace of the $(M + Na)^+$ of an oligosaccharide, measured at high energy. It allows one to determine the positional isomers of all its glycosidic bonds. (Reproduced (modified) from Ref. 214 with permission)

molecular species; the various first-generation fragments observed are identified not only on the basis of their mass but also of their fragmentation; and filiation of the fragments can be established experimentally.

7.4.3 Degradation of oligosaccharides coupled with mass spectrometry

There are a variety of chemical or enzymatic degradation methods of oligosaccharides that have been used in conjunction with mass spectrometry.[221,222] Three of these methods are described in detail below.

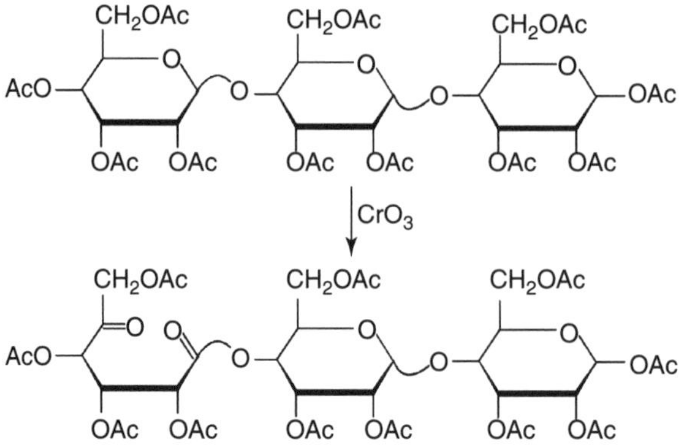

Figure 7.51
Chromium oxide action on the α- and β-anomers of the glycosidic bonds

The first method is based on the selective oxidation by chromium trioxide of the β-anomer of derivatized hexoses in order to yield a ketoester, as shown in Figure 7.51. Thus, this method allows the determination of the anomeric configuration of the various glycosidic bonds.[223] The oligosaccharide mass difference observed before and after oxidation allows one to determine the number of β-bonds that are present: an $N \times 14$ Da increment corresponds to the presence of N β-bonds. The oxidized bond positions may be detected using fragmentations of the glycosidic bonds. Finally, a soft methanolysis at room temperature may counter a possible low intensity of the ions derived from cleavage of the glycosidic bond, because it cleaves preferentially the ester bonds formed at the oxidized β-anomers.

The second method relies on enzymatic degradation of oligosaccharides by endoglucosidases. These enzymes catalyze selectively the hydrolysis of glucosidc bonds of monosaccharides located at the non-reducing end. The specificity of the different endoglucosidase concerns the nature of the monosaccharides at the reducing end or the configuration of the anomeric carbon atom involved in the glycosidic bond. It can also be sensitive to the nature of the monosaccharides at the penultimate monosaccharide. As an example, the β-galactosidase of *Streptococcus pneumoniae* cleaves the bonds Gal(β1-4)Glc or Gal(β1-4)GlcNAc whereas the α-L-fucosidase from bovine epididyme cleaves only the Fuc(α1–6)x bond.

The full structural determination of the structure of an oligosaccharide by this method includes complete sequencing by successive enzymatic digests using specific exoglycosidases and analysis of the products of each hydrolysis by mass spectrometry to observe the effectiveness of the hydrolysis by the decrease of molecular weight.[224]

A combination of the specific enzymes with the sensitivity, speed and accuracy of mass spectrometry allows complete sequence determination, including the anomeric configuration. According to the specificity of the enzymes, positional isomers of the glycosidic bonds can be recognized.

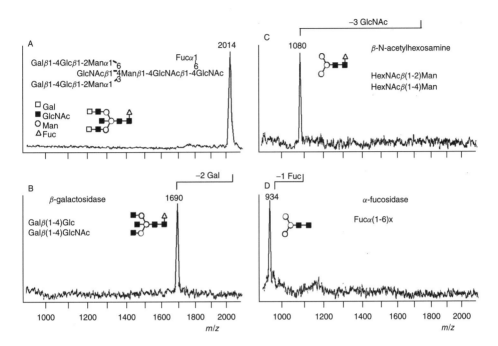

Figure 7.52
The MALDI spectrum from successive enzymatic digests of a native oligosaccharide. The structural information obtained from the specificity of the exoglucosidases used is indicated by italics in the complete structure in A. (Reproduced (modified) from Ref. 225 with permission)

An application of this method is displayed in Figure 7.52.[225] The sequential treatment of the studied oligosaccharide with three specific exoglycosidases allows two Gal, three GlcNac and one Fuc residue to be removed successively, as shown by the detected ions of the molecular species.

In the case of an oligosaccharide of totally unknown structure, this method can be long and costly due to the need to use many different enzymes. However, it has the advantage that it can be applied to complex mixtures of oligosaccharides.[226]

The last method described here for the degradation of oligosaccharides is a modification of the methylation analysis described in section 7.4.2.[227]

It is based on a derivatization process that starts with ethanolysis of the permethylated oligosaccharide (Figure 7.53). The partly methylated ethylglycosides are deuteromethylated on the hydroxyl groups made free by the ethanolysis step. The fingerprint of the glycosidic positions is held by deuteromethylation of these positions. The ethyl group specific to the 1 position will allow alcohol elimination from this position to be distinguished from the other alcohol eliminations (loss of methanol or deuterated methanol). The different monosaccharides derivatives then are analyzed by chemical ionization GC/MS/MS.

Gas chromatographic analysis allows the determination of the monosaccharides present by comparing the retention times with those of reference compounds. It is noteworthy to mention that the chromatographic profile is quite simple, because the

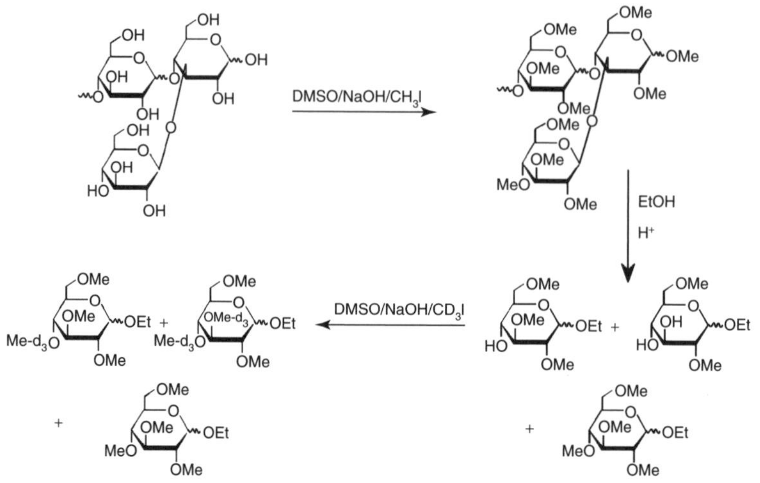

Figure 7.53
Sequence of reactions allowing permethylated ethylglycosides to be selectively deuteromethylated

deuterated compounds are not separated from the non-deuterated, even if a small difference in retention time can be evidenced by comparison of specific ion chromatograms. Each monosaccharide can give a maximum of four peaks, corresponding to the α/β and the pyran/furan isomers.

Analysis by positive ion chemical ionization of the permethylated ethylglycosides leads to spontaneous fragmentation of the molecular species. As shown in Figure 7.54, this fragmentation produces oxonium ions (resulting from the loss of ethanol) at m/z 219 (for non-deuterated hexoses). The oxonium ions then can lose a methanol molecule to yield a fragment 32 mass units lower (m/z 187 for the non-deuterated hexoses).

The masses of the oxonium ions will allow the number of deuteromethyl groups present to be determined and so yield information on the position of the monosaccharide in the chain. Indeed, the oxonium ion will appear at m/z 219 (d$_0$) for a hexose in a non-reducing terminal position, at m/z 222 (d$_3$) for a hexose in an internal position and at m/z 225 (d$_6$) or 228 (d$_9$) for a hexose at a branching point. This same information can be obtained from the oxonium ions of N-acetylhexoses at m/z

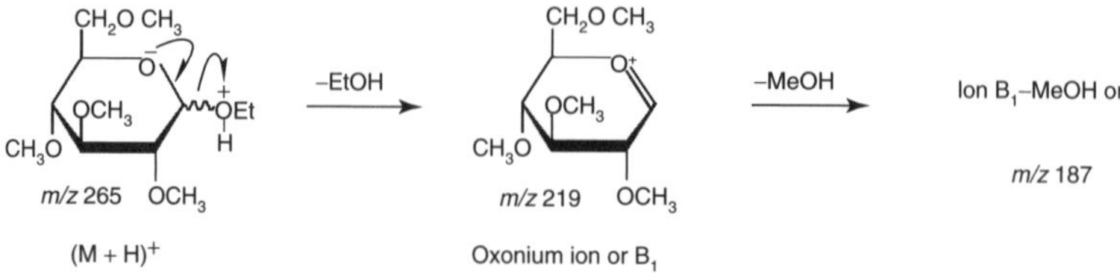

Figure 7.54
Fragmentation scheme of ethylglycosides produced in a CI source

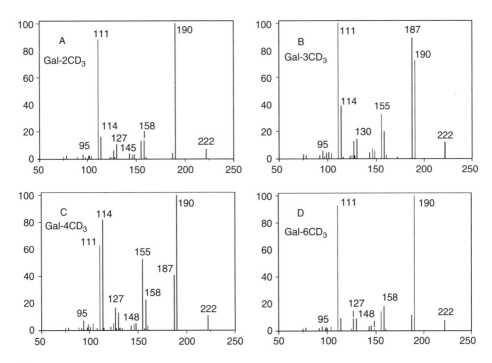

Figure 7.55
The CI/MS/MS mass spectra of oxoniums ions from galactose: (A) Gal-2CD$_3$; (B) Gal-3CD$_3$; (C) Gal-4CD$_3$; (D) Gal-6CD$_3$. (© E. de Hoffmann)

260 (d$_0$), 263 (d$_3$) and 266 (d$_6$) or for a deoxyhexose at m/z 189 (d$_0$), 192 (d$_3$) and 195 (d$_6$), respectively.

Subsequent analysis of the fragments of the oxonium ions by CID/MS/MS at low energy allows the positions of the deuterated methyl groups to be assigned corresponding to the branching positions. This is illustrated in Figure 7.55, displaying the product ion spectra obtained by MS/MS of the oxonium ions of Gal derivatives deuteromethylated on the 2, 3, 4 or 6 position, respectively. It is possible to differentiate the various positions of the CD$_3$.

This differentiation is based mainly on the ratios of the intensities of the two couples of ions at m/z 187 and 190 and m/z 111 and 114, respectively. The monosaccharides substituted on the 2 and 6 positions display the same spectrum. They can, however, be distinguished clearly by the fragmentation spectrum of the (oxonium-MeOH) ions. A systematic study has shown that the MS/MS spectra are predictable and largely independent of the nature of the monosaccharide.

This method alone does not allow the determination of the structure of oligosaccharides. If, however, it is used in conjunction with an MS/MS analysis of the whole permethylated oligosaccharides, the complete structure can be deduced.

Figure 7.56 describes the application of this method to the LNF-II oligosaccharide. The mixture of derivatized monosaccharides obtained after permethylation, ethanolysis and deuteropermethylation is analyzed by chemical ionization GC/MS/MS. The GC chromatogram obtained is displayed along with the ion chromatograms

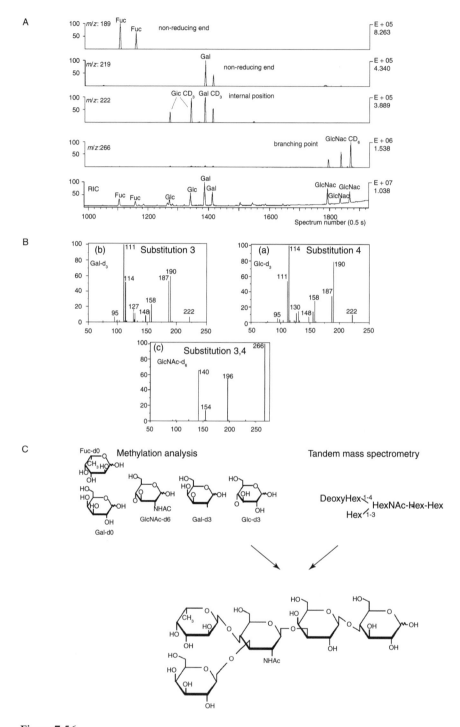

Figure 7.56
(A) The GC chromatogram of a mixture of derivatized monosaccharides obtained from the LNF-II oligosaccharide. (B) The MS/MS product ion spectra of the oxonium ions: (a) at *m/z* 222, monosubstituted Glc; (b) at *m/z* 222, monosubstituted Gal; (c) at *m/z* 266, disubstituted GlcNAc. (C) Deduction of the structure of LNF-II. (© E. de Hoffmann)

of different oxonium ions. The multiplicity of the chromatographic traces results from the presence of the α/β and pyran/furan isomers for each glycoside.

From the retention times, compared with those of standards, the presence of Fuc, Glc, GlcNac and two Gal residues is clearly established. The masses detected for these different monosaccharides allow the positions of these residues to be assigned. The Fuc and one of the Gal residues are located at the non-reducing ends, based on the mass of their oxonium ions at m/z 189 and 219, respectively. Indeed, they are not deuterated. The second Gal residue and the Glc residue are positioned inside the oligosaccharide, because they include one deuteromethyl group as shown by the masses of the oxonium ions at m/z 222. Finally, the oxonium ion of GlcNac appears at m/z 266, corresponding to the incorporation of six deuterium atoms. This GlcNac is thus located at a branching point.

The next step is to assign the positions of the deuteromethyl groups and thus of the glycosidic bonds. This is deduced from MS/MS fragmentation spectra of the corresponding oxonium ions. The spectra obtained are displayed in Figure 7.56 and lead to the following conclusions: Glc is deuterated at the 4 position, Gal at the 3 position and GlcNac at the 3 and 4 positions.

This information together with that from the MS/MS fragmentation spectrum of the whole permethylated oligosaccharide (see Figure 7.48), allows the complete structure of LNF-II to be deducted, except for the anomeric configuration of the glycosidic bonds.

7.5 Lipids

Lipids are made up of many classes of very different molecules that all show solubility properties in organic solvents. Mass spectrometry plays a key role in the biochemistry of lipids. Indeed, mass spectrometry allows not only the detection and determination of the structure of these molecules but also their quantification. For practical reasons, only the fatty acids, acylglycerols and bile acids are discussed here, although other types of lipids such as phospholipids,[228-231] steroids,[232-234] prostaglandins,[235] ceramides,[236,237] sphingolipids[238,239] and leukotrienes[240,241] have been analyzed successfully by mass spectrometry. Moreover, the described methods will be limited to those that are based only on mass spectrometry, even if the majority of these methods generally are coupled directly or indirectly with separation techniques such as GC or HPLC. A book on the mass spectrometry of lipids was published in 1993.[242]

7.5.1 Fatty acids

Fatty acids are a class of molecules that are made up of a long hydrocarbon chain, of varying length and varying degrees of unsaturation, terminated by a carboxylic group. Some organisms such as bacteria, sponges and some plants can produce modified fatty acids: branchings; introduction of a cyclopropane, cyclopropene or epoxy ring; hydroxylations or alkoxylations; or even unusual unsaturations. These fatty acids are interesting not only for their particular biological activities but also for their rarity, which allows one to classify and identify the organisms that produce them much faster than by the classical characterization techniques.

The classical analysis method consists of extracting the lipids from the biological material and then characterizing the previously purified fatty acids by comparing their retention times with those of reference fatty acids, by degradative analyses and by using a series of spectroscopic methods. Numerous chromatographic techniques, including thin-layer chromatography (TLC), HPLC and GC, are used in the purification.

Because fatty acids derived from natural sources are present in a mixture, an ideal analysis method for these molecules should be applicable to mixtures without requiring a prior separation or derivatization. Mass spectrometry is an excellent tool for determining the structure of fatty acids present in a mixture. It is possible to determine not only the molecular weight and thus the elemental composition but also, in most cases, the nature and position of the branching and the other substituents on the carbon chain.[234,244] Furthermore, such an analysis requires low quantities ranging from 10 pg to 100 ng of total lipid, depending upon the analyzed sample, the ionization method used and the configuration of the spectrometer.[245,246]

A general analysis method for fatty acids is based on the high-energy tandem mass spectrometry of carboxylate anions. Indeed, FAB, desorption chemical ionization (DCI), ESI or atmospheric pressure chemical ionization (APCI) spectra measured in the negative mode are characterized by the single presence of molecular ion species and their isotopic clusters and the absence of fragments.[247–249] This characteristic is used to facilitate the determination of the molecular weights of fatty acids even when they are present in a complex mixture, but it does not provide information concerning structure. This inconvenience may be overcome by transferring the extra energy necessary for fragmentation to the stable ions produced during ionization by using MS/MS.

The high-energy tandem mass spectrometric analysis of carboxylate ions from fatty acids yields a series of homologous fragments all separated by 14 Th. These fragments correspond to the formal loss of alkane: CH_4, C_2H_6, ..., C_nH_{2n+2}. They are produced by a highly specific 1,4-elimination mechanism (charge remote fragmentation) of a hydrogen molecule, as shown in Figure 7.57.[250] This mechanism has been confirmed by neutral fragment reionization, demonstrating that the lost neutral is indeed an alkene.[251] However, other mechanisms compatible with the experimental data, implying the homolytic rupture of C–C or C–H bonds, also were proposed.[252,253]

This loss of C_nH_{2n+2} starts at the terminal alkyl and progresses along the hydrocarbon chain. In the case of saturated fatty acids, the abundance of these fragments yields a characteristic spectrum such as that shown in Figure 7.58.[254,255]

The presence of unsaturation or substituents disrupts or disturbs the characteristics of the spectrum. The nature of the disruption allows one to distinguish the type of structural modification whereas the localization on the chain can be determined from the point where the perturbation occurs.

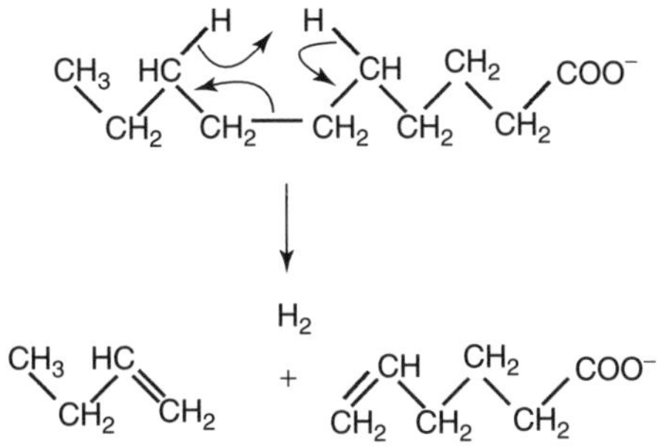

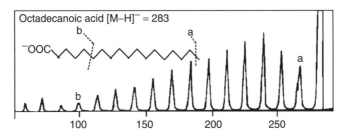

Figure 7.57
The 1,4-elimination mechanism of H_2 occurring along the fatty acid chain in order to yield fragments corresponding to the loss of an alkene

Octadecanoic acid $[M–H]^- = 283$

Figure 7.58
Collision-induced dissociation FAB/MS/MS trace of octadecanoic acid acquired in the negative ion mode at high energy. (Reproduced (modified) from Ref. 255 with permission)

The presence of branching on the hydrocarbon chain is indicated by the suppression of fragmentation at the branching point and by accentuation of fragmentation of the bond adjacent to the branching on the alkyl side. Thus the spectra of a branched fatty acid are characterized by a two-methylene 'hole', i.e. by two peaks in the series separated by 28 Th. The CID/MS/MS spectra of two branched fatty acids, presented in Figure 7.59, show how easy it is to localize the branching.[256]

The presence of an unsaturation within a fatty acid is indicated and its position is established by the absence of fragments derived from cleavages of this unsaturated bond and the adjacent ones. This corresponds, on the spectrum, to a four-carbon atom 'hole', i.e. by two peaks in the series separated by 54 Th (Figure 7.60).[257] This absence of ions results from the fact that cleaving a vinylic bond or a double bond is not energetically favored. The localization of double bonds in unsaturated fatty acids is made more difficult and even impossible as the number of unsaturations increases because the process of losing alkanes is hidden by the loss of 45 Da ($\cdot$COOH).

Other modifications of the hydrocarbon chain of fatty acids, such as hydroxylation or introduction of a cyclopropane, cyclopropene or epoxy ring, could be identified

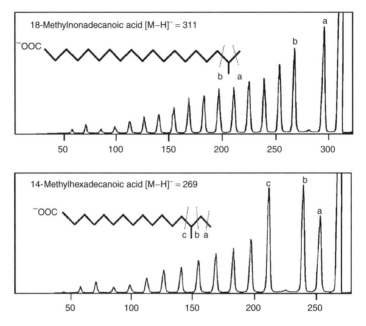

Figure 7.59
Collision-induced dissociation FAB/MS/MS traces of 18-methylnona-decanoic acid and 14-methylhexadecanoic acid, acquired in the negative ion mode at high energy. (Reproduced (modified) from Ref. 256 with permission)

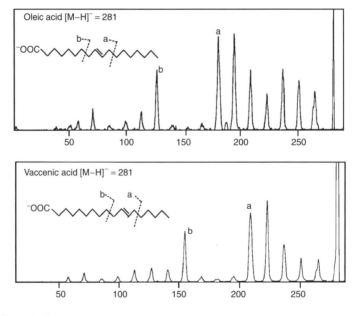

Figure 7.60
Collision-induced dissociation FAB/MS/MS traces of oleic acid and vaccenic acid, measured in the negative mode at high energy. (Reproduced (modified) from Ref. 257 with permission)

and located by this method.[258] High-energy MS/MS applied to $(M + Li)^+$ or $(M - H + 2Li)^+$ corresponding to fatty acids cationized by Li^+ (preferred over other alkali metals because they bind more strongly to carboxylic groups) yields results that are comparable with those obtained in negative ion mode. The only important difference lies with polyunsaturated fatty acids, which yield ions in the positive mode that are interpreted more easily.

An alternative method for the analysis of fatty acids by mass spectrometry is based on the preparation of picolinic esters.[259,260] These derivatives allow the GC separation of fatty acids and their EI spectra display abundant diagnostic fragment ions. These fragments correspond to C−C bond cleavages all along the hydrocarbon chain. The proposed mechanism starts with the production by EI of a radical-cation located at the nitrogen atom. This radical-cation rearranges to yield a distonic radical-cation by abstraction of a hydrogen atom from the chain. The C−C bonds then are broken by a reaction initiated at the radical site to yield an alkene and an alkyl radical. The presence of a modification along the hydrocarbon chain will cause an alteration of the mass spectrum. Figure 7.61 shows the mass spectra of two ramified fatty

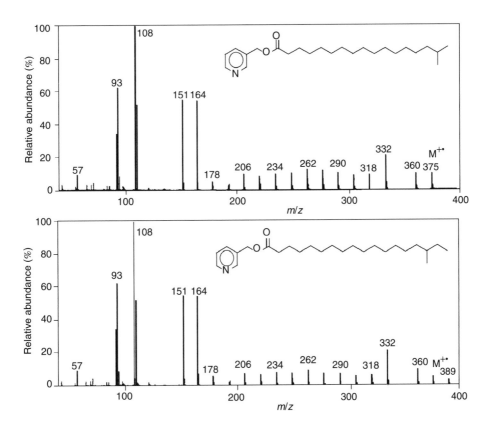

Figure 7.61
Electron ionization (25 eV) mass spectra of picolinic esters from *iso*-octadenoic acid and *ante*-isononadecanoic acid. The fragment ion at m/z 151 corresponds to a McLafferty rearrangement product. (Reproduced (modified) from Ref. 259 with permission)

acid picolinic esters as examples. This method based on GC/MS does not require high-energy MS/MS.

Low-energy CID of the molecular ions of fatty acid methyl esters obtained by EI (70 eV) also has been studied.[261] Such fatty acids methyl esters decompose in the tandem quadrupole mass spectrometer to yield a regular homologous series of carbomethoxy ions. Decomposition of the molecular ions of several methyl-branched fatty acid methyl esters, such as phytanic acid shown in Figure 7.62, reveals enhanced radical site cleavage at the alkyl branching positions. This technique of low-energy CID of molecular ions generated by EI provides a sensitive, powerful and simple approach to the determination of methyl- or alkyl-branched saturated fatty acids without the need for special derivatization or high-energy MS/MS.

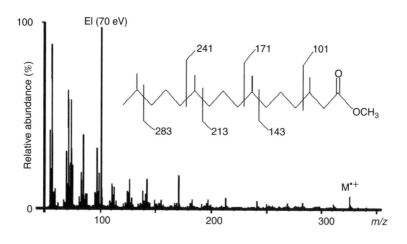

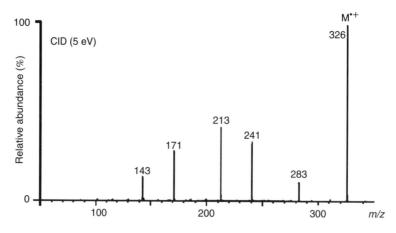

Figure 7.62
The EI/MS/MS spectrum of methyl 3,7,11,15-tetramethylhenadecanoate (methyl phytanate) and the collision-induced dissociation EI/MS/MS spectrum of the molecular ion (m/z 326) at low energy (5 eV). (Reproduced modified from Ref. 261 with permission)

7.5.2 Acylglycerols

Acylglycerols are the fatty acid esters of glycerol. Because every alcohol function of glycerol can be esterified, the molecules can contain one, two, or three fatty acids, termed mono-, di- or triacylglycerols, respectively. Characterization of an acylglycerol molecule requires not only the identification of its component fatty acids but also their positions in the glycerol molecule (positional isomer).

The general procedure allowing the characterization of an acylglycerol consists of hydrolyzing the acylglycerol that has been purified previously by TLC or HPLC in order to analyze the fatty acids obtained by the methods described earlier in this chapter. For determination of the positions of the various component fatty acids in an acylglycerol, no simple method exists that does not use mass spectrometry, in spite of the progress in acylglycerol analysis in the last few years using high-resolution GC or reverse-polarity HPLC. The most frequently used method consists of treating the purified acylglycerol with a lipase that specifically cleaves the central position ester bond and then characterizing the fatty acid that is liberated. Obviously the complete identification of acylglycerols using classical methods is long and difficult and requires a large quantity of sample.[262]

At present, mass spectrometry allows the analysis of an acylglycerol by identifying its various component fatty acids and their positions in the glycerol without necessitating prior chromatographic separation. The quantity of sample that is used during a mass spectrometric analysis is about 1 picomole.[263]

Various ionization techniques, such as EI,[264] CI,[265] DCI,[266] PD,[267] FAB,[268] APCI[269] and ESI,[270] have been used successfully. Overall, the acylglycerol spectra obtained in the positive mode contain as predominant ions, in addition to the molecular ion $M^{\bullet+}$ in EI or the molecular species ions $(M + H)^+$, $(M + Na)^+$ or $(M + NH_4)^+$ in the case of other ionization techniques, the ions issued from the loss of every acyloxy group present in the molecule, labeled $(M - R_nCOO)^+$, and the corresponding acylium ions, labeled $(R_nCO)^+$. The labeled ions $(M - R_nCOO)^+$ have been referred to as 'diglyceride-type ions'. In the negative mode the predominant ions are the molecular species ion $(M - H)^-$ and the ion corresponding to the fatty acids that are present, labeled $(R_nCOO)^-$, as shown in Figure 7.63. The relative importance of these various ions depends, of course, on the ionization technique used. For example, in the positive mode, the intensity of the molecular species ion is weak in EI and FAB and stronger in CI and DCI, and is the only ion present in ESI.

Results obtained by CID/MS/MS product ion spectra of triacylglycerols in positive ion mode appear to be quite independent of the type of selected precursor as well as the ionization mode used to obtain the precursor ion.[268,270,271] These fragmentation spectra contain acylium ions of the fatty acids and, more abundantly, diacylglycerol-type ions.

Useful qualitative and quantitative information could be obtained from mass spectrometry. Starting from the measured mass of each acylglycerol, the elemental composition and the number of unsaturations can be determined. A chromatographic separation is no longer necessary because this method based on mass spectrometry allows one to determine the molecular weights of a mixture of various components. Hence an analysis using this method requires minimal preparation and achieves

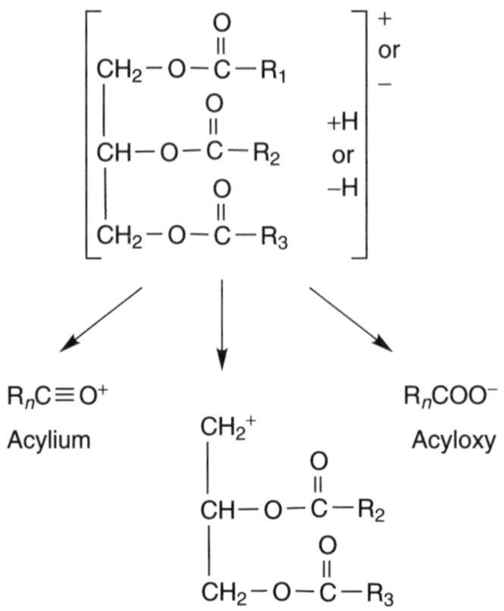

Figure 7.63
General fragmentation diagram of acylglycerols

important savings in time. However, more information is necessary to determine the fatty acid composition of an acylglycerol. A triacylglycerol containing three C_{16} fatty acids has the same molecular weight as a triacylglycerol containing C_{14}, C_{16} and C_{18} fatty acids. This extra information is supplied by the fragments derived from cleavage of the fatty acid chains, which allow one to determine the molecular mass of each fatty acid linked to the glycerol. This fatty acid composition can be determined in the presence of other acylglycerols only if MS/MS is used, because all the acylglycerols in the mixture undergo fragmentation and yield ions that may overlap one another, thereby preventing correct interpretation of the spectrum.

It should be noted that only the elemental composition of each fatty acid is obtained, without other structural information on double bond position and isomerism, branching, etc. However, high-energy CID of the anions of the fatty acids allows this information to be obtained.[272] For mono- and diacylglycerols, structural information can be obtained after derivatization with nicotinic acid.[273]

An unambiguous, rapid and sensitive method based on MS/MS of the deprotonated molecular ions of acylglycerols obtained in negative mode by DCI was developed in order to determine the position of each fatty acid linked to the glycerol molecule, i.e. to determine the positional isomers.[274]

The low-energy tandem mass spectra of the deprotonated molecular ions of acylglycerol contain a type of ion whose formal mass-based composition corresponds to a ketone obtained by the combination of two fatty acid chains with a carbonyl group, minus a proton. The ketone contains mainly the chains of the central fatty acid combined with one of the two external fatty acids, even if the ketone containing the two external fatty acids is present with a much weaker intensity. The formation of these ions may be explained by an internal Claisen condensation followed by a

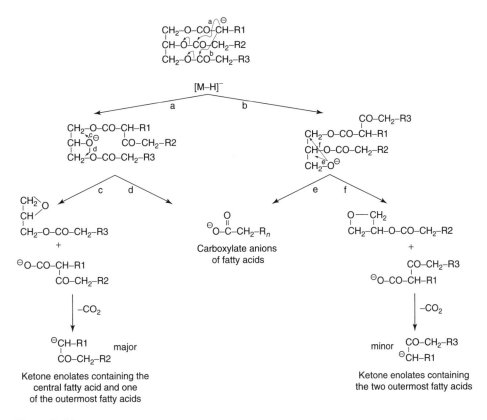

Figure 7.64
Claisen condensation mechanism responsible for the formation of ketones, allowing the determination of positional isomers

fragmentation induced by a nucleophile substitution and then by a decarboxylation, as shown in Figure 7.64.

This mechanism of formation explains the sensitivity of these ions for the positional isomers. Indeed, expulsion of the ketone containing the hydrocarbon chains of the central fatty acid combined with one of the outermost fatty acids (Figure 7.64, pathway c) requires the formation of a neutral epoxide, which is faster than the neutral oxetane formation necessary for expulsion of the ketone resulting from the condensation between the two outermost fatty acids (Figure 7.64, pathway f). This initial reaction thus occurs preferentially.

This method can be applied directly to acylglycerols present in a mixture, as is illustrated by the mass spectrometric analysis of natural cocoa butter.[274] This analysis (Figure 7.65) allowed the determination of the complete structure of the predominant acylglycerols in the cocoa (Table 7.11).

7.5.3 Bile acids

Bile acids are a family of molecules derived from cholesterol. They are characterized by a 5β steroid ring made up of four fused cycles bearing a side chain attached to

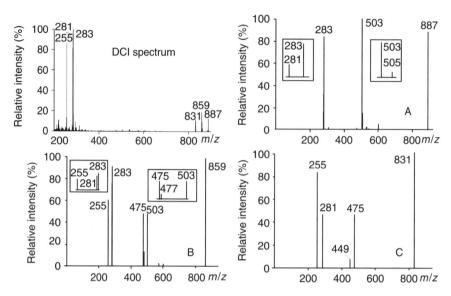

Figure 7.65
Desorption chemical ionization (DCI) mass spectrum of natural cocoa butter and collision-induced dissociation DCI/MS/MS traces of deprotonated molecular ions of 887 Th (A), 859 Th (B) and 831 Th (C). (Reproduced (modified) from Ref. 274 with permission)

Table 7.11 Masses observed in the various fragment spectra of the three deprotonated molecular ions present in the DCI spectrum of natural cocoa butter (P = palmitic acid, O = oleic acid, S = stearic acid)

$(M - H)^-$	R_nCOO^-	$R_nCOR'_n$	Deduced structure
831	255 (P), 281 (O)	475 (PO), *449*	POP
859	255 (P), 281 (O), 283 (S)	475 (PO), 503 (OS), *477*	POS
887	281 (O), 283 (S)	503 (SO), *505*	SOS

Numbers in italic indicate ions of low abundance.

the C-17 carbon atom of the cycle, terminated by a carboxylic group. They differ from each other by the number and position of hydroxyl or keto groups and by the presence of unsaturations in the steroid cycle. Furthermore, these bile acids can exist as free carboxylic acids or as amide conjugates of the carboxylic groups with glycine ($NH_2CH_2CO_2H$) or taurine ($NH_2CH_2CH_2SO_3H$). Table 7.12 lists the structure of the bile acids encountered most often in the literature.

Mass spectrometry has become an indispensable method for the analysis of bile acids by virtue of its power to identify, assign structure and quantify free or conjugated bile acids, either pure or in mixtures. It is useful not only to study the metabolism of bile acids but also for the detection and diagnosis of metabolic diseases. Indeed, numerous metabolic diseases resulting from an alteration of the

Table 7.12 Structure and nomenclature of various bile acids

X = OH free bile acid
= NHCH$_2$CO$_2$H glycinoconjugated
= NHCH$_2$CH$_2$SO$_3$H tauroconjugated

Trivial name	Chemical name	R^1	R^2	R^3	R^4
Cholanic acid	5β-Cholan-24-oic acid	H	H	H	H
Lithocholic acid	5β-Cholan-24-oic-3α-ol acid	αOH	H	H	H
Hyodeoxycholic acid	5β-Cholan-24-oic-3α,6α-diol acid	αOH	αOH	H	H
Murocholic acid	5β-Cholan-24-oic-3α,6β-diol acid	αOH	βOH	H	H
Chenodeoxycholic acid	5β-Cholan-24-oic-3α,7α-diol acid	αOH	H	αOH	H
Ursodeoxycholic acid	5β-Cholan-24-oic-3α,7β-diol acid	αOH	H	βOH	H
Deoxycholic acid	5β-Cholan-24-oic-3α,12α-diol acid	αOH	H	H	αOH
Cholic acid	5β-Cholan-24-oic-3α,7α,12α-triol acid	αOH	H	α OH	αOH
Hyocholic acid	5β-Cholan-24-oic-3α,6α,7α-triol acid	αOH	αOH	αOH	H
Dehydrocholic acid	5β-Cholan-24-oic-3,7,12-trione acid	O	H	O	O

conversion of cholesterol to bile acids have been described, including peroxisomal disorders resulting in a block of β-oxidation of the lateral chain and other enzyme deficiencies interfering with the biochemistry of the side chain or the steroid nucleus.

Before the advent of soft ionization techniques, the analysis of bile acids was long and tedious and needed large sample quantities. First, the bile acids had to be extracted from the biological fluid and separated by lipophilic ion exchange chromatography into four classes: unconjugated, tauro- or glycinoconjugated and sulfated. Next, each fraction separately was hydrolyzed, extracted and derivatized. The bile acids then were identified and quantitated in four GC/MS runs.

The emergence of soft ionization techniques such as FAB, thermospray (TSP), APCI and ESI, combined with MS/MS, will considerably simplify these analysis. On a much smaller sample quantity, bile acids can be analyzed without prior separation or derivatization. Soft ionization techniques are well suited for such polar, non-volatile thermolabile compounds.

The FAB, TSP, APCI and ESI mass spectra in the positive ionization mode display the protonated molecular ion, generally accompanied by other adducts of the molecular species and fragments resulting from the loss of one or several water molecules originating from the ring hydroxyl groups.[275–277] As an acidic function is always present in these compounds, they also yield intense ions in the negative ion mode.[278-280] The spectra contain mainly deprotonated molecular ions. The presence of some adducts and a weak fragmentation sometimes also are observed. The relative importance of these ions depends on the sample preparation, the ionization method used and the experimental conditions. Generally, ESI spectra are simpler than FAB or TSP spectra.

The spectra thus contain few structural information and the use of MS/MS is very useful. The analysis of free or conjugated bile acids by high-energy MS/MS allows the observation of ions resulting from charge remote fragmentation (CRF) of the steroid cycle and the side chain.[281-283] Stereoisomers yield very similar spectra that do not allow differentiation between them in mixtures. But CRF provides information on the nature and position of substituents in the steroid ring. The position of double bonds also can be deduced by the presence of specific fragments or by the absence of some fragmentations.

The low-energy CID of bile acids is characterized by the presence of charge-induced fragmentations.[284] These fragmentations are often of little analytical interest because they result mostly in the loss of small molecules, such as water, formic acid,

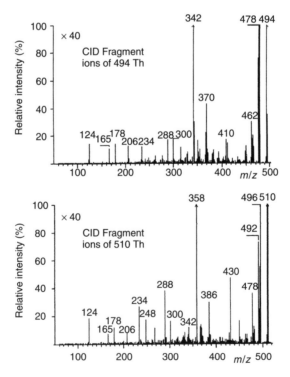

Figure 7.66
Collision-induced dissociation FAB/MS/MS traces of the taurine conjugates of 7α-hydroxy-3-oxochol-4-en-24-oic acid (*top*) and 7α, 12α-dihydro-3-oxochol-4-en-24-oic acid (*bottom*). (Reproduced (modified) from Ref. 286 with permission)

carbon dioxide, etc. However, the fragmentation of taurine conjugates in negative ion mode at low energy displays both charge-induced fragmentations CRF, yielding structural information on the ring substitution.[285]

This experimental observation is illustrated in Figure 7.66 by the tandem mass spectra of two bile acids of the Δ^4-3-oxo class: the taurine conjugates of 7α-hydroxy-3-oxochol-4-en-24-oic acid and 7α, 12α-dihydro-3-oxochol-4-en-24-oic acid.[286] Similar spectra were obtained for the Δ^4-3-hydroxy derivatives. The general fragmentation diagram deduced from these spectra is shown in Figure 7.67.

Knowing the fragmentation diagram of the bile acids, i.e. the structures of the fragments produced, allows one to determine the complete structure of the molecule. For example, the masses of fragments A and B indicate the presence or the absence of a hydroxyl group at position C-12.

The usefulness of low-energy MS/MS has been shown in the diagnosis of metabolic diseases by analysis of underivatized bile acids, conjugated or not, in complex biological samples such as urine or serum. A series of neutral loss and precursor ion scans was developed in order to allow the selective detection of molecular ion species of some bile acid classes present within a complex mixture. These scans mainly include: the scan of the precursor ions of 124 Th (anions of taurine), to detect all the taurine conjugates; the scan of neutral losses of 152 and 154 Da (mass of fragment A), to characterize the Δ^4-3-oxo and Δ^4-3-hydroxy taurine conjugates; and the scan of the neutral loss of 62 Da (corresponding to the loss of $CO_2 + H_2O$), to detect the free hydroxylated C-12 bile acids and their glycine conjugates. This method is rapid because it does not require prior chromatographic separation, degradation or derivatization.

The example in Figure 7.68 shows the analysis of a mixture containing 11 bile acids.[286] The FAB spectrum containing the molecular ion species of all of the bile acids present in the mixture (more intense than the background noise) is not sufficient to determine the composition of the mixture. Only the various specific scans allow

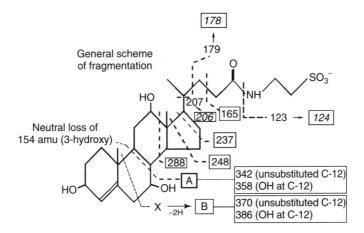

Figure 7.67
General fragmentation diagram of the Δ^4-cholenoic taurine conjugates[285]

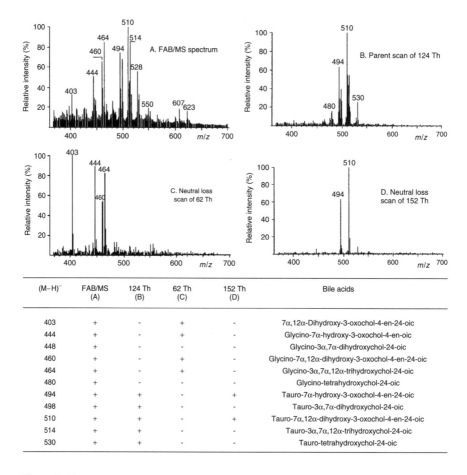

(M–H)⁻	FAB/MS (A)	124 Th (B)	62 Th (C)	152 Th (D)	Bile acids
403	+	-	+	-	7α,12α-Dihydroxy-3-oxochol-4-en-24-oic
444	+	-	+	-	Glycino-7α-hydroxy-3-oxochol-4-en-oic
448	+	-	-	-	Glycino-3α,7α-dihydroxychol-24-oic
460	+	-	+	-	Glycino-7α,12α-dihydroxy-3-oxochol-4-en-24-oic
464	+	-	+	-	Glycino-3α,7α,12α-trihydroxychol-24-oic
480	+	-	-	-	Glycino-tetrahydroxychol-24-oic
494	+	+	-	+	Tauro-7α-hydroxy-3-oxochol-4-en-24-oic
498	+	+	-	-	Tauro-3α,7α-dihydroxychol-24-oic
510	+	+	-	+	Tauro-7α,12α-dihydroxy-3-oxochol-4-en-24-oic
514	+	+	-	-	Tauro-3α,7α,12α-trihydroxychol-24-oic
530	+	+	-	-	Tauro-tetrahydroxychol-24-oic

Figure 7.68
(A) The FAB spectrum of a mixture of 11 bile acids from a patient's urine, measured in negative mode. (B) Spectrum of the 124 Th precursors, allowing the selective detection of taurine conjugates. (C) Spectrum of 62 Da neutral losses. (D) Spectrum of 152 Da neutral losses. (Reproduced (modified) from Ref. 286 with permission)

the determination of the nature of all the bile acids present in the mixture, without however determining their stereochemistry.

7.6 References

1. Bleakney W., *Phys. Rev.*, **34**, 157 (1929).
2. Munson M.S.B. and Field F.H., *J. Am. Chem. Soc.*, **88**, 2621 (1966).
3. Beckey H.D., *Z. Anal. Chem.*, **197**, 80 (1963).
4. Morris H.R., *Soft Ionization Biological Mass Spectrometry*, Heyden, London, 1980.
5. Thorgersen D.F., Skowronsky R.P. and Macfarlane R.D., *Biochem. Biophys. Res. Commun.*, **60**, 616 (1974).
6. Harkansson P., Kamenski I., Sundquist B., *et al.*, *J. Am. Chem. Soc.*, **104**, 2498 (1982).

7.   Barber M., Bordoli R.S., Sedwick R.D., *et al.*, *J. Chem. Soc. Chem. Commun.*, **7**, 325 (1981).
8.   Cotter R.J., *Anal. Chem.*, **56**, 485A (1984).
9.   Fenn J.B., Mann M., Meng C.K., *et al.*, *Science*, **246**, 64 (1989).
10. Karas M. and Hillenkamp F., *Anal. Chem.*, **60**, 2299 (1988).
11. Krishna R. and Wold F., *Protein Structure–A Practical Approach*, 2nd edition, ed. by T.E. Creighton, Oxford University Press, New York, 1997, pp. 391–402.
12. McCloskey J.A., *Methods in Enzymology*, Vol. **193**, Academic Press, New York, 1990, pp. 351–538.
13. Cotter J.C., *Anal. Chem.*, **64**, 1027 (1992).
14. Roepstorff P., *Trends Anal. Chem.*, **12**, 413 (1993).
15. Eckart K., *Mass Spectrom. Rev.*, **13**, 23 (1994).
16. Siuzdak G., *Proc. Natl. Acad. Sci. USA*, **91**, 11290 (1994).
17. Feistner G.J., Faull K.F., Barofsky D.F., *et al.*, *J. Mass Spectrom.*, **30**, 519 (1995).
18. Burlingame A.L., Boyd R.K. and Gaskell S.J., *Anal. Chem.*, **68**(12), 599R (1996).
19. Kuster B. and Mann M., *Curr. Opin. Struc. Biol.*, **8**(3), 393 (1998).
20. Mann M. and Talbo G., *Curr. Opin. Biotechnol.*, **7**(1), 11 (1996).
21. Vorm O., Roepstorff P. and Mann M., *Anal. Chem.*, **66**, 3281 (1994).
22. Valaskovic G.A., Kelleher N.L. and McLafferty F.W., *Science*, **273**(5279), 1199 (1996).
23. Martin S.E., Shabanowitz J., Hunt D., *et al.*, *Anal. Chem.*, **72**, 4266 (2000).
24. Cohen S.L. and Chait B.T., *Anal. Chem.*, **68**(1), 31 (1996).
25. Naylor S.A., Findeis A.F., Gibson B.W., *et al.*, *J. Am. Chem. Soc.*, **108**, 6359 (1986).
26. Beavis R.C. and Chait B.T., *Anal. Chem.*, **62**, 1836 (1990).
27. Wilm M., Shevchenko A., Houthaeve T., *et al.*, *Nature*, **379**(6564), 466 (1996).
28. Davis M.T., Stahl D.C. and Lee T.D., *J. Am. Soc. Mass Spectrom.*, **5**, 571 (1994).
29. Emmet M.R. and Caprioli R.M., *J. Am. Soc. Mass Spectrom.*, **5**, 605 (1994).
30. Wahl J.H., Goodlett D.R., Udseth H.R., *et al.*, *Electrophoresis*, **14**, 448 (1993).
31. Bean M.F., Carr S.A., Thorne G.C., *et al.*, *Anal. Chem.*, **63**, 1473 (1991).
32. Papayannopoulos I.A., *Mass Spectrom. Rev.*, **14**, 49 (1995).
33. Roepstorff P. and Fohlman J., *Biomed. Mass Spectrom.*, **11**, 601 (1984); **12**, 631 (1985).
34. Biemann K., *Biomed. Environ. Mass Spectrom.*, **16**, 99 (1988).
35. Hunt D.F., Yates J.R., Shabanowitz J., *et al.*, *Proc. Natl. Acad. Sci. USA*, **83**, 6233 (1986).
36. Johnson R.S. and Biemann K., *Biomed. Environ. Mass Spectrom.*, **18**, 945 (1989).
37. Heerma W. and Kulik W., *Biomed. Environ. Mass Spectrom.*, **16**, 155 (1988).
38. Falik A.M., Hines W.M., Medzihradsky K.F., *et al.*, *J. Am. Soc. Mass Spectrom.*, **4**, 882 (1993).
39. Johnson R.S., Martin S.A., Biemann K., *et al.*, *Anal. Chem.*, **59**, 2621 (1987).
40. Tomer K.B., Crow F.W. and Gross M.L., *J. Am. Chem. Soc.*, **105**, 5487 (1983).
41. Johnson R.S., Martin S.A. and Biemann K., *Int. J. Mass Spectrom. Ion Processes*, **86**, 137 (1988).
42. Alexander A.J. and Boyd R.K., *Int. J. Mass Spectrom. Ion Processes*, **90**, 211 (1989).
43. Tang X.J., Thibault P. and Boyd R.K., *Anal. Chem.*, **65**, 2824 (1993).
44. Downard K.M. and Biemann K., *J. Am. Soc. Mass Spectrom.*, **5**, 966 (1994).
45. Covey T.R., Huang E.C. and Henion J.D., *Anal. Chem.*, **63**, 1193 (1991).
46. Watkins P.F.E., Jardine I. and Zhou J.X.G., *Biochem. Soc. Trans.*, **19**, 957 (1991).
47. Kaufmann R., Spengler B. and Lutzenkirchen, *Rapid Commun. Mass Spectrom.*, **7**, 902 (1993).
48. Huberty M.C., Vath J.E., Yu W., *et al.*, *Anal. Chem.*, **65**, 2791 (1993).
49. Rouse J.C., Yu W. and Martin S.A., *J. Am. Mass Spectrom.*, **6**(9), 822 (1996).
50. vanDongen W.D., Ruijters H.F.M., Luinge H.J., *et al.*, *J. Mass Spectrom.*, **31**(10), 1156 (1996).
51. Biemann K., *Methods Enzymol.*, **193**, 455 (1990).

52. Martin S.A., Johnson R.S., Costello C.E., *et al.*, in *The Analysis of Peptides and Proteins by Mass Spectrometry*, ed. by C.J. McNeal, Wiley, New York, 1988, pp. 135–150.
53. Cardenas M.S., van der Heeft E. and de Jong A.P.J.M., *Rapid Commun. Mass Spectrom.*, **11**(12), 1271 (1997).
54. Hendrickson R.C., Skipper J.C., Shabanowitz J., *et al.*, *Immunology Methods Manual*, ed. by I. Lefkovits, Academic Press, San Diego, CA, 1997.
55. Shevchenko A., Chernushevich I., Ens W., *et al.*, *Rapid Commun. Mass Spectrom.*, **11**(9), 1015 (1997).
56. Vath J.E. and Biemann K., *Int. J. Mass Spectrom. Ion Processes*, **100**, 287 (1990).
57. Hulst A.G. and Kientz C.E., *J. Mass Spectrom.*, **31**(10), 1188 (1996).
58. Scoble H.A., Biller J.E. and Biemann K., *Fresenius Z. Anal. Chem.*, **327**, 239 (1987).
59. Johnson R.S. and Biemann K., *Biomed. Environ. Mass Spectrom.*, **18**, 945 (1989).
60. Zidarov D., Thibault P., Evans M.J., *et al.*, *Biomed. Environ. Mass Spectrom.*, **19**, 13 (1990).
61. Hines W.M., Faliek A.M., Burlingame A.L., *et al.*, *J. Am. Mass Spectrom.*, **3**, 326 (1992).
62. Eng J.K., McCormack A.L. and Yates J.R., *J. Am. Soc. Mass Spectrom.*, **5**, 976 (1994).
63. Yates J.R., Eng J.K. and McCormack A.L., *Anal. Chem.*, **67**, 1426 (1995).
64. Yates J.R., Eng J.K., Clauser K.R., *et al.*, *J. Am. Soc. Mass Spectrom.*, **7**(11), 1089 (1996).
65. Roepstorff P., *Curr. Opin. Biotechnol.*, **8**(1), 6 (1997).
66. Yates J.R., *J. Mass Spectrom.*, **33**(1), 1 (1998).
67. Jungblut P. and Thiede B., *Mass Spectrom. Rev.*, **16**(3), 145 (1997).
68. Deissler H., Wilm M, Genc B., *et al.*, *J. Biol. Chem.*, **272**(27), 16761 (1997).
69. Henzel W.J., Billeci T.M., Stults J.T., *et al.*, *Proc. Natl. Acad. Sci. USA*, **90**, 5011 (1993).
70. Mann M., Hojrup P. and Roepstorff P., *Biol. Mass Spectrom.*, **22**, 338 (1993).
71. Pappin D., Hojrup P. and Bleasby A.J., *Curr. Biol.*, **3**, 327 (1993).
72. Yates J.R., Speicher S., Griffin P.R., *et al.*, *Anal. Biochem.*, **214**, 397 (1993).
73. Mann M. and Wilm M., *Anal. Chem.*, **66**, 4390 (1994).
74. Mann M., *Trends Biochem. Sci.*, **21**(12), 494 (1996).
75. McCormack A.L., Schieltz D.M., Goode B., *et al.*, *Anal. Chem.*, **69**(4), 767 (1997).
76. Shevchenko A., Wilm M., Vorm O., *et al.*, *Anal Chem.*, **68**(5), 850 (1996).
77. Eckerskorn C. and Grimm R., *Electrophoresis*, **17**(5), 899 (1996).
78. Wilm M., Shevchenko A., Houthaeve T., *et al.*, *Nature*, **379**(6564), 466 (1996).
79. Shevchenko A., Jensen O.N., Podtelejniikov A.V., *et al.*, *Proc. Natl. Acad. Sc. USA*, **93**(25), 14440 (1996).
80. Wada Y., *Biol. Mass Spectrom.*, **21**, 617 (1992).
81. Li M.X., Wu J.T. and Liu L., *Rapid Commun. Mass Spectrom.*, **11**(1), 99 (1997).
82. Lewis J.K., Krone J.R. and Nelson R.W., *BioTechniques*, **24**(1), 102 (1998).
83. Wada Y., Matsuo T. and Sukarai T., *Mass Spectrom. Rev.*, **8**, 379 (1988).
84. Witkowska H.E., Bitsch F. and Shackleton C.H., *Hemoglobin*, **17**, 227 (1993).
85. Carver M.F.H. and Huisman T.H.J., *Hemoglobin*, **21**(6), 505 (1997).
86. Wada Y., Fujita T., Hayashi A., *et al.*, *Biomed. Environ. Mass Spectrom.*, **18**, 563 (1989).
87. Baczynskyj L. and Bronson G.E., *Rapid Commun. Mass Spectrom.*, **4**, 533 (1990).
88. Nelson R.W., *Mass Spectrom. Rev.*, **16**(6), 353 (1998).
89. Johnson R.S. and Biemann K., *Biochemistry*, **26**, 1209 (1987).
90. Biemann K. and Scoble A., *Science*, **237**, 992 (1987).
91. Annan R.S. and Carr S.A., *J. Protein Chem.*, **16**(5), 391 (1997).
92. Kuster B. and Mann M., *Curr. Opin. Struct. Biol.*, **8**(3), 393 (1998).
93. Smith D.L. and Zhou Z., *Methods Enzymol.*, **193**, 374 (1990).
94. Akashi S., Hirayama K., Seino T., *et al.*, *Biomed. Environ. Mass Spectrom.*, **15**, 541 (1988).
95. Zhang W., Czernik A.J., Yungwirth T., *et al.*, *Protein Sci.*, **3**, 677 (1994).
96. Burlingame A.L., *Curr. Opin. Biotechnol.*, **7**(1), 4 (1996).

97. Wilm M., Neubauer G. and Mann M., *Anal. Chem.*, **68**(3), 527 (1996).
98. Neubauer G. and Mann M., *J. Mass Spectrom.*, **32**(1), 94 (1997).
99. Annan R.S. and Carr S.A., *J. Protein Chem.*, **16**(5), 391 (1997).
100. Jedrzejewski P.T. and Lehmann W.D., *Anal. Chem.*, **69**(3), 294 (1997).
101. Gibson B.W. and Cohen P., *Methods Enzymol.*, **193**, 501 (1990).
102. Papayannopoulos I.A. and Biemann K., *Peptide Res.*, **5**, 83 (1987).
103. Andersen J.S., Svensson B. and Roepstorff P., *Nat. Biotechnol.*, **14**(4), 449 (1996).
104. Clerc F.F., Monegier B., Faucher D., *et al.*, *J. Chromatogr.*, **662**, 245 (1994).
105. Andersen J.S., *Biochem. Soc. Trans.*, **23**, 917 (1995).
106. Van Dorsselaer A., Bitsch F., Green B., *et al.*, *Biomed. Environ. Mass Spectrom.*, **19**, 692 (1990).
107. Chait B.T.,Wang R., Beavis R.C., *et al.*, *Science*, **262**, 89 (1993).
108. Bartlet-Jones M., Jeffery W.A., Hansen H.F., *et al.*, *Rapid Commun. Mass Spectrom.*, **8**, 737 (1994).
109. Thiede B., Salnikow J. and WittmannLiebold B., *Eur. J. Biochem.*, **244**(3), 750 (1997).
110. Patterson D.H., Tarr G.E., Renier F.E., *et al.*, *Anal. Chem.*, **67**, 3971 (1995).
111. Gamen B., Li Y.T. and Henion J.D., *J. Am. Chem. Soc.*, **113**, 7818 (1991).
112. Katta V. and Chait B.T., *J. Am. Chem. Soc.*, **113**, 8534 (1991).
113. Smith R.D. and Light-Wahl K.J., *Biol. Mass Spectrom.*, **22**, 493 (1993).
114. Smith R.D. and Zhang Z., *Mass Spectrom. Rev.*, **13**, 411 (1994).
115. Przybylski M. and Glocker M.O., *Angew. Chem. Int. Ed. Engl.*, **35**(8), 807 (1996).
116. Loo J.A., *Mass Spectrom. Rev.*, **16**(1), 1 (1997).
117. Winston R.A. and Fitzgerald M.C., *Mass Spectrom. Rev.*, **16**(4), 165 (1997).
118. Loo J.A., *J. Mass Spectrom.*, **30**, 180 (1995).
119. Fan X. and Beavis R.C., *Org. Mass Spectrom.*, **28**, 1424 (1993).
120. Glocker M.O., Bauer S.H.J., Kast J., *et al.*, *J. Mass Spectrom.*, **31**(11), 1221 (1996).
121. Loo J.A., Edmonds C.G., Udseth H.R., *et al.*, *Anal. Chem.*, **62**, 693 (1990).
122. Chowdhury S.K., Katta V. and Chait B.T., *J. Am. Chem. Soc.*, **112**, 9012 (1990).
123. Konermann L., Rosell F.I., Mauk A.G., *et al.*, *Biochemistry*, **34**(18), 5554 (1997).
124. Katta V. and Chait B.T., *Rapid Commun. Mass Spectrom.*, **5**, 214 (1991).
125. Katta V. and Chait B.T., *J. Am. Chem. Soc.*, **115**, 6317 (1993).
126. Zhang Z. and Smith D.L., *Protein Sci.*, **5**, 1282 (1993).
127. Suckeau D., Mak M. and Przybylski M., *Proc. Natl. Acad. Sci. USA*, **89**, 5630 (1992).
128. Glocker M.O., Borchers C., Fielder W., *et al.*, *Bioconjugate Chem.*, **5**, 583 (1994).
129. Massotte D., Yamamoto M., Scianimanico S., *et al.*, *Biochemistry*, **32**, 13787 (1993).
130. Cohen S.L., Ferré-D'Amaré A.R., Burley S.K., *et al.*, *Protein Sci.*, **4**, 1088 (1995).
131. Suckau D., Kohl J., Shneider K., *et al.*, *Proc. Natl. Acad. Sci. USA*, **87**, 9848 (1990).
132. Zhao Y.M., Muir T.W., Kent S.B.H., *et al.*, *Proc. Natl. Acad. Sci. USA*, **93**(9), 4020 (1996).
133. Murray K.K., *et al.*, *J. Mass Spectrom.*, **31**(11), 1203 (1996).
134. Limbach P.A., *Mass Spectrom. Rev.*, **15**(5), 297 (1996).
135. Nordhoff E., Kirpekar F. and Roepstorff P., *Mass Spectrom. Rev.*, **15**(2), 67 (1996).
136. Crain P.F. and McCloskey J.A., *Curr. Opin. Biotechnol.*, **9**(1), 25 (1998).
137. Stults J.T. and Masters J.C., *Rapid Commun. Mass Spectrom.*, **5**, 359 (1991).
138. Wu K.J., Shaler T.A. and Becker C.H., *Anal. Chem.*, **66**, 1637 (1994).
139. Kirpekar F., Nordhoff E., Kristiansen K., *et al.*, *Nucleic Acids Res.*, **22**, 3688 (1994).
140. Wu K.J., Stedding A. and Becker C.H., *Rapid Commun. Mass Spectrom.*, **7**, 142 (1993).
141. Fitzgerald M.C., Zhu L. and Smith L.M., *Rapid Commun. Mass Spectrom.*, **7**, 895 (1993).
142. Fenn J.B., Mann M., Meng C.K., *et al.*, *Mass Spectrom. Rev.*, **9**, 37 (1990).
143. Limbach P., Crain P.F. and Mc Closkey J.A., *J. Am. Soc. Mass Spectrom.*, **6**, 27 (1995).
144. Little D.P., Chorus R.A., Speir J.P., *et al.*, *J. Am. Chem. Soc.*, **116**, 4893 (1994).

145. Little D.P., Thannhauser T.W. and McLafferty F.W., *Proc. Natl. Acad. Sci. USA*, **92**, 2318 (1995).
146. Ball R.W. and Packman L.C., *Anal. Biochem.*, **246**(2), 185 (1998).
147. Van Ausdall D.V. and Marshall W.S., *Anal. Biochem.*, **256**(2), 220 (1998).
148. Lui Y.H., Bai J., Lubman D.M., *et al.*, *Anal. Chem.*, **67**, 3482 (1995).
149. Doktycz M.J., Hurst G.B., Goudarzi S.H., *et al.*, *Anal. Biochem.*, **230**, 205 (1995).
150. Chang L., Tang K., Shell M., *et al.*, *Rapid Commun. Mass Spectrom.*, **9**, 772 (1995).
151. Pomerantz S.C., Kowalak J.A. and McCloskey J.A., *J. Am. Soc. Mass Spectrom.*, **4**, 204 (1993).
152. Pieles U., Zürcher W., Shär M., *et al.*, *Nucleic Acids Res.*, **21**, 3191 (1993).
153. Limbach P.A., McCloskey J.A. and Crain P.F., *Nucleic Acids Symp. Ser.*, **31**, 127 (1994).
154. Smirnov I.P., Roskey M.T., Juhasz P., *et al.*, *Anal. Biochem.*, **238**(1), 19 (1996).
155. Keoug T., Baker T.R., Dobson R.L.M., *et al.*, *Rapid Commun. Mass Spectrom.*, **7**, 195 (1993).
156. Shaler T.A., Tan Y., Wickham J.N., *et al.*, *Rapid Commun. Mass Spectrom.*, **9**, 942 (1995).
157. Goodlett D.R., Camp II D.G., Hardin C.C., *et al.*, *Biol. Mass Spectrom.*, **22**, 181 (1993).
158. Pocsfalvi G., Dilanda G., Ferranti P., *et al.*, *Rapid Commun. Mass Spectrom.*, **11**(3), 265 (1997).
159. Cheng X.H., Morin P.E., Harms A.C., *et al.*, *Anal. Biochem.*, **239**(1), 35 (1996).
160. Triolo A., Arcamone F.M., Raffaeli A., *et al.*, *J. Mass Spectrom.*, **32**(11), 1186 (1997).
161. Bayer E., Bauer T., Schmeer K., *et al.*, *Anal. Chem.*, **66**, 3858 (1994).
162. Grotjahn L., in *Mass Spectrometry in Biomedical Research*, ed. by S.J. Gaskell, Wiley, New York, 1986, pp.215–234.
163. Nordhoff E., Karas M., Cramer R., *et al.*, *J. Mass Spectrom.*, **30**, 99 (1995).
164. Juhasz P., Roskey M.T., Smirnov I.P., *et al.*, *Anal. Chem.*, **68**(6), 941 (1996).
165. Stemmler E.A., Buchanan M.V., Hurst G.B., *et al.*, *Anal. Chem.*, **67**, 2924 (1995).
166. Little D.P., Chorush R.A., Spier J.P., *et al.*, *J. Am. Chem. Soc.*, **116**, 4893 (1994).
167. Boschenok J. and Sheil M.M., *Rapid Commun. Mass Spectrom.*, **10**(1), 144 (1996).
168. Little D.P., Aaserud D.J., Valaskoviic G.A., *et al.*, *J. Am. Chem. Soc.*, **118**(39), 9352 (1996).
169. McLuckey S.A., Van Berkel G.J. and Glish G.L., *J.Am. Soc. Mass Spectrom.*, **3**, 60 (1992).
170. McLuckey S.A. and Goudarzi H.S., *J. Am. Chem. Soc.*, **115**, 12085 (1993).
171. Viari A., Ballini J.P., Vigny P., *et al.*, *Biomed. Environ. Mass Spectrom.*, **16**, 225 (1987).
172. Barry J.P., Vouros P., Schepdael A.V., *et al.*, *J. Mass Spectrom.*, **30**, 993 (1995).
173. McLuckey S.A. and Goudarzi H.S., *J. Am. Chem. Soc.*, **115**, 12085 (1993).
174. Bartlett M.G., McCloskey J.A., Manalili S., *et al.*, *J. Mass Spectrom.*, **31**(11), 1277 (1996).
175. Hettich R.L. and Stemmler E.A., *Rapid Commun. Mass Spectrom.*, **10**(3), 321 (1996).
176. Ni J.S., Pomerantz S.C., Rozenski J., *et al.*, *Anal. Chem.*, **68**(13), 1989 (1996).
177. Rozenski J., Crain P.F. and McCloskey J.A., *Nucleic Acids Res.*, **27**(3), 196 (1999).
178. Chiarelli M.P. and Lay O.J., *Mass Spectrom. Rev.*, **11**, 447 (1992).
179. McCloskey J.A. and Crain P.F., *Int. J. Mass Spectrom. Ion Processes*, **118/119**, 593 (1992).
180. Crain P.F. and McCloskey J.A., in *Biological Mass Spectrometry*, ed. by T. Matsuo, R.M. Caprioli and M.L., Gross, Wiley, New York, 1994, pp. 509–537.
181. Biemann K. and McCloskey J.A., *J. Am. Chem. Soc.*, **84**, 2005 (1962).
182. Wilson M.S. and McCloskey J.A., *J. Am. Chem. Soc.*, **97**, 3436 (1975).
183. Crow F.W., Tomer K.B., Gross M.L., *et al.*, *Anal. Biochem.*, **139**, 243 (1984).
184. Chaudhary A.K., Nokubo M., Oglesby T.D., *et al.*, *J. Mass Spectrom.*, **30**, 1157 (1995).
185. Nelson C.C. and McCloskey J.A., *Adv. Mass Spectrom.*, **11A**, 260 (1989).

186. Dizdaroglu M., *Methods Enzymol.*, **193**, 842 (1990).
187. Crain P.F., Hashizume T., Nelson C.C., *et al.*, in *Biological Mass Spectrometry*, ed. by A.L. Burlingame and J.A. McCloskey, Elsevier, Amsterdam, 1990, pp. 509–525.
188. Pomerantz S.C. and McCloskey J.A., *Methods Enzymol.*, **193**, 796 (1990).
189. Apruzzese W.A. and Vouros P., *J. Chromatogr.*, **794**(1/2), 97 (1998).
190. Esmans E.L., Broes D., Hoes I., *et al.*, *J. Chromatogr.*, **794**(1/2), 109 (1998).
191. Kowalak J.A., Pomerantz S.C., Crain P.F., *et al.*, *Nucleic Acids Res.*, **21**, 4577 (1993).
192. Iden C.R. and Rieger R.A., *Biomed. Environ. Mass Spectrom.*, **18**, 617 (1989).
193. Reinhold V.N., Reinhold B.B. and Chan S., in *Biological Mass Spectrometry Present and Future*, ed. by T. Matsuo, R.M. Caprioli, M.L. Gross and Y. Seyama, Wiley, New York, 1994, pp. 403–462.
194. Reinhold V.N., Reinhold B.B. and Costello C.E., *Anal. Chem.*, **67**, 1772 (1995).
195. Harvey D.J., Naven T.J.P. and Kuster B., *Biochem. Soc. Trans.*, **24**(3), 905 (1996).
196. Merkle R.K. and Poppe I., *Methods in Enzymology*, Vol. 230, ed. by J.K. Lennarz and G.W. Hart, Academic Press, New York, 1994, pp. 1–15.
197. Hellerqvist C.G., *Methods in Enzymology*, Vol. 193, ed. by J.A. McCloskey, Academic Press, New York, 1990, pp. 554–573.
198. Bahr U., Pfenninger A., Karas M., *et al.*, *Anal. Chem.*, **69**(22), 4530 (1997).
199. Dell A., *Methods in Enzymology*, Vol. **193**, ed. by J.A. McCloskey, Academic Press, New York, 1990, pp. 647–660.
200. Gillece-Castro B.L. and Burlingame A.L., *Biological Mass Spectrometry*, ed. by A.L. Burlingame and J.A. McCloskey, Elsevier, Amsterdam, 1990, pp. 411–436.
201. Garozzo D., Giuffrida M., Impallomeni G., *et al.*, *Anal. Chem.*, **62**, 279 (1990).
202. Carr S.A., Reinhold V.N., Green B.N., *et al.*, *Biomed. Mass Spectrom.*, **12**, 288 (1985).
203. Bosso C., Heyraud A. and Patron L., *Org. Mass Spectrom.*, **26**, 321 (1991).
204. Domon B., Muller D.R. and Richter W.J., *Org. Mass Spectrom.*, **24**, 357 (1989).
205. Orlando R., Bush A. and Fenselau C., *Biomed. Environ. Mass Spectrom.*, **19**, 747 (1990).
206. Duffin K.L., Welply J.K., Huang E., *et al.*, *Anal. Chem.*, **64**, 1440 (1992).
207. Phillips N.J., Apicella M.A., Griffiss J.M., *et al.*, *Biochemistry.*, **32**, 2003 (1993).
208. Linsley K., Chan S-Y, Chan S., *et al.*, *Anal. Biochem.*, **219**, 207 (1994).
209. Reinhold B.B., Chan S-Y, Chan S., *et al.*, *Org. Mass Spectrom.*, **29**, 736 (1995).
210. Spengler B., Kirch D., Kaufmann R., *et al.*, *Org. Mass Spectrom.*, **29**, 782 (1994).
211. Perreault H. and Costello C.E., *Org. Mass Spectrom.*, **29**, 720 (1994).
212. Cancilla M.T., Penn S.G., Caroll J.A., *et al.*, *J. Am. Chem. Soc.*, **118**(28), 6736 (1996).
213. Domon B. and Costello C.E., *Glycoconj. J.*, **5**, 397 (1988).
214. Lemoine J., Fournet B., Despeyroux D., *et al.*, *J. Am. Soc. Mass Spectrom.*, **4**, 197 (1993).
215. Domon B., Muller D.R. and Richter W.J., *Org. Mass Spectrom.*, **29**, 713 (1994).
216. Lemoine J., Chirat F. and Domon B., *J. Mass Spectrom.*, **31**(3), 908 (1996).
217. Domon B., Muller D.R. and Richter W.J., *Biomed. Environ. Mass Spectrom.*, **19**, 390 (1990).
218. Viseux N., de Hoffmann E. and Domon B., *Anal. Chem.*, **69**(16), 3193 (1997).
219. Solouki T., Reinhold B.B., Costello C.E., *et al.*, *Anal. Chem.*, **70**(5), 857 (1998).
220. Sheeley D.M. and Reinhold V.N., *Anal. Chem.*, **70**(14), 3053 (1998).
221. Angel A.S. and Nilsson B., *Methods in Enzymology*, Vol. 193, ed. by J.A. McCloskey, Academic Press, New York, 1990, pp. 721–730.
222. Cancilla M.T., Penn S.G. and Lebrilla C.B., *Anal. Chem.*, **70**(4), 663 (1998).
223. Khoo K.H. and Dell A., *Glycobiology*, **1**, 83 (1990).
224. Sutton C.W., O'Neil J.A. and Cottrell J.S., *Anal. Biochem.*, **218**, 34 (1994).
225. Kuster B., Naven T.J.P. and Harvey D.J., *J. Mass Spectrom.*, **31**(10), 1131 (1996).
226. Harvey D.J., Rudd P.M., Bateman R.H., *et al.*, *Org. Mass Spectrom.*, **29**, 753 (1994).
227. Viseux N., de Hoffmann E. and Domon B., *Glycobiology*, **6**(7), 121 (1996).
228. Jensen N.J. and Gross M.L., *Mass Spectrom. Rev.*, **7**, 41 (1988).

229. Gage D.A., Huang Z.H. and Sweeley C.C., in *Mass Spectrometry Clinical and Biomedical Applications*, ed. by M.D. Desiderato, Plenum Press, New York, 1994, pp. 53–87.

230. Kim H.Y., Wang T.C.L. and Ma Y.C., *Anal. Chem.*, **66**, 3977 (1994).

231. Murphy R.C. and Harrson K.A., *Mass Spectrom. Rev.*, **13**, 57 (1994).

232. Tomer K.B. and Gross M.L., *Biomed. Environ. Mass Spectrom.*, **15**, 89 (1989).

233. Tabet J.C., in *Applications of Mass Spectrometry to Organic Stereochemistry*, ed. by J.S. Splitter and F. Turecek, VCH, New York, 1994, pp. 543–591.

234. Bean K.A. and Henion J.B., *J. Chromatogr. B*, **690**(1/2), 65 (1997).

235. Zirrolli J.A., Davoli E., Bettazzoli L., *et al.*, *J. Am. Soc. Mass Spectrom.*, **1**, 325 (1990).

236. Ann Q. and Adams J., *Anal. Chem.*, **65**, 7 (1993).

237. Gu M., Kerwin J.L. and Watts J.D., *Anal. Biochem.*, **244**(2), 347 (1997).

238. Adams J. and Ann Q., *Mass Spectrom. Rev.*, **12**, 51 (1993).

239. Mano N., Oda Y. and Yamada K., *Anal. Biochem.*, **244**(3), 291 (1997).

240. Wheelan P., Zirrolli J.A. and Murphy R.C., *J. Am. Soc. Mass Spectrom.*, **7**(2), 140 (1996).

241. Murphy R.C., *J. Mass Spectrom.*, **30**, 5 (1995).

242. Murphy R.C., *Mass Spectrometry of Lipids, Handbook of Lipid Research*, Vol. 7, Plenum Press, New York, 1993.

243. Jensen N.J. and Gross M.L., *Mass Spectrom. Rev.*, **6**, 497 (1987).

244. Kuksis A. and Myher J.J., *J. Chromatogr.*, **671**, 35 (1995).

245. Tomer K.B., Jensen N.J. and Gross M.L., *Anal. Chem.*, **58**, 2429 (1986).

246. Kerwin J.L. and Torvik J.J., *Anal. Biochem.*, **237**(1), 56 (1996).

247. Jensen N.J., Tomer K.B. and Gross M.L., *J. Am. Chem. Soc.*, **105**, 5487 (1985).

248. Bambagiotti M.A., Coran S.A., Vincieri F.F., *et al.*, *Org. Mass Spectrom.*, **21**, 485 (1986).

249. Kerwin J.L., Wiens A.M. and Ericsson L.H., *J. Mass Spectrom.*, **31**, 184 (1996).

250. Adams J., *Mass Spectrom. Rev.*, **9**, 141 (1990).

251. Cordero M.M. and Wesdemiotis C., *Anal. Chem.*, **66**, 861 (1994).

252. Nizigiyimana L., Van den Heuvel H. and Claeys M., *J. Mass Spectrom.*, **32**(3), 277 (1997).

253. Cheng C.F., Pittenauer E. and Gross M.L., *J. Am. Soc. Mass Spectrom.*, **9**(8), 840 (1998).

254. Jensen N.J., Tomer K.B. and Gross M.L., *J. Am. Chem. Soc.*, **107**, 1863 (1985).

255. Jensen N.J., Tomer K.B. and Gross M.L., *Anal. Chem.*, **57**, 2018 (1985).

256. Jensen N.J. and Gross M.L., *Lipids*, **21**, 362 (1986).

257. Tomer K.B., Crow F.W. and Gross M.L., *J. Am. Chem. Soc.*, **105**, 5487 (1983).

258. Tomer K.B., Jensen N.J. and Gross M.L, *Anal. Chem.*, **59**, 1576 (1987).

259. Harvey D.J., *Biomed. Mass Spectrom.*, **9**, 33 (1982).

260. Dobson G. and Christie W.W., *Trends Anal. Chem.*, **15**(3), 130 (1996).

261. Zirrolli J.A. and Murphy R.C., *J. Am. Soc. Mass Spectrom.*, **4**, 223 (1993).

262. Ruizgutierrez V. and Barron L.J.R., *J. Chromatogr.*, **671**, 113 (1995).

263. Laakso P., *Food Rev. Int.*, **12**(2), 199 (1996).

264. Hites R.A., *Anal. Chem.*, **42**, 1736 (1970).

265. Murata T. and Takahashi S., *Anal. Chem.*, **49**, 728 (1977).

266. Schulte E., Hohn M. and Rapp U., *Fresenius Z. Anal. Chem.*, **307**, 115 (1981).

267. Showell J.S., Fales H.M. and Sokoloski E.A., *Org. Mass Spectrom.*, **24**, 632 (1989).

268. Evans C., Traldi P., Bambagiotti-Alberti M., *et al.*, *Biol. Mass Spectrom.*, **20**, 351 (1991).

269. Neff W.E. and Byrdwell W.C., *J. Am. Oil Chem. Soc.*, **72**, 5487 (1995).

270. Duffin K.L., Henion J.D. and Shieh J.J., *Anal. Chem.*, **63**, 1781 (1991).

271. Anderson M.A., Collier L., Dilliplane R., *et al.*, *J. Am. Oil Chem. Soc.*, **70**, 905 (1993).

272. Bambagiotti M.A., Coran S.A., Vincieri F.F., *et al.*, *Org. Mass Spectrom.*, **21**, 485 (1986).

273. Keusgen M., Curtis J.M. and Ayer S.W., *Lipids*, **31**(2), 231 (1996).

274. Stroobant V., Rozenberg R., Bouabsa E.M., *et al.*, *J. Am. Soc. Mass Spectrom.*, **6**, 498 (1995).

275. Ballatore A.M., Beckner C.F., Caprioli R.M., *et al.*, *Steroids*, **41**, 197 (1983).

276. Ito Y., Takeuchi T. Ishii D., Goto M., *et al.*, *J. Chromatogr.* **358**, 201 (1986).
277. Roda A., Giacchini A.M. and Baraldini M., *J. Chromatogr.* **665**, 281 (1995).
278. Evans J.E., Ghosh A., Evans B.A., *et al.*, *Biol. Mass Spectrom.*, **22**, 331 (1993).
279. Scalia S. and Games D.E., *Org. Mass Spectrom.*, **27**, 1266 (1992).
280. Warrack B.M. and Didonato G.C., *Biol. Mass Spectrom.*, **22**, 101 (1993).
281. Lier J.G., Kingston E.E. and Beynon J.H., *Biomed. Mass Spectrom.*, **12**, 95 (1985).
282. Tomer K.B., Jensen N.J. and Gross M.L., *Biomed. Mass Spectrom.*, **13**, 265 (1986).
283. Griffiths W.J., Egestad B. and Sjovall J., *Rapid Commun. Mass Spectrom.*, **7**, 235 (1993).
284. Eckers C., New A.P., East P.B., *et al.*, *Rapid Commun. Mass Spectrom.*, **4**, 449 (1990).
285. Stroobant V., Libert R., Van Hoof F., *et al.*, *J. Am. Soc. Mass Spectrom.*, **6**, 588 (1995).
286. Libert R., Hermans D., Draye J.P., *et al.*, *Clin. Chem.*, **37**, 2102 (1991).

8

Exercises

Questions

8.1. Calculate the isotopic distribution of:

    (a) $CH_2=CHCl$
    (b) CH_2BrCl

8.2. The molecular ion of a product occurs at 59.9670 Th. Identify the product and calculate the relative abundance of the ion with a nominal mass of 62 Th.

8.3. Which resolution is required to distinguish the molecular ions of:

    (1) CO and C_2H_4
    (2) $C_{10}H_{21}CHO$ and $C_{12}H_{26}$
    (3) The monoisotopic peaks of Question 8.1

8.4. Spectrum A in Figure 8.1 is the electron ionization (EI) mass spectrum of S_8. Spectra B and C are the tandem mass spectrometry (MS/MS) fragmentation spectra of S_8 for the 256 Th and 258 Th fragments, respectively. Explain the observed isotope ratios.

8.5. The electrospray negative ion spectrum of an oligonucleotide displays the following ions: 2903.8 (40%), 2580.9 (67%), 2323.7 (100%), 2111.4 (71%) and 1935.7 (38%). What is the molecular weight of this oligonucleotide derivative?

8.6. The specific interaction of Cu ions with a 26-residue peptide present at the surface of a human protein has been studied by mass spectrometry. (Hutchens T.W., Nelson R.W., Allen M.H., *et al.*, *Biol. Mass Spectrom.*, **21**, 151 (1992)). The positive ion electrospray spectra A and B (Figure 8.2) have been obtained, respectively, before and after the addition of Cu(II).

    What is the molecular weight of this peptide?
    What is the maximum number of Cu atoms included in the peptide?
    Which is the oxidation stage of the Cu atoms?

8.7. How can you distinguish between the spectrum of *n*-butyraldehyde and that of isobutyraldehyde?

8.8. The spectra displayed in Figure 8.3 originate from isomeric ketones analyzed by electron ionization. They comprise only linear saturated chains. Give the developed formula of each of these ketones. Justify your answer by the interpretation of at least three fragments, one of them resulting from a rearrangement.

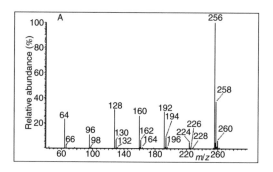

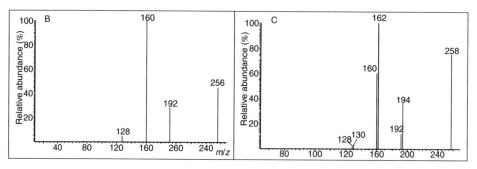

Figure 8.1
The EI mass spectrum (A) and MS/MS spectra (B,C) of S_8 for the 256 Th (B) and 258 Th (C) processors

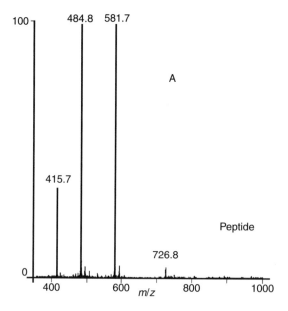

Figure 8.2
Positive ion electrospray spectra of a 26-residue peptide: (A) free peptide ; (B) in the presence of Cu(II) ions

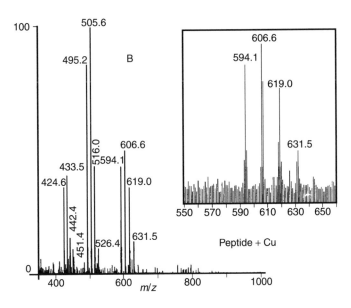

Figure 8.2
(*continued*)

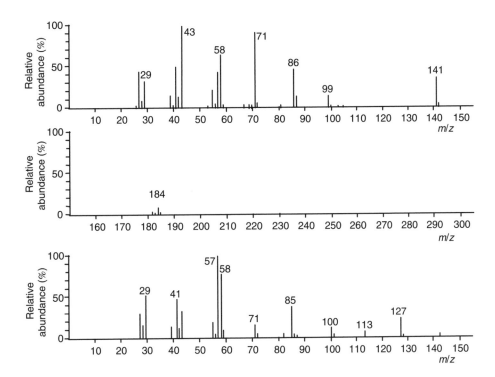

Figure 8.3
Spectra of three isomass ketones of molecular weight 184 Da

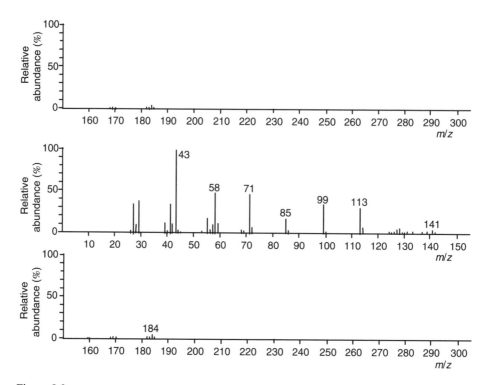

Figure 8.3
(*continued*)

8.9. A compound has a molecular peak at 115 Th. Where is the peak corresponding to the metastable loss of CH₃ in the spectrum obtained with a magnetic instrument using an EB configuration?

8.10. Explain the following: 'MIKE spectroscopy' and 'B/E linked scan'.

8.11. Interpret the main peaks in the following spectra:

8.11.1. Methylene chloride   **8.11.2.** Acetone

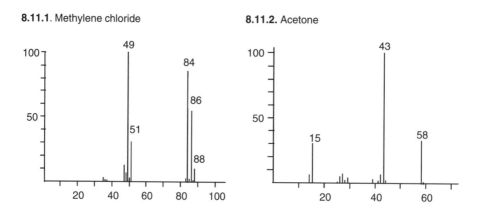

8.11.3. Ethylamine

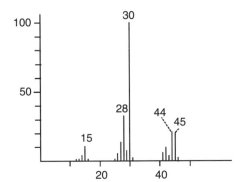

8.11.4. 1-Butene

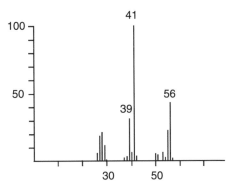

8.11.5. Ethylbenzene

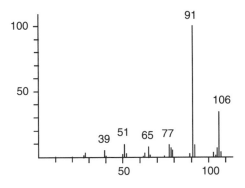

8.11.6. Tetralin

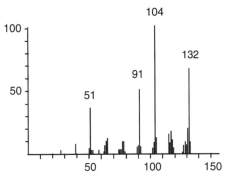

8.11.7. *n*-Butyraldehyde

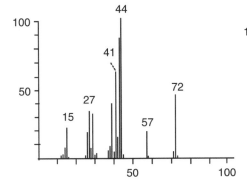

8.11.8. Cyclohexanol

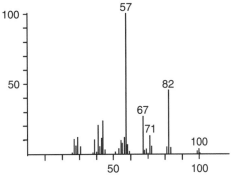

8.11.9. 2-Dodecanone

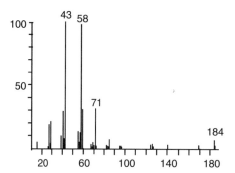

8.11.10. *n*-Dodecane

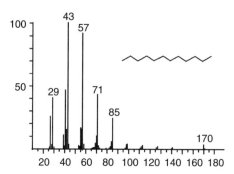

8.11.11. 1-Phenyl-*n*-hexane

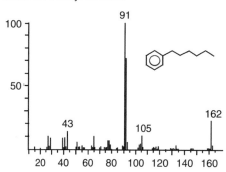

8.11.12. Cyclohexane

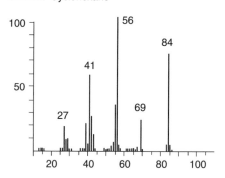

8.12. Identify the product and interpret the main peaks in the following:

8.12.1.

m/z	Relative abundance (%)
14	3.4
15	20
16	0.2
19	2
31	10
32	9.8
33	94
34	100
35	1.1

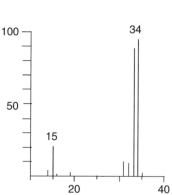

(a)

8.12.2.

m/z	Relative abundance (%)
12	3.2
13	4.4
14	4.6
16	1.5
28	32
29	100
30	88
31	1.2

8.12.3.

m/z	Relative abundance (%)
15	15.3
27	37
28	32
29	44
39	12
41	27
42	12
43	100
44	3.2
57	2.3
58	12
59	0.5

8.12.4.

m/z	Relative abundance (%)
26	38
27	74
28	12
29	4.4
44	14
45	32
55	74
56	2.5
57	0.2
71	4.2
72	100
73	3.5
74	0.5

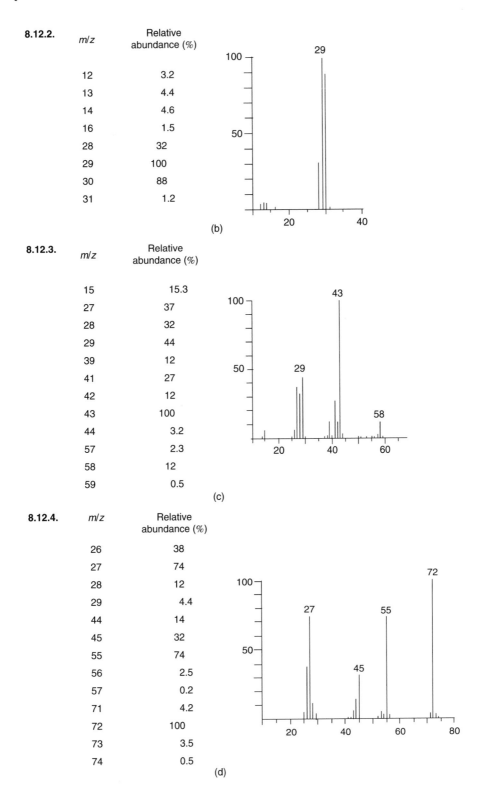

(b)

(c)

(d)

8.12.5.

m/z	Relative abundance (%)
14	5.1
19	8.3
33	36
52	100
53	0.5
71	30

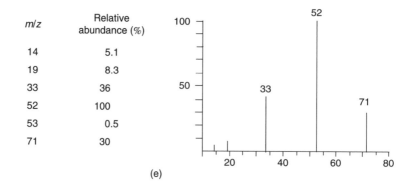

(e)

8.12.6.

m/z	Relative abundance (%)
19	0.2
31	1.7
35	3.1
37	1.2
50	6.4
69	100
70	1.2
85	18
86	0.2
87	5.8
104	0.7
106	0.2

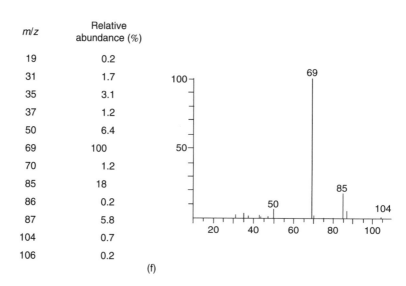

(f)

8.13. Identify the product that yields the following mass spectrum (Silverstein R.M., Bassler G.C. and Morril T.C., *Spectrometric Identification of Organic Compounds*, 5th edition, Wiley, New York, 1991):

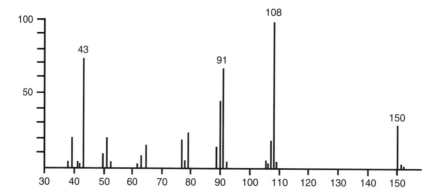

In the IR spectrum the main peaks appear at 3058, 2941, 1745, 1385, 1225, 1026, 749 and 697 cm^{-1}. In ^{1}H-NMR:

δ	Intensity	Multiplicity
7.22	5	Singlet
5.00	2	Singlet
1.96	3	Singlet

8.14. Identify the product that yields the following mass spectrum (Silverstein R.M., Bassler G.C. and Morril T.C., *Spectrometric Identification of Organic Compounds*, 5th edition, Wiley, New York, 1991):

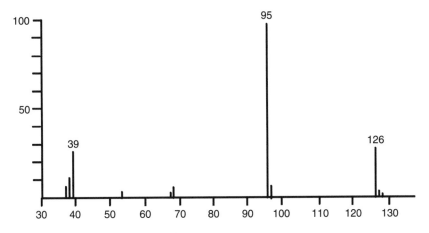

In the IR spectrum the main peaks appear at 3106, 2941, 1730, 1587, 1479, 1449, 1393, 1299, 1205, 1121 and 758 cm^{-1}. In ^{1}H-NMR:

δ	Intensity	Multiplicity
7.51	1	Quadruplet
7.02	1	Quadruplet
6.45	1	Quadruplet
3.80	3	Singlet

In off-resonance decoupled ^{13}C-NMR:

δ	Multiplicity
160	Singlet
146	Doublet
144	Singlet
118	Doublet
112	Doublet
51	Quadruplet

8.15. Identify the product that yields the following mass spectrum (Silverstein R.M., Bassler G.C. and Morril T.C., *Spectrometric Identification of Organic Compounds*,

5th edition, Wiley, New York, 1991):

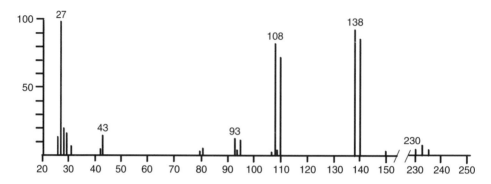

The IR spectrum shows the following main peaks: 2970, 2850, 1425, 1360, 1279, 1120 and 630 cm^{-1}. In ^{1}H-NMR:

δ	Intensity	Multiplicity
3.40	1	Multiplet
3.88	1	Multiplet

8.16. Identify the product that yields the following mass spectrum (Silverstein R.M., Bassler G.C. and Morril T.C., *Spectrometric Identification of Organic Compounds*, 5th edition, Wiley, New York, 1991):

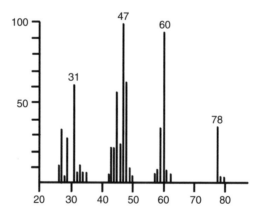

In the IR spectrum the main peaks appear at 3367, 3030, 2558, 1429, 1298, 1050 and 1020 cm^{-1}. In off-resonance decoupled ^{13}C-NMR:

δ	Multiplicity
28	Triplet
64	Triplet

8.17. A 429 Da peptide gives the MS/MS product ion spectrum from its protonated molecular ion displayed in Figure 8.4. What is the sequence of this peptide?

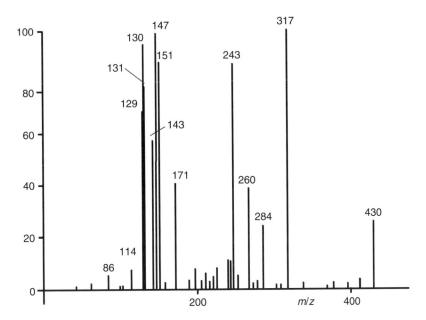

Figure 8.4
Fragment ion spectrum of a protonated peptide

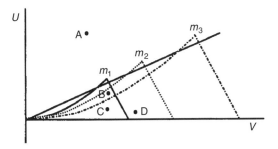

Figure 8.5
Stability diagram for three ions of mass m_1, m_2 and m_3 Th in U, V coordinates

8.18. Consider Figure 8.5, representing the stability diagram for three ions of mass m_1, m_2 and m_3 Th in U, V coordinates. Explain which ions will be observed if the U, V values are adjusted successively to the values indicated by points A–D in the figure.

8.19. A quadrupole with $r_0 = 1$ cm operating at an RF frequency of 0.5 MHz has a stability diagram with an apex at $a = 0.22$ and $q = 0.72$ for an ion of mass 2000 u. What are the corresponding U and V values?

8.20. The same quadrupole as in Question 8.19 operating in the 'RF-only' mode has values of $a = 0$ and $q = 0.91$. Which is the minimum V_{rf} voltage allowing all ions of less than 100 Th to be eliminated?

8.21. If a V_{rf} of 2 kV is applied to the same quadrupole as in Question 8.19, from which m/z value will the ions be unstable?

8.22. In an ideal ion trap, i.e. $r_0^2 = 2z_0^2$, with $r_0 = 1$ cm at $v = 1$ MHz, calculate the a and q values at the apex of the stability diagram.

8.23. In an ideal ion trap, i.e. $r_0^2 = 2z_0^2$, with $r_0 = 1$ cm at $v = 1$ MHz, $U = 4$ V and $V = 400$ V for a nitrogen ion, calculate the corresponding a and q values.

8.24. In an ideal ion trap, i.e. $r_0^2 = 2z_0^2$, with $r_0 = 1$ cm at $v = 1.1$ MHz, $U = 4$ V and $V = 500$ V, what is the resonant frequency that should be applied to expel an ion of 100 Th?

8.25. A time-of-flight (TOF) analyzer has the following characteristics: $V_s = -3$ kV, $V_R = 130$ V, $D = 0.522$ m, $L_1 = 1$ m.

    (a)   At which distance L_2 does it focalize ions having the same $m/z(m_p)$ of 578 Th?
    (b)   What is the focal distance for post-source decay ions with $m/z = 560$ Th?
    (c)   At which value should V_R be reduced in order to focalize the fragment ions at the same distance as the precursors?

Answers
8.1. Calculate the isotopic distribution of:

    (a) $CH_2=CHCl$

Answer:

	m/z nominal	m/z exact	Relative abundance (%)
$^{12}C_2H_3{}^{35}Cl$	62	61.99232	74.11
$^{12}C^{13}CH_3{}^{35}Cl$	63	62.99568	1.65
$^{12}C_2H_3{}^{37}Cl$	64	63.98862	23.7
$^{13}C_2H_3{}^{35}Cl$	64	63.99903	0.009
$^{13}C^{12}CH_3{}^{37}Cl$	65	64.99198	0.53
$^{13}C_2H_3{}^{37}Cl$	66	65.99533	0.003

    (b) CH_2BrCl

Answer:

	m/z nominal	m/z exact	Relative abundance (%)
$^{12}CH_2{}^{79}Br^{35}Cl$	128	127.90283	37.99
$^{13}CH_2{}^{79}Br^{35}Cl$	129	128.90618	0.42
$^{12}CH_2{}^{79}Br^{37}Cl$	130	129.89913	12.15
$^{12}CH_2{}^{81}Br^{35}Cl$	130	129.90080	36.95
$^{13}CH_2{}^{79}Br^{37}Cl$	131	130.90248	0.14
$^{13}CH_2{}^{81}Br^{35}Cl$	131	130.90415	0.41
$^{12}CH_2{}^{81}Br^{37}Cl$	132	131.89710	11.82
$^{13}CH_2{}^{81}Br^{37}Cl$	133	132.90046	0.13

8.2. The molecular ion of a product occurs at 59.9670 Th. Identify the compound and calculate the relative abundance of the ion with a nominal mass of 62 Th.

Answer:
It is COS; the ion at 62 Th due to $CO^{34}S$ has a relative abundance of 4.44%.

8.3. Which resolution is required to distinguish the molecular ions of:

    (1)   CO and C_2H_4
    (2) $C_{10}H_{21}CHO$ and $C_{12}H_{26}$
    (3)   The monoisotopic peaks of Exercise 8.1

Answer:
    (1)   770
    (2)   4800
    (3)   6200(a) and
         78 000(b)

8.4. Spectrum A in Figure 8.1 is the electron ionization (EI) mass spectrum of S_8. Spectra B and C are the tandem mass spectrometry (MS/MS) fragmentation spectra of S_8 for the 256 Th and 258 Th fragments, respectively. Explain the observed isotope ratios.

Answer:
The two main isotopes of sulfur are ^{32}S and ^{34}S. The latter is present with a natural abundance of 4.44%, taking ^{32}S as 100%. At m/z 256 the formula is S_8. The 258 Th fragment should have a relative intensity of $8 \times 0.0444 = 0.355$ or 35.5%, and the 260 Th fragment corresponds to the probability of having two ^{34}S atoms together, thus $(8 \times 0.044)(7 \times$

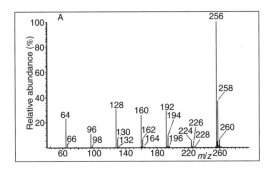

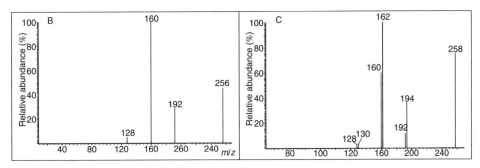

Figure 8.1

0.044) $= 0.3552 \times 0.3108 = 0.1104$ or 11.04%. The peak at 192 Th has the formula S_6. The 194 Th fragment will have an abundance of $6 \times 0.044 = 0.266$ or 26.6% of the abundance at m/z 192. The 196 Th fragment will have an abundance corresponding to one out of six and one out of the remaining five being ^{34}S. This is thus $(6 \times 0.0444)(5 \times 0.044) = 0.2664 \times 0.222 = 0.059$ or 5.9% of the abundance at m/z 192.

Spectrum B is the MS/MS product ion spectrum of 256 Th, thus isotopically pure $^{32}S_8$. Its fragments contain only ^{32}S and there are no isotope peaks.

Spectrum C is the MS/MS product ion spectrum of 258 Th, thus having the formula $^{34}S^{32}S_7$. One out of eight sulfur atoms is now ^{34}S. The fragment at m/z 160–162 has the formula S_5 and results from the loss of three sulfur atoms. The probability is thus 5/8 that the ^{34}S is saved, producing the m/z 162 fragment, and 3/8 that it is lost, which leads to the 160 Th fragment. The ratio of the relative abundances '160/162' is thus $(3/8)/(5/8) = 3/5$ or 60%. At 128 and 130 Th, the formula is S_4 and thus the probability is equal for holding or losing the ^{34}S, leading to two peaks of equivalent abundance.

8.5. The electrospray negative ion spectrum of an oligonucleotide displays the following ions: 2903.8 (40%), 2580.9 (67%), 2323.7 (100%), 2111.4 (71%) and 1935.7 (38%). What is the molecular weight of this oligonucleotide derivative?

Answer:
Using the formula for multiply charged negative ions $z_1 = j \, (m_2 + m_p)/(m_2 - m_1)$, where m_p is the proton mass and for the extreme ions $j = 4$, one obtains $z_1 = 12$. The formula $M = z_i \, (m_i + m_p)$ yields $M = 23240.5, 23236.5, 23247.1, 23237.2$ and 23238.5 Da for the five ions. The average value is $M = 23240 \pm 3$ Da.

8.6. The specific interaction of Cu ions with a 26-residue peptide present at the surface of a human protein has been studied by mass spectrometry. The positive ion electrospray spectra obtained before and after addition of Cu(II) are displayed in Figure 8.2.

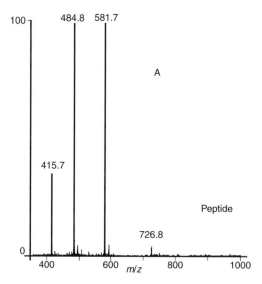

Figure 8.2

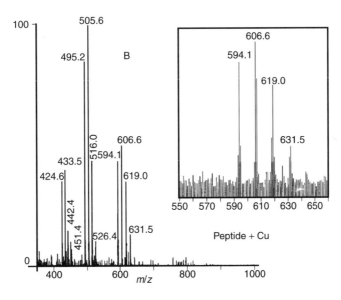

Figure 8.2 (*continued*)

What is the molecular weight of this peptide?
What is the maximum number of Cu atoms included in the peptide?
Which is the oxidation stage of the Cu atoms?

Answer:
The molecular weight of the peptide is deduced from its electrospray spectrum, (Figure 8.2A). If we suppose that the observed peaks in this spectrum correspond to different charge states of the peptide, than we can calculate the number of charges z_1 for the peak detected at m_1 : $z_1 = j\ (m_2 - 1)/(m_2 - m_1)$, where j corresponds to the number of peaks $+1$ separating m_1 and m_2; z_1 has to be rounded to the nearest integer. The molecular weight is then $M = z_1(m_1 - 1)$.

For spectrum A, applying the formula with $m_1 = 484.8$, $m_2 = 581.7$ and $j = 1$ one obtains a charge state of $+6$ for the number of charges on the ions at m/z 484. From this charge state and the m/z value, one obtains 2902.8 Da for the molecular weight.

Table 8.1 reports the molecular weights deduced from every observed mass and their more accurate average value.

Table 8.1 Molecular weights calculated from each observed mass

Number of charges	Detected m/z	Molecular weight
7+	415.7	2903.4
6+	484.8	2902.8
5+	581.7	2903.5
4+	726.8	2903.2
		Average = 2903.2

Table 8.2 Comparison of expected and observed molecular weight (MW) for different numbers of Cu atoms in various charge states

Number of Cu atoms	MW if Cu(0) $[M + n\text{Cu}]$	MW if Cu(I) $[M + n\text{Cu} - n\text{H}]$	MW if Cu(II) $[M + n\text{Cu} - 2n\text{H}]$	Observed MW	Difference
0	2903.02	2903.02	2903.02	2903.2	62.1
1	2966.52	2965.52	2964.52	2965.3	62.4
2	3030.02	3028.02	3026.02	3027.7	62.3
3	3093.52	3090.52	3087.52	3090.0	62.6
4	3157.02	3153.02	3149.02	3152.6	Average = 62.35

After addition of a Cu(II) salt to the solution, new peaks are observed as displayed in Figure 8.2B. Four new peaks appear at every charge state. The molecular weight determination of these peaks allows one to conclude that a maximum of four Cu atoms are incorporated. Table 8.2 summarizes the expected molecular weights for different numbers of Cu atoms in various charge states and compares them with the experimental molecular weights. The copper is thus present as Cu(I). Indeed, for each Cu atom added the molecular weight increases by 62.35 Da, corresponding to the atomic weight of Cu (63.5) minus a proton. Because the observed charge state does not change with the addition of Cu atoms, it must be concluded that it is monocharged. In conclusion, the ions observed in spectrum B correspond to the formula $[M + x\text{H} + n\text{Cu} - n\text{H}]^{x+}$, where x is the number of charges (varying between 4 and 7) and n is the number of Cu atoms (varying from 1 to 4).

8.7. How can you distinguish between the spectrum of n-butyraldehyde and that of isobutyraldehyde?

Answer:
n-Butyraldehyde can undergo a McLafferty rearrangement to yield an ion at m/z 44 Th through rHi followed by αi, whereas isobutyraldehyde cannot undergo such a rearrangement.

8.8. The spectra displayed in Figure 8.3 originate from isomeric ketones analyzed by electron ionization. They comprise only linear saturated chains. Give the developed formula of each of these ketones. Justify your answer by the interpretation of at least three fragments, one of them resulting from a rearrangement.

Answer:
The molecular weight of these ketones is 184. The weight of the sum of the aliphatic chains is thus $184 - 28 = 156$, i.e. $C_{11}H_{24}$. Acylium ions are quite stable and α cleavage often occurs adjacent to the carbonyl group.

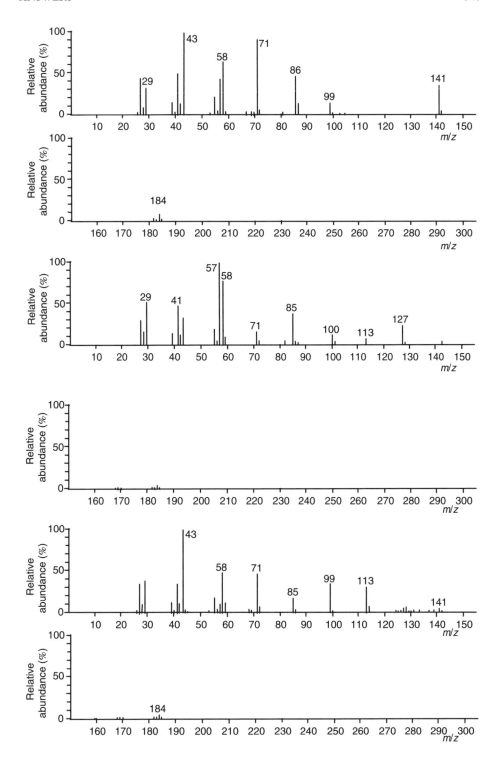

Figure 8.3

The first important observed fragments at the higher mass side appear at 141, 127 and 113 u for the three ketones, respectively. We can speculate that this corresponds to the higher mass acylium ion in each case.

For the top spectrum $184 - 141 = 43$ indicates that a C_3H_7 radical is lost. We then should expect the other acylium ion to result from the loss of C_8H_{17} and to appear at m/z 71. This is indeed an intense peak and, as expected, more abundant than the 141 Th peak because the lost alkyl radical is larger. This first spectrum could correspond to a C_7H_{15}-CO-C_3H_7 ketone. We will now try to confirm this by looking at the rearrangement peaks. The McLafferty rearrangement is indeed often important for ketones. The three spectra show a peak at m/z 58. The even mass indicates a rearrangement, but this one will not give information because it is the same for the three compounds.

In the first spectrum, another rearrangement peak appears at m/z 86. The following scheme shows the corresponding McLafferty rearrangement:

m/z 86

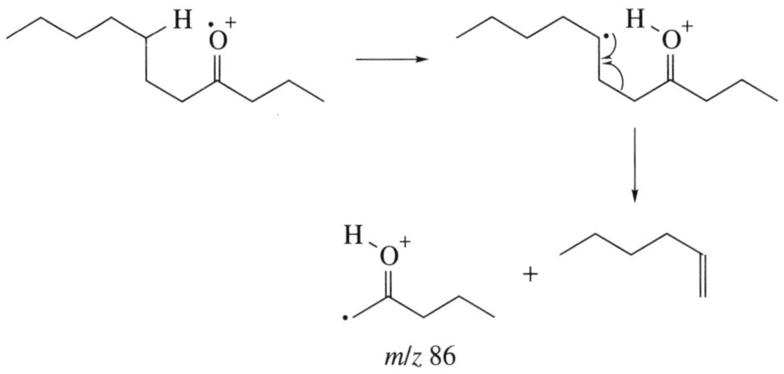

This confirms the proposed structure. The m/z 58, common to the three ketones, results from a second rearrangement and fragmentation of this m/z 86 ion:

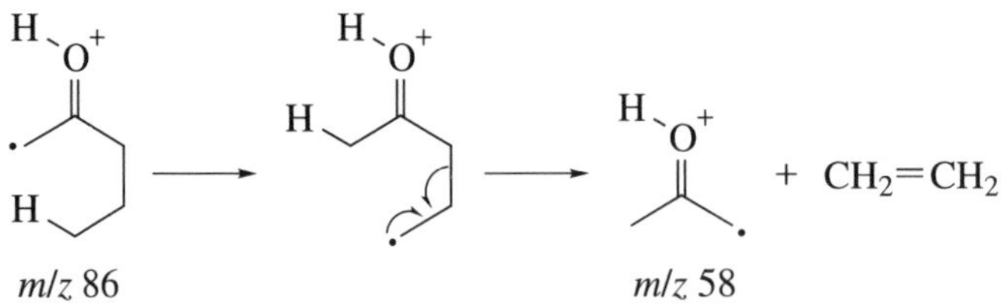

m/z 86 m/z 58

By the same reasoning, the two other ketones can be shown to be $C_6H_{13}-CO-C_4H_9$ and $C_5H_{11}-CO-C_5H_{11}$, respectively. For the latter, the McLafferty rearrangement gives a weak peak at m/z 114.

8.9. A compound has a molecular peak at 115 Th. Where is the peak corresponding to the metastable loss of CH_3 in the spectrum obtained with a magnetic instrument using an EB configuration?

Answer:
The metastable occurs at $100^2/115$ Th $= 86.9$ Th.

8.10. Explain the following: 'MIKE spectroscopy' and 'B/E linked scan'.

Answer:
MIKE stands for 'Mass-analyzed ion kinetic energy' and is a tandem mass spectrometry method that allows product ion spectra to be recorded with a magnetic instrument of 'inverse' or BE geometry. The parent ion m_p is selected with the magnetic analyzer B and the metastable fragment ions m_f are analyzed by scanning the electric sector. The relevant equation is $E_p/E_f = m_p/m_f$. If kinetic energy E_k' is released during the fragmentation process, the kinetic energy of the fragments will be between $(E_{kf} + E_k')$ and $(E_{kf} - E_k')$. A broadening of the peak will result, which allows direct measurement of the released kinetic energy.

In B/E linked scan, the precursor ion is first focalized by the appropriate values of both B_p and E_p. Then, B and E are scanned together in such a way that the ratio B/E remains constant. A fragment produced between the source and the analyzers, either in EB or BE configuration, will be focalized at values of B_f and E_f such that $B_p/E_p = B_f/E_f$. Because $B_p/B_f = m_p/m_f$, the mass of the fragment can be determined. This method gives a better resolution than the MIKE method but does not allow measurement of the kinetic energy released.

8.11. Interpret the main peaks of the following spectra:

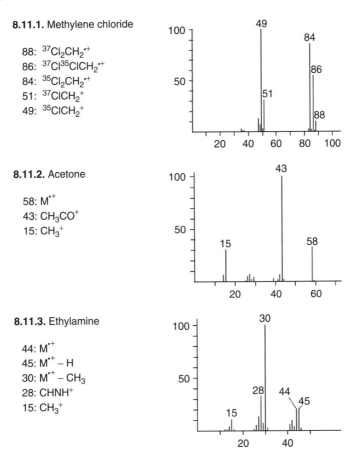

8.11.1. Methylene chloride

88: $^{37}Cl_2CH_2^{•+}$
86: $^{37}Cl^{35}ClCH_2^{•+}$
84: $^{35}Cl_2CH_2^{•+}$
51: $^{37}ClCH_2^+$
49: $^{35}ClCH_2^+$

8.11.2. Acetone

58: $M^{•+}$
43: CH_3CO^+
15: CH_3^+

8.11.3. Ethylamine

44: $M^{•+}$
45: $M^{•+} - H$
30: $M^{•+} - CH_3$
28: $CHNH^+$
15: CH_3^+

8.11.4. 1-Butene

56: $M^{•+}$
55: $M^{•+} - H$
41: $M^{•+} - CH_3$
39: $C_3H_3^+$

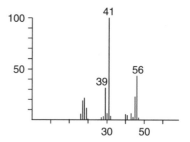

8.11.5. Ethylbenzene

106: $M^{•+}$
91: tropylium ion
77: phenyl
65: retro-Diels–Alder of 91
51: retro-Diels–Alder of 77
39: cyclopropenium

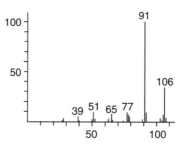

8.11.6: Tetralin

132: $M^{•+}$
104: $M^{•+} - C_2H_4$ retro-Diels–Alder
91: Tropylium ion
51: $C_4H_3^+$

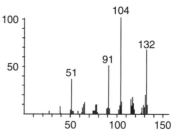

8.11.7. n-Butyraldehyde

72: $M^{•+}$
57: $M^{•+} - CH_3^•$
44: $C_2H_4O^{•+}$ McLafferty
43: $C_3H_7^+$ σ cleavage
41: $C_3H_5^+$
39: $C_3H_3^+$
29: $C_2H_5^+$
15: CH_3^+

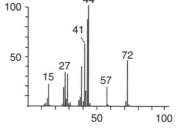

8.11.8. Cyclohexanol

100: $M^{•+}$
82: $M^{•+} - H_2O$
71: $M^{•+} - CHO^•$
67: $M^{•+} - H_2O - CH_3^•$
57: $M^{•+} - C_2H_3O^•$

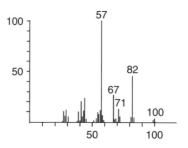

8.11.9. 2-Dodecanone

184: $M^{\bullet+}$

85: $C_6H_{13}^+$ through σ cleavage
71: $C_5H_{11}^+$
58: $C_3H_6O^{\bullet+}$ McLafferty
43: $C_3H_7^+$ through σ cleavage or
    $C_2H_3O^+$ through α cleavage

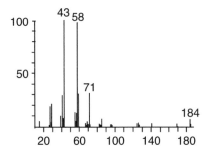

8.11.10. *n*-Dodecane

The main peaks are explained by σ cleavage

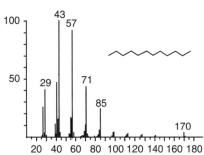

8.11.11. 1-Phenyl-*n*-hexane

162: $M^{\bullet+}$

105: $M^{\bullet+} - C_4H_9^{\bullet}$

91: tropylium ion

43: $C_3H_7^{\bullet+}$

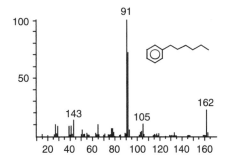

8.7.12: Cyclohexane

84: molecular ion

69: ring opening,
    hydrogen rearrangement,
    then cleavage

56: α cleavage after ring
    opening

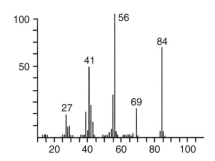

8.12. Identify the product and interpret the main peaks in the following:

8.12.1.

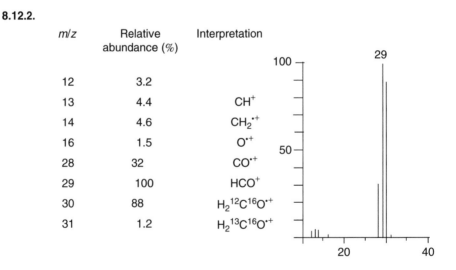

m/z	Relative abundance (%)	Interpretation
14	3.4	
15	20	CH_3^+
16	0.2	
19	2	F^+
31	10	CF^+
32	9.8	$CHF^{\cdot+}$
33	94	CH_2F^+
34	100	$^{12}CH_3F^{\cdot+}$
35	1.1	$^{13}CH_3F^{\cdot+}$

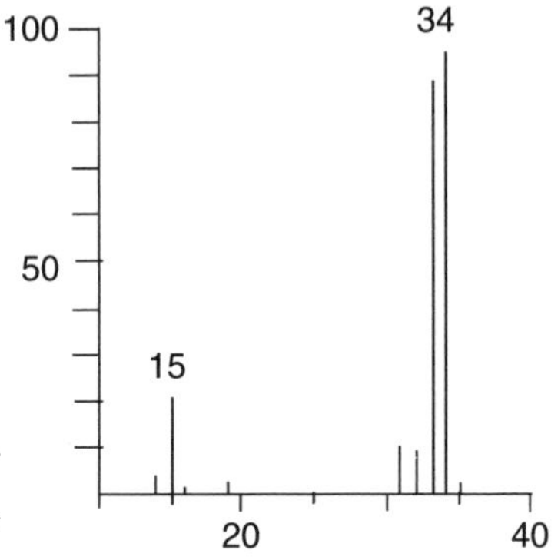

The intensity ratio of the peaks 35/34 indicates the presence of one C atom and the peak at 15 Th is ascribed to CH_3; if the molecular peak occurs at 34 Th, the $^{\cdot}CH_3$ loss leaves 19, which is due to fluorine; the formula proposed is thus CH_3F (methyl fluoride).

8.12.2.

m/z	Relative abundance (%)	Interpretation
12	3.2	
13	4.4	CH^+
14	4.6	$CH_2^{\cdot+}$
16	1.5	$O^{\cdot+}$
28	32	$CO^{\cdot+}$
29	100	HCO^+
30	88	$H_2{}^{12}C^{16}O^{\cdot+}$
31	1.2	$H_2{}^{13}C^{16}O^{\cdot+}$

The intensity ratio of the peaks 31/30 indicates the presence of only one C and the remainder (18) is due to H_2O. The structure suggested is thus $CH_2{=}O$ (formaldehyde).

8.12.3.

m/z	Relative abundance (%)	Interpretation
15	15.3	CH_3^+
27	37	$C_2H_3^+$
28	32	$C_2H_4^{•+}$
29	44	$C_2H_5^+$
39	12	$C_3H_3^+$
41	27	$C_3H_5^+$
42	12	$C_3H_6^{•+}$
43	100	$C_3H_7^+$
44	3.2	$^{13}C^{12}C_2H_7^+$
57	2.3	$C_4H_9^+$
58	12	$C_4H_{10}^+$

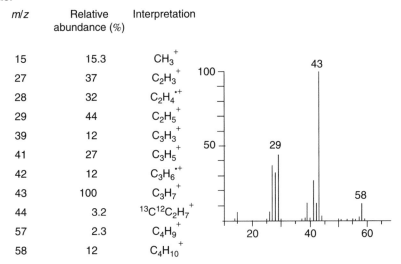

The intensity ratio of the peaks 59/58 indicates C_4; the intensity ratio 44/43 indicates C_3; if the molecular peak occurs at 58 Th, peak 43 indicates a $^•CH_3$ loss; the crude formula C_4H_{10} is thus proposed. The peak at 29 Th due to $^+C_2H_5$ allows one to exclude isobutane; the structure proposed is thus n-butane. The main peaks that are observed are due to σ cleavages.

8.12.4.

m/z	Relative abundance (%)	Interpretation
26	38	$C_2H_2^{•+}$
27	74	$C_2H_3^{•+}$
28	12	$CO^{•+}$
29	4.4	
44	14	$COO^{•+}$
45	32	$COOH^{•+}$
55	74	$M^{•+} - {}^•OH$
56	2.5	
57	0.2	
71	4.2	$M^{•+} - H^•$
72	100	$^{12}C_3H_4{}^{16}O_2^{•+}$
73	3.5	
74	0.5	

The intensity ratio 73/72 points to the presence of C_3; the intensity ratio 74/72 points to the presence of O_2; the residual mass can be justified only by four H atoms, which leads to the crude formula $C_3H_4O_2$. If the molecular peak occurs at 72 Th, the peak at 55 Th can be due to an $^\bullet$OH loss; the intensity ratio 57/56/55 confirms the existence of C_3 and O in this ion with a residual mass of 15 due to CH_3, which leads to the same crude formula. The number of unsaturations is $N = 2$. The peak at 45 Th due to $COOH^+$ leads us to suggest the following structure: $CH_2{=}CH{-}COOH$ (acrylic acid).

8.12.5.

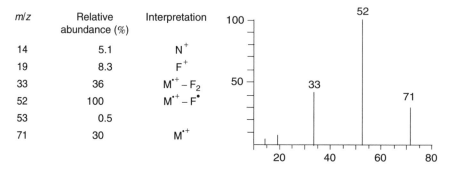

m/z	Relative abundance (%)	Interpretation
14	5.1	N^+
19	8.3	F^+
33	36	$M^{\bullet+} - F_2$
52	100	$M^{\bullet+} - F^\bullet$
53	0.5	
71	30	$M^{\bullet+}$

Because the molecular ion has an odd mass, it must contain an odd number of nitrogen atoms. With one nitrogen, the residual mass 57 can be ascribed to F_3; the presence of one nitrogen is confirmed by the intensity ratio 53/52; the formula proposed, NF_3 (nitrogen trifluoride), is confirmed by interpretation of the peaks that are observed.

8.12.6.

m/z	Relative abundance (%)	Interpretation
19	0.2	F^+
31	1.7	CF^+
35	3.1	Cl^+
37	1.2	
50	6.4	$CF_2^{\bullet+}$
69	100	CF_3^+
70	1.2	
85	18	$CClF_2^+$
86	0.2	
87	5.8	
104	0.7	$C^{35}Cl^{19}F_3^{\bullet+}$
106	0.2	

The intensity ratios 106/104 and 87/85 point to the presence of Cl with an $F^\bullet$ loss between the two; a 35 u loss leads to the 70/69 cluster, whose intensity ratio points to the presence of one carbon atom. If the molecular peak occurs at 104 Th, the residual mass from CCl is 57, which can be explained by F_3. The formula that we suggest is thus $CClF_3$ (chlorotrifluoromethane).

8.13. Identify the product that yields the following mass spectrum:

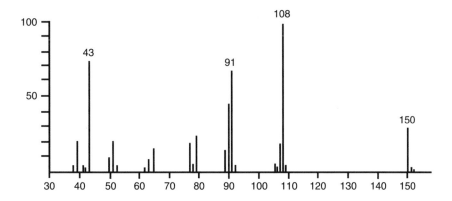

In the IR spectrum the main peaks occur at 3058, 2941, 1745, 1385, 1225, 1026, 749 and 697 cm^{-1}. In ^{1}H-NMR:

δ	Intensity	Multiplicity
7.22	5	Singlet
5.00	2	Singlet
1.96	3	Singlet

Answer:

In the mass spectrum, the peak at m/z 150 can be the molecular peak (even number of nitrogen atoms). The isotopic cluster suggests $C_9H_{10}O_2$ as the most probable formula, with five rings and/or double bonds.

In the IR spectrum, the peak at 1745 cm^{-1} corresponds to the unconjugated C=O stretch, the peak at 1225 cm^{-1} is the C−O−C stretch of an acetal, the peak at 1026 cm^{-1} can be the asymmetric stretch of COC and the peaks at 749 and 697 cm^{-1} point to a singly substituted benzene.

All this leads us to suggest the formula (benzyl acetate):

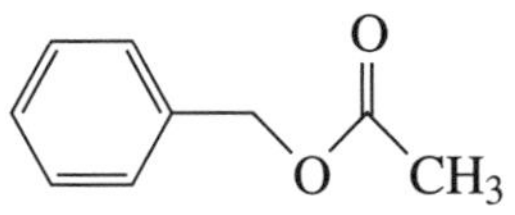

$$\text{benzyl acetate structure}$$

This is confirmed by the ^{1}H-NMR spectrum:

δ	Intensity	Assignment
7.22	5	Benzene protons
5.00	2	CH_2
1.96	3	CH_3

The main peaks in the mass spectrum are easily ascribed to the suggested formula:

150:	molecular peak	
108:	$COCH_3$ loss with H transfer	
91:	tropylium ion	
77,78:	benzene ions	
43:	CH_3CO^+	

8.14. Identify the product that yields the following mass spectrum:

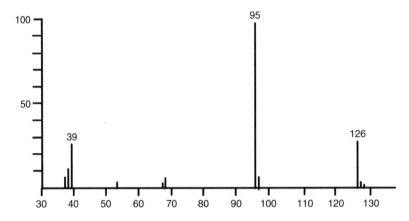

In the IR spectrum the main peaks that are observed occur at 3106, 2941, 1730, 1587, 1479, 1449, 1393, 1299, 1205, 1121 and 758 cm^{-1}. In ^{1}H-NMR:

δ	Intensity	Multiplicity
7.51	1	Quadruplet
7.02	1	Quadruplet
6.45	1	Quadruplet
3.80	3	Singlet

In off-resonance decoupled ^{13}C-NMR:

δ	Multiplicity
160	Singlet
146	Doublet
144	Singlet
118	Doublet
112	Doublet
51	Quadruplet

Answer:
The mass spectrum shows a peak at m/z 126 that can be the molecular peak (even number of nitrogen atoms); the isotopic cluster corresponds to $C_6H_6O_3$ (the exact mass can be

confirmed by high resolution) with four rings and two double bonds. The base peak at m/z 95 corresponds to M − OCH$_3$.

The IR spectrum shows a peak at 1730 cm^{-1} that can be ascribed to a conjugated C=O and peaks at 1205 and 1121 cm^{-1} that can be ascribed to an ester C−O stretch; the peak at 758 cm^{-1} indicates the presence of an aromatic ring and the same is true for the peaks at 3106, 1587 and 1479 cm^{-1}.

The ^{1}H-NMR spectrum confirms the presence of six protons and the ^{13}C-NMR confirms the presence of six carbons.

All this leads us to suggest the formula (carboxymethylfuran):

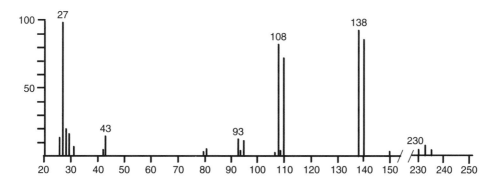

This structure is confirmed by the decoupled ^{13}C-NMR spectrum:

δ	Multiplicity	Number of H	Assignment
160	Singlet	0	C-5
146	Doublet	1	C-4
144	Singlet	0	C-1
118	Doublet	1	C-2
112	Doublet	1	C-3
51	Quadruplet	3	C-6

8.15. Identify the product that yields the following mass spectrum:

The IR spectrum shows the following main peaks: 2970, 2850, 1425, 1360, 1279, 1120 and 630 cm^{-1}. In ^{1}H-NMR:

δ	Intensity	Multiplicity
3.40	1	Multiplet
3.88	1	Multiplet

Answer:
The mass spectrum with the isotopic cluster at m/z 230, 232 and 234 indicates the presence of two bromine atoms; the remainder (72) can be ascribed to C_4H_8O. The symmetry of the ^{1}H-NMR spectrum indicates an AA'BB'-type symmetric structure, so we can suggest that the compound is di(bromoethyl) ether:

$$BrCH_2CH_2OCH_2CH_2Br$$

The IR spectrum confirms this structure:

2970 and 2850 cm^{-1}:	methylene stretch
1360 and 1279 cm^{-1}:	CH_2Br wagging
1120 cm^{-1}:	aliphatic ether
630 cm^{-1}:	CBr stretch

The main peaks in the mass spectrum are identified with this formula:

230:	molecular peak
138,140:	CH_2Br loss and H transfer
108,110:	OCH_2CH_2Br loss and H transfer
93, 95:	CH_2Br^+
27:	$C_2H_3^+$

8.16. Identify the product that yields the following mass spectrum:

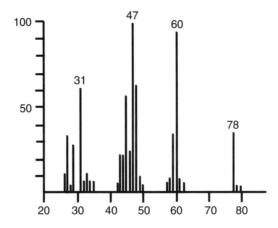

In the IR spectrum, the main peaks appear at 3367, 3030, 2558, 1429, 1298, 1050 and 1020 cm^{-1}. In off-resonance decoupled ^{13}C-NMR:

δ	Multiplicity
28	Triplet
64	Triplet

Answer:
The molecular peak at m/z 78 indicates a molecule with an even number of nitrogen atoms. The intensity of the $M + 2$ peak indicates the presence of S; the remainder (46) could be due to C_2H_6O.

The crude formula that is suggested is thus C_2H_6OS, with no ring or double bond. The decoupled ^{13}C-NMR spectrum confirms the presence of two C; in the off-resonance spectrum, the triplets indicate that four out of the six hydrogens belong to neighboring methylene groups. The structure that is proposed is thus 2-hydroxyethanethiol:

$$HOCH_2CH_2SH$$

The IR spectrum confirms such a structure:

3367 cm^{-1}:	OH stretch
2558 cm^{-1}:	SH stretch
1050 cm^{-1}:	CH$_2$OH primary alcohol

The main peaks in the mass spectrum are easily assigned:

78:	molecular peak
60:	H$_2$O loss
47:	CH$_2$SH$^+$
31:	CH$_2$OH$^+$

8.17. A 429 Da peptide gives the MS/MS product ion spectrum from its protonated molecular ion displayed in Figure 8.4. What is the sequence of this peptide?

Answer:
The difference between b_n and a_n fragments is 28 u. This applies to fragments 114/86 and 143/171. Thus $b_1 = 114$ and $b_2 = 171$ u. The N-terminal sequence is deduced: Lxx-Gly ... , with Lxx either Leu or Ile.

Similarly, y_n and z_n fragments differ by 17 Da, which applies to ion couples 130/147 and 243/260. Thus, $y_1 = 147$ and $y_2 = 260$ u. The C-terminal sequence is deduced: Lxx-Lys.

Combining this N- and C-terminal information, one gets the sequence Lxx-Gly-Lxx-Lys. The molecular weight is indeed 429 Da. This sequence is further confirmed by the observation of $y_3 = 317$ and $b_3 = 284$ u.

8.18. Consider Figure 8.5, representing the stability diagram for three ions of mass m_1, m_2 and m_3 Th in U, V coordinates. Explain which ions will be observed if the U, V values are adjusted successively to the values indicated by points A–D in the figure.

Answer:
At A, none of these ions will be observed because the point is outside the stability curves. At B, both ions m_1 and m_2 will be observed because the point is inside their stability diagrams; m_3 will not be observed. At C, the three ions will be observed because this point is within the three stability diagrams. Point D is inside the stability diagrams of m_2 and m_3, which will be observed; m_1 will not be detected.

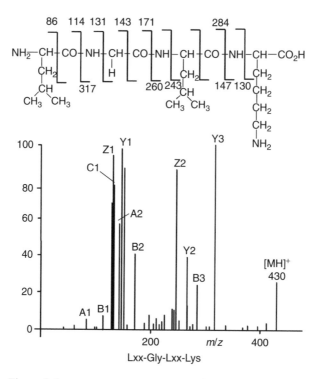

Figure 8.4
Fragment ion spectrum of a protonated peptide and (*top*)
the sequence of ions

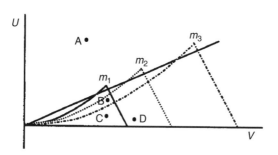

Figure 8.5

8.19. A quadrupole with $r_0 = 1$ cm operating at an RF frequency of 0.5 MHz has a stability diagram with an apex at $a = 0.22$ and $q = 0.72$ for an ion of mass 2000 u. What are the corresponding U and V values?

Answer:
Applying the formula for a quadrupole:

$$a_u = \frac{8eU}{mr_0{}^2\omega^2} \text{ and } q_u = \frac{4eV}{mr_0{}^2\omega^2}$$

with $e = 1.6 \times 10^{-19}$ C, $m = 2000 \times 1.66 \times 10^{-27} = 3.32 \times 10^{-24}$ kg, $r_0 = 0.01$ m and $\omega = 2\pi f = 2 \times 3.14 \times 500\,000 = 3.14 \times 10^6$ rad/s^{-1}, one finds: $U = 563$ V and $V = 3.68$ kV.

8.20. The same quadrupole as in Question 8.19 operating in the 'RF-only' mode has values of $a = 0$ and $q = 0.91$. Which is the minimum V_{rf} voltage allowing eliminate all ions of less than 100 Th to be eliminated?

Answer:
$V_{min} = 233$ V.

8.21. If a V_{rf} of 2 kV is applied to the same quadrupole as in Question 8.19, from which m/z value will the ions be unstable?

Answer:
$m_{min} = 859$ Da.

8.22. In an ideal ion trap, i.e. $r_0^2 = 2z_0^2$, with $r_0 = 1$ cm at $\nu = 1$ MHz, calculate the a and q values at the apex of the stability diagram.

Answer:
Referring to the values for U and V given in the figure below, one finds: $a_z = -0.67$ and $q_z = -1.24$ for the lower apex and $a_z = 0.15$ and $q_z = -0.78$ for the upper apex.

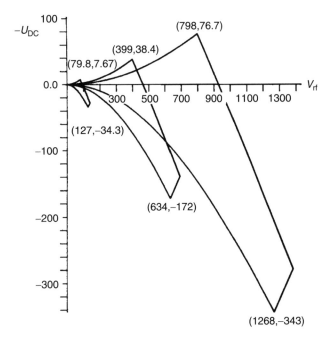

8.23. In an ideal ion trap, i.e. $r_0^2 = 2z_0^2$, with $r_0 = 1$ cm at $\nu = 1$ MHz, $U = 4$ V and $V = 400$ V for a nitrogen ion $N_2{}^{\bullet+}$, calculate the corresponding a and q values.

Answer:
Using the appropriate formula for an ideal ion trap

$$a_u = a_z = -2a_r = \frac{-16zeU}{m\left(r_0^2 + 2z_0^2\right)\omega^2}, \quad q_u = q_z = -2q_r = \frac{8zeV}{m\left(r_0^2 + 2z_0^2\right)\omega^2}$$

the results are:

$$a_z = -0.0109 \quad a_r = 0.0055$$

$$q_z = -0.55 \quad q_r = 0.27$$

8.24. In an ideal ion trap, i.e. $r_0^2 = 2z_0^2$, with $r_0 = 1$ cm at $v = 1.1$ MHz, $U = 4$ V and $V = 500$ V, what is the resonant frequency that should be applied to expel an ion of 100 Th?

Answer:
Applying the formulae $\beta_z = (a_z + q_z^2/2)^{1/2}$ and $v_z = \beta_z v/2$, one finds $v = 153$ kHz.

8.25. A time-of-flight (TOF) analyzer has the following characteristics: $V_s = -3$ kV, $V_R = 130$ V, $D = 0.522$ m, $L_1 = 1$ m.

   (a)   At which distance L_2 does it focalize ions having the same $m/z(m_p) = 578$ Th?
   (b)   What is the focal distance for post-source decay ions with $m/z = 560$ Th?
   (c)   At which value should V_R be reduced in order to focalize the fragment ions at the same distance as the precursors?

Answer:
   (a) $x_p = V_s D/(V_R - V_s) = 0.5$ m; $L_{2p} = 4x - L_1 = 1$ m
   (b) $x_f = x_p m_f/m_p = 0.48$ m; $L_{2f} = 0.94$ m
   (c) $V_{rf} - V_s = (V_{rp} - V_s)m_f/m_p = 3025$ V; $V_{rf} = 25$ V

Appendices

1 Nomenclature

A report (Price P., *J. Am. Soc. Mass. Spectrum.*, **2**, 336 (1991), about standard definitions of terms relating to mass spectrometry were presented by the committee on measurements and standards of the American Society for Mass Spectrometry.

Units

1. The mass unit (u) is also called the dalton (Da). It is defined as 1/12 of the mass of a carbon-12 atom:
$$1 \text{ u} = 1 \text{ Da} = 1.660540 \times 10^{-27} \text{ kg}$$

2. The thomson (Th) is the m/z unit.

3. The charge unit is the charge of the electron and is an absolute value:
$$e = 1.602177 \times 10^{-19} \text{ C}$$

Definitions

1. Average mass or chemical mass: mass calculated using a weighted average of the natural isotopes for the atomic mass of each element. Example: average mass of $CH_3Br = (12.01115 + 3 \times 1.00797 + 79.904)\text{u} = 94.93906$ u. This is the mass that a chemist normally uses in stoichiometric calculations.

2. Nominal ion mass: mass calculated using the integer mass of the predominant isotope of each element. Example: nominal mass of $CH_3OH = {}^{12}C^1H_3^{16}O^1H = (12 + 4 \times 1 + 16)\text{u} = 32$ u.

3. Monoisotopic ion mass: mass calculated using the exact mass of the predominant isotope of each element, which takes into account the mass defects. Example: $CH_3Br = {}^{12}C^1H_3^{79}Br = (12.0000 + 3 \times 1.007825 + 78.918336)\text{u} = 93.941011$ u.

4. m/z: the ratio of the mass number to the charge number. This is an abstract, unitless number that allows us to talk about, for example, $m/z = 60$. If we define m/z as the mass-to-charge ratio, m is given in dalton and the charge is given in e, thus m/z is in thomson (Th). For example, an ion with mass $m = 200$ u and with charge $2e$ has $m/z = 100$ Th.

5. Mass spectrum: plot of ion abundance versus mass-to-charge ratio normalized to the most abundant.

6. Base peak: the most intense peak in the spectrum.

7. Isotopic peak: peak due to other isotopes of the same chemical but different isotopic composition.

8. Relative abundance: normalization relative to the base peak.

9. Relative intensity: ratio of the peak intensity to that of the base peak.

10. Percentage of the total intensity: abundance of an ion over the total abundance of all ions within a specified mass region.

Analyzers

1. Mass spectrometer: an instrument that analyzes ions according to their m/z ratio and measures electrically the number (abundance) of the ions.

2. Mass spectrograph: an instrument that separates the ion beams according to their m/z ratio and records the deflection and intensity of the beams directly on a photographic plate or film.

3. Single-focusing spectrometer: an instrument that focuses ions of a given m/z even though the initial directions of the ions diverge.

4. Double-focusing spectrometer: an instrument that focuses ions with a given m/z ratio even though their initial directions diverge and their kinetic energies are different.

5. Nier–Johnson geometry: double-focusing spectrometer in which a $\pi/2$ rad deflection in a radial electrostatic field is followed by a $\pi/3$ rad magnetic deflection.

6. Mattauch–Herzorg geometry: contrivance for a double-focusing spectrometer where a $\pi/4\sqrt{2}$ rad deflection in a radial electrostatic field is followed by a magnetic deflection of $\pi/2$ rad.

7. Static field spectrometer: a spectrometer that separates the ion beams using fields that remain constant in time.

8. Dynamic field spectrometer: a spectrometer in which the separation of ion beams depends on fields that vary in time.

9. Electrostatic analyzer: a kinetic energy focusing instrument that produces an electrostatic field that is perpendicular to the ions' displacement direction. Its effect is to bring all ions of the same kinetic energy to a single focus.

10. Magnetic analyzer: an instrument that causes a direction focusing produced by a magnetic field that is perpendicular to the ions' displacement direction. Its effect is to bring to a common focus all the ions of a given momentum. This can be converted to m/z ratios if all the ions have the same kinetic energy.

11. Quadrupole analyzer: a mass filter that produces a quadrupolar field with a DC component and an AC component such that only ions with a given m/z pass through.

12. Time-of-flight analyzer: an instrument that measures the time of flight of the ions over a known fixed distance.

13. Reflectron (also known as an ion mirror): a device used in a time-of-flight mass spectrometer that retards and then reverses ion velocities in order to correct for the flight times of ions having different kinetic energies

14. Cyclotron resonance analyzer (also known as a Penning ion trap): an instrument that confines ions by placing them in a static magnetic field. Inside the field, the ions are subject to the Lorentzian force that causes ions of a particular m/z to cyclotron at a specific frequency (cyclotron frequency).

15. Ion trap analyzer (also known as a Paul ion trap): a type of mass analyzer in which ions are confined in space by means of a three-dimensional, rotationally symmetric quadrupolar electric field capable of storing ions at selected m/z ratios.

Detection

1. Electron multiplier: device that multiplies an electronic current by accelerating the electrons on the surface of an electrode. The collision yields a number of secondary electrons higher than the number of incident electrons. The secondary electrons are accelerated towards another electrode (or another part of the same electrode), which in turn gives off secondary electrons, thereby continuing the process.

2. Detection limit: the smallest sample quantity that yields a signal that can be distinguished from the background noise. See also *sensitivity*.

3. Resolution: the ratio $m/\delta m$ where m and $m + \delta m$ are the mass numbers of the two ions that yield neighboring peaks with a valley depth $x\%$ of the weakest peak's intensity. Hence, to interpret the value of the resolution, one has to consider together the value of m and x. In the commercial description of mass spectrometers, $x = 10$ is normally used for magnetic instruments and $x = 50$ is normally used for quadrupole instruments. Another definition entails using for δm the width of an isolated peak at $x\%$ of its maximum.

4. Sensitivity: the sensitivity of an instrument is the ratio of the ionic current change to the sample flux change in the source (in Cb μg^{-1}). The analytical sensitivity is the smallest quantity of compound yielding a definite signal-to-noise-ratio, often 10:1.

Ionization

1. Ionization: a process that yields an ion from an atom or from a neutral molecule.

2. Electron ionization (EI): ionization of the sample by a beam of electrons most often accelerated by a potential of about 70 eV.

3. Field ionization (FI): removal of an electron from the sample molecule by the application of an intense electric field.

4. Field desorption (FD): production of ions from the sample laid on a solid surface and submitted to an intense electric field.

5. Chemi-ionization: production of an ion through reaction of the sample molecule with another molecule that was excited beforehand:

$$M + A^* \rightarrow MA^+ + e^-$$

6. Chemical ionization (CI): formation of a new ionized species when the sample molecule reacts with an ion.

7. Laser ionization (LI): production of ions by irradiation of the sample with a laser beam.

8. Photoionization (PI): ionization using photons.

9. Thermal ionization: occurs when an atom or a molecule is ionized through interaction with a heated solid surface or through its presence in a high-temperature gaseous environment.

10. Electron attachment: a resonance process whereby an electron is incorporated into an atomic or molecular orbital.

11. Ion pair formation: ionization process where the only products are a positive ion fragment and a negative ion fragment.

12. Vertical ionization: a process in which an electron is removed from a molecule so fast that a positive ion is produced without any change in position or momentum of the atoms. The ion that results is often in an excited state.

13. Adiabatic ionization: the electron is removed from the molecule in its fundamental state, which produces an ion in its fundamental state.

14. Dissociative ionization: a process in which a gas molecule decomposes in order to form products, one of which is an ion.

Ion types

1. Molecular ion: ion derived from the neutral molecule by loss or gain of an electron or other simple unit.

2. Pseudomolecular ion: ion produced from the molecule by abstraction of a proton or a hydride, or by the addition of a proton or other ion. It allows one to deduce the molecular mass. It is also called 'molecular species ion'.

3. Precursor ion: ion that decomposes or changes its charge, yielding a product ion.

4. Fragment ion: ion derived from the decomposition of a parent ion.

5. Metastable ion: ion that fragments spontaneously in a region where there is no deflection field, usually between the source and the first sector.

6. Radical ion (synonymous with odd-electron ion): ion containing an unpaired electron.

7. Distonic ion: radical ion in which the charge and radical sites are formally located on different atoms in the molecule.

8. Isotopic ion: any ion containing one or more of the less abundant naturally occurring isotopes of the elements that make up its structure.

9. Isobaric ion: ion of the same normal mass (integral).

10. Adduct: ion formed from interaction between two species, usually an ion and a molecule, containing all the atoms of one species plus one or several atoms of the other.

Ion–molecule reaction

1. Collision-activated or collision-induced dissociation (CAD or CID): the rapidly moving projectile ion dissociates through interaction with a neutral immobile target molecule.

2. Collisional activation (CA): the rapidly moving projectile ion is excited by interaction with an immobile neutral target molecule.

3. Elastic collision: ion–neutral interaction in which the total kinetic energy of the collision partners remains unchanged.

4. Inelastic collision: ion–neutral interaction in which the total kinetic energy (and thus the internal energy) of the collision partners changes.

5. Charge stripping: ion–molecule reaction that increases the ion positive charge.

Fragmentation

1. Single electron transfer: half-arrow ⤴

2. Electron pair transfer: full arrow ⟶

3. McLafferty rearrangement: $\xrightarrow{\text{rH}}$

2 Abbreviations

AC	alternating current
ADC	analog-to-digital converter
AE	appearance energy
AED	atomic emission detector
AMS	accelerator mass spectrometry
API	atmospheric pressure ionization
APCI	atmospheric pressure chemical ionization
CA	collisional activation
CAD	collision-activated dissociation
CAR	collision-activated reaction
CE	capillary electrophoresis
CE/MS	capillary electrophoresis/mass spectrometry
CEM	channel electron multiplier
CF	continuous flow
CF-FAB	continuous-flow fast atom bombardment
CI	chemical ionization
CID	collision-induced dissociation
CID/MS/MS	collision-induced dissociation/tandem mass spectrometry
CRF	charge remote fragmentation
CS	charge stripping
CZE	capillary zone electrophoresis
DAC	digital-to-analog converter
DADI	direct analysis of daughter ion
DC	direct current
DCI	desorption chemical ionization

DE	delayed extraction
DEI	desorption electron ionization
DI	desorption ionization
DLI	direct liquid introduction
DMS	dynamic mass spectrometry
EA	electron affinity
ECCI	electron capture chemical ionization
EE	even electron number
EHD	electron hydrodynamic desorption
EI	electron ionization (formerly electron impact)
ESI	electrospray ionization
ESI/MS	electrospray ionization/mass spectrometry
ESI/FTICR	electrospray ionization/fourier transform ion cyclotron resonance
FA	flowing afterglow
FAB	fast atom bombardment
FAB/MS	fast atom bombardment/mass spectrometry
FD	field desorption
FFR	field-free region
FFT	fast fourier transform
FI	field ionization
FIB	fast ion bombardment
FID	flame ionization detection
FTMS	Fourier transform mass spectrometry
FTNMR	Fourier transform nuclear magnetic resonance
FTICR	Fourier transform ion cyclotron resonance
FVP	flash vacuum pyrolysis
FWHM	full width at half-maximum
GC	gas chromatography
GC/MS	gas chromatography/mass spectrometry
GDMS	glow discharge mass spectrometry
HPLC	high-performance liquid chromatography
HPLC/APCI	high-performance liquid chromatography/atmospheric pressure chemical ionization
HPLC/MS	high-performance liquid chromatography/mass spectrometry
HPMS	high-pressure pulsed mass spectrometry
HR	high resolution
IC	ion chromatography
ICP	inductively coupled plasma
ICPMS	inductively coupled plasma mass spectrometry
ICR	ion cyclotron resonance
ICR/FTMS	ion cyclotron resonance/Fourier transform mass spectrometry
IE	ionization energy (formerly ionization potential)
IKES	ion kinetic energy spectrometry
IR	infrared
IRMPD	infrared multiple photon dissociation
ISP	ionspray
ITD	ion trap detector
ITMS	ion trap mass spectrometry
ITR	integrating transient recorder

LC	liquid chromatography
LC/MS	liquid chromatography/mass spectrometry
LD	laser desorption
LDLPMS	laser desorption laser photoionization mass spectrometry
LI	laser ionization
LIMS	laser ionization mass spectrometry
LIF	laser-induced fluorescence
LPCI	low-pressure chemical ionization
LSIMS	liquid secondary ion mass spectrometry
MALDI	matrix-assisted laser desorption/ionization
MALDI/TOF	matrix-assisted laser desorption/ionization time-of-flight
MIKES	mass-analyzed ion kinetic energy spectroscopy
MIMS	membrane introduction mass spectrometry
MPI	multiphoton ionization
MRM	metastable reaction monitoring
MS	mass spectrometry
MS/MS	tandem mass spectrometry
MS^n	general designation of mass spectrometry to the nth degree
$N_f R$	neutral fragment reionization
NMR	nuclear magnetic resonance
NICI	negative ion chemical ionization
NR	neutralization reionization
NRMS	neutralization reionization mass spectrometry
OE	odd electron number
PA	proton affinity
PB	particle beam
PBM	probability-based matching
PCR	polymerase chain reaction
PD	plasma desorption
PDMS	plasma desorption mass spectrometry
PI	photoionization
PID	photoinduced dissociation
PSD	post-source decay
QET	quasi-equilibrium theory
RA	relative abundance
RDA	retro-diels−alder
REMPI	resonance-enhanced multiphoton ionization
RF	radiofrequency
RIC	reconstructed ion chromatogram
RIMS	resonance ionization mass spectrometry
RR	reaction region
RTOF	reflectron time-of-flight
SFC	supercritical fluid chromatography
SID	surface-induced dissociation
SIFT	selected-ion flow tube
SIM	selected-ion monitoring
SIMS	secondary ion mass spectrometry
SIR	selected-ion recording
SIS	selected-ion storage

SORI	sustained off-resonance irradiation
SRM	selected-reaction monitoring
SSMS	spark source mass spectrometry
SWIFT	selected waveform inverse Fourier transform
TCC	time-compressed chromatography
TDC	time-to-digital conversion
TIC	total ion current
TIMS	thermal ionization mass spectrometry
TLC	thin-layer chromatography
TLF	time-lag focusing
TOF	time-of-flight
TSP	thermospray
UV	ultraviolet

3 Fundamental Physical Constants

Quantity	Symbol	Value	Units
Speed of light in vacuum	c	299 792 458	m s^{-1}
Gravitation constant	G	6.67259	10^{-11} m^3 kg^{-1} s^{-2}
Planck constant	h	6.6260755	10^{-34} J s
Elementary charge	e	1.60217733	10^{-19} C
Electron mass	m_e	9.1093897	10^{-31} kg
Proton mass	m_p	1.6726231	10^{-27} kg
Neutron mass	m_n	1.6749286	10^{-27} kg
Atomic mass unit, dalton	u, Da	1.6605402	10^{-27} kg
Boltzmann constant	k	1.380658	10^{-23} J K^{-1}
		1.60217733	10^{-19} J
Electronvolt	eV	23.06054	kcal mol^{-1}
		96.4853	kJ mol^{-1}
Avogadro constant	N_A	6.0221367	10^{23} mol^{-1}
Molar gas constant	R	8.314510	J mol^{-1} K^{-1}

Prefix	Symbol	Factor	Prefix	Symbol	Factor
deca	da	10^1	deci	d	10^{-1}
hecto	h	10^2	centi	c	10^{-2}
kilo	k	10^3	milli	m	10^{-3}
mega	M	10^6	micro	µ	10^{-6}
giga	G	10^9	nano	n	10^{-9}
tera	T	10^{12}	pico	p	10^{-12}
peta	P	10^{15}	femto	f	10^{-15}
exa	E	10^{18}	atto	a	10^{-18}
			zepto	z	10^{-21}

These tables are from the *Handbook of Chemistry and Physics* (71st edition, CRC Press, Boca Raton, FL, 1990) and the Internet site of the NIST Reference on Constants, Units and Uncertainty (http://physics.nist.gov/cuu/Constants/index.html).

4A Table of Isotopes in Ascending Mass Order

Z	Symbol	Nominal mass	%	Rel. %	Isotopic mass	Average mass
0	n	1	—		1.008665	—
1	H	1	99.985	100	1.007825	1.00794
	D	2	0.015	0.015	2.014	
2	He	3	0.000137	0.000137	3.016030	4.00260
		4	≈100	100	4.00260	
3	Li	6	7.5	8.0108	6.015121	6.941
		7	92.5	100	7.016003	
4	Be	9	100	100	9.012182	9.012182
5	B	10	19.9	24.84	10.012937	10.811
		11	80.1	100	11.009305	
6	C	12	98.90	100	12.000000	12.011
		13	1.10	1.112	13.003355	
7	N	14	99.63	100	14.003074	14.00674
		15	0.37	0.37	15.000108	
8	O	16	99.76	100	15.994915	15.9994
		17	0.04	0.04	16.999133	
		18	0.20	0.20	17.999160	
9	F	19	100	100	18.998403	18.9984
10	Ne	20	90.48	100	19.992435	20.1797
		21	0.27	0.298	20.993843	
		22	9.25	10.22	21.991264	
11	Na	23	100	100	22.989768	22.9898
12	Mg	24	78.99	100	23.985042	24.3050
		25	10.00	12.66	24.985837	
		26	11.01	13.94	25.982593	
13	Al	27	100	100	26.981539	26.9815
14	Si	28	92.21	100	27.976927	28.0855
		29	4.67	5.065	28.976495	
		30	3.10	3.336	29.973770	
15	P	31	100	100	30.973762	30.9738
16	S	32	95.03	100	31.972070	32.066
		33	0.75	0.789	32.971456	
		34	4.22	4.44	33.967866	
		36	0.02	0.021	35.967080	
17	Cl	35	75.77	100	34.968852	35.453
		37	24.23	31.98	36.965903	
18	Ar	36	0.337	0.338	35.967545	39.948
		38	0.063	0.0633	37.962732	
		40	99.600	100	39.962384	
19	K	39	93.2581	100	38.963707	39.0983
		40	0.0117	0.0125	39.963999	
		41	6.7302	7.22	40.961825	

(continued overleaf)

Z	Symbol	Nominal mass	%	Rel. %	Isotopic mass	Average mass
20	Ca	40	96.941	100	39.962591	40.078
		42	0.647	0.66742	41.958618	
		43	0.135	0.139	42.958766	
		44	2.086	2.152	43.955480	
		46	0.004	0.004	45.953689	
		48	0.187	0.193	47.952533	
21	Sc	45	100	100	44.955911	44.956
22	Ti	46	8.00	10.84	45.952629	47.88
		47	7.3	9.892	46.951764	
		48	73.8	100	47.947947	
		49	5.51	7.466	48.947871	
		50	5.4	7.317	49.944792	
23	V	50	0.25	0.251	49.947161	50.9415
		51	99.75	100	50.943962	
24	Cr	50	4.345	5.185	49.946046	51.9961
		52	83.79	100	51.940509	
		53	9.50	11.34	52.940651	
		54	2.365	2.82	53.938882	
25	Mn	55	100	100	54.938046	54.9380
26	Fe	54	5.9	6.43	53.939612	55.847
		56	91.72	100	55.934939	
		57	2.1	2.29	56.935396	
		58	0.28	0.305	57.933277	
27	Co	59	100	100	58.933198	58.9332
28	Ni	58	68.27	100	57.935346	58.6934
		60	26.10	38.23	59.930788	
		61	1.13	1.66	60.931058	
		62	3.59	5.26	61.928346	
		64	0.91	1.33	63.927968	
29	Cu	63	69.17	100	62.929598	63.546
		65	30.83	44.57	64.927765	
30	Zn	64	48.6	100	63.929145	65.39
		66	27.9	57.41	65.926034	
		67	4.1	8.44	66.927129	
		68	18.8	38.68	67.924846	
		70	0.6	1.23	69.925325	
31	Ga	69	60.108	100	68.925580	69.723
		71	39.892	66.37	70.924700	
32	Ge	70	20.5	56.16	69.924250	72.61
		72	27.4	75.07	71.922079	
		73	7.8	21.37	72.923463	
		74	36.5	100	73.921177	
		76	7.8	21.37	75.921401	
33	As	75	100	100	74.921594	74.9216

Z	Symbol	Nominal mass	%	Rel. %	Isotopic mass	Average mass
34	Se	74	0.9	1.80	73.922475	78.96
		76	9.1	18.24	75.919212	
		77	7.6	15.23	76.919912	
		78	23.6	47.29	77.917309	
		80	49.9	100	79.916520	
		82	8.9	17.84	81.916698	
35	Br	79	50.69	100	78.918336	79.904
		81	49.31	97.28	80.916289	
36	Kr	78	0.35	0.614	77.920401	83.80
		80	2.25	3.947	79.916380	
		82	11.6	20.35	81.913482	
		83	11.5	20.175	82.914135	
		84	57.0	100	83.911507	
		86	17.3	30.35	85.910610	
37	Rb	85	72.17	100	84.911794	85.4678
		87	27.83	38.562	86.909187	
38	Sr	84	0.56	0.68	83.913431	87.62
		86	9.86	11.94	85.909267	
		87	7.00	8.5	86.908884	
		88	82.58	100	87.905619	
39	Y	89	100	100	88.905849	88.906
40	Zr	90	51.45	100	89.904703	91.224
		91	11.22	21.73	90.905643	
		92	17.15	33.33	91.905039	
		94	17.38	33.78	93.906314	
		96	2.80	5.44	95.908275	
41	Nb	93	100	100	92.906377	92.906
42	Mo	92	14.84	61.50	91.906808	95.94
		94	9.25	38.33	93.905085	
		95	15.92	65.98	94.905840	
		96	16.68	69.13	95.904678	
		97	9.55	39.58	96.906020	
		98	24.13	100	97.905406	
		100	9.63	39.91	99.907477	
43	Tc			100		
44	Ru	96	5.54	17.53	95.907599	101.07
		98	1.86	5.89	97.905267	
		99	12.7	40.19	98.905939	
		100	12.6	38.87	99.904219	
		101	17.1	54.11	100.905582	
		102	31.6	100	101.904348	
		104	18.6	58.86	103.905424	
45	Rh	103	100	100	102.905500	102.905

(*continued overleaf*)

Z	Symbol	Nominal mass	%	Rel. %	Isotopic mass	Average mass
46	Pd	102	1.02	3.73	101.905634	106.42
		104	11.14	40.76	103.904029	
		105	22.33	81.71	104.905079	
		106	27.33	100	105.903478	
		108	26.46	96.82	107.903895	
		110	11.72	42.88	109.905167	
47	Ag	107	51.839	100	106.905092	107.868
		109	48.161	94.90	108.904757	
48	Cd	106	1.25	4.35	105.906461	112.411
		108	0.89	3.10	107.904176	
		110	12.49	43.47	109.903005	
		111	12.80	44.55	110.904182	
		112	24.13	83.99	111.902758	
		113	12.22	42.53	112.904400	
		114	28.73	100	113.903357	
		116	7.49	26.07	115.904754	
49	In	113	4.3	4.49	112.904061	114.82
		115	95.7	100	114.903880	
50	Sn	112	0.97	2.98	111.904826	118.710
		114	0.65	1.99	113.902784	
		115	0.36	1.10	114.903348	
		116	14.53	43.58	115.901747	
		117	7.68	23.57	116.902956	
		118	24.22	73.32	117.901609	
		119	8.58	26.33	118.903310	
		120	32.59	100	119.902200	
		122	4.63	14.21	121.903440	
		124	5.79	17.77	123.905274	
51	Sb	121	57.4	100	120.903821	121.752
		123	42.6	74.22	122.904216	
52	Te	120	0.095	0.28	119.904048	127.60
		122	2.59	7.65	121.903054	
		123	0.905	2.67	122.904271	
		124	4.79	14.14	123.902823	
		125	7.12	21.02	124.904433	
		126	18.93	55.89	125.903314	
		128	31.70	93.59	127.904463	
		130	33.87	100	129.906229	
53	I	127	100	100	126.904476	126.9045
54	Xe	124	0.10	0.37	123.905894	131.29
		126	0.09	0.33	125.904281	
		128	1.91	7.10	127.903531	
		129	26.4	98.14	128.904780	
		130	4.1	15.24	129.903509	
		131	21.2	78.81	130.905072	
		132	26.9	100	131.904144	
		134	10.4	38.866	133.905395	
		136	8.9	33.09	135.907214	

Z	Symbol	Nominal mass	%	Rel. %	Isotopic mass	Average mass
55	Cs	133	100	100	132.905429	132.905
56	Ba	130	1.101	1.536	129.906284	137.34
		132	0.097	0.135	131.905045	
		134	2.42	3.77	133.904493	
		135	6.59	9.2	134.905671	
		136	7.81	10.9	135.904559	
		137	11.32	15.8	136.905815	
		138	71.66	100	137.905235	
57	La	138	0.090	0.09	137.90711	138.91
		139	99.91	100	138.906347	
72	Hf	174	0.162	0.46	173.940044	178.49
		176	5.206	14.83	175.941406	
		177	18.606	53.01	176.943217	
		178	27.297	77.77	177.943696	
		179	13.629	38.83	178.945812	
		180	35.100	100	179.946545	
73	Ta	180	0.012	0.012	179.947462	180.948
		181	99.988	100	180.947992	
74	W	180	0.12	0.39	179.946701	183.85
		182	26.3	85.67	181.948202	
		183	14.28	46.51	182.950220	
		184	30.7	100	183.950928	
		186	28.6	93.16	185.954357	
75	Re	185	37.40	59.74	184.952951	186.207
		187	62.60	100	186.955744	
76	Os	184	0.02	0.05	183.952488	190.2
		186	1.58	3.85	185.953830	
		187	1.6	3.90	186.955741	
		188	13.3	32.44	187.955860	
		189	16.1	39.27	188.958137	
		190	26.4	64.39	189.958436	
		192	41.0	100	191.961467	
77	Ir	191	37.3	59.49	190.960584	192.22
		193	62.7	100	192.962917	
78	Pt	190	0.01	0.03	189.959917	195.08
		192	0.79	2.34	191.961019	
		194	32.9	97.34	193.962655	
		195	33.8	100	194.964766	
		196	25.3	74.85	195.964926	
		198	7.2	21.30	197.967869	
79	Au	197	100	100	196.966543	196.967
80	Hg	196	0.15	0.50	195.965807	200.59
		198	10.0	33.56	197.966743	
		199	16.9	56.71	198.968254	
		200	23.1	77.52	199.968300	
		201	13.2	44.30	200.970277	
		202	29.8	100	201.970617	
		204	6.85	22.99	203.973467	

(continued overleaf)

Z	Symbol	Nominal mass	%	Rel. %	Isotopic mass	Average mass
81	Tl	203	29.524	41.89	202.972320	204.383
		205	70.476	100	204.974401	
82	Pb	204	1.4	2.67	203.973020	207.2
		206	24.1	45.99	205.974440	
		207	22.1	42.18	206.975872	
		208	52.4	100	207.976627	
83	Bi	209	100	100	208.980374	208.980
90	Th	232	100	100	232.038054	232.038
92	U	234	0.0055	0.0055	234.040946	238.03
		235	0.720	0.725	235.043924	
		238	99.2745	100	238.050784	

Source: Handbook of Chemistry and Physics, 71st edition, CRC Press, Boca Raton, FL, 1990.

4B Table of Isotopes in Alphabetical Order

Z	Symbol	Nominal mass	%	Rel. %	Isotopic mass	Average mass
47	Ag	107	51.839	100	106.905092	107.868
		109	48.161	94.90	108.904757	
13	Al	27	100	100	26.981539	26.9815
18	Ar	36	0.337	0.338	35.967545	39.948
		38	0.063	0.0633	37.962732	
		40	99.600	100	39.962384	
33	As	75	100	100	74.921594	74.9216
79	Au	197	100	100	196.966543	196.967
5	B	10	19.9	24.84	10.012937	10.811
		11	80.1	100	11.009305	
56	Ba	130	1.101	1.536	129.906284	137.34
		132	0.097	0.135	131.905045	
		134	2.42	3.77	133.904493	
		135	6.59	9.2	134.905671	
		136	7.81	10.9	135.904559	
		137	11.32	15.8	136.905815	
		138	71.66	100	137.905235	
4	Be	9	100	100	9.012182	9.012182
83	Bi	209	100	100	208.980374	208.980
35	Br	79	50.69	100	78.918336	79.904
		81	49.31	97.28	80.916289	
6	C	12	98.90	100	12.000000	12.011
		13	1.10	1.112	13.003355	

Z	Symbol	Nominal mass	%	Rel. %	Isotopic mass	Average mass
20	Ca	40	96.941	100	39.962591	40.078
		42	0.647	0.66742	41.958618	
		43	0.135	0.139	42.958766	
		44	2.086	2.152	43.955480	
		46	0.004	0.004	45.953689	
		48	0.187	0.193	47.952533	
48	Cd	106	1.25	4.35	105.906461	112.411
		108	0.89	3.10	107.904176	
		110	12.49	43.47	109.903005	
		111	12.80	44.55	110.904182	
		112	24.13	83.99	111.902758	
		113	12.22	42.53	112.904400	
		114	28.73	100	113.903357	
		116	7.49	26.07	115.904754	
17	Cl	35	75.77	100	34.968852	35.453
		37	24.23	31.98	36.965903	
27	Co	59	100	100	58.933198	58.9332
24	Cr	50	4.345	5.185	49.946046	51.9961
		52	83.79	100	51.940509	
		53	9.50	11.34	52.940651	
		54	2.365	2.82	53.938882	
55	Cs	133	100	100	132.905429	132.905
29	Cu	63	69.17	100	62.929598	63.546
		65	30.83	44.57	64.927765	
9	F	19	100	100	18.998403	18.9984
26	Fe	54	5.9	6.43	53.939612	55.847
		56	91.72	100	55.934939	
		57	2.1	2.29	56.935396	
		58	0.28	0.305	57.933277	
31	Ga	69	60.108	100	68.925580	69.723
		71	39.892	66.37	70.924700	
32	Ge	70	20.5	56.16	69.924250	72.61
		72	27.4	75.07	71.922079	
		73	7.8	21.37	72.923463	
		74	36.5	100	73.921177	
		76	7.8	21.37	75.921401	
1	H	1	99.985	100	1.007825	1.00794
	D	2	0.015	0.015	2.014	
2	He	3	0.000137	0.000137	3.016030	4.0026
		4	≈100	100	4.00260	
72	Hf	174	0.162	0.46	173.940044	178.49
		176	5.206	14.83	175.941406	
		177	18.606	53.01	176.943217	
		178	27.297	77.77	177.943696	
		179	13.629	38.83	178.945812	
		180	35.100	100	179.946545	

(continued overleaf)

Z	Symbol	Nominal mass	%	Rel. %	Isotopic mass	Average mass
80	Hg	196	0.15	0.50	195.965807	200.59
		198	10.0	33.56	197.966743	
		199	16.9	56.71	198.968254	
		200	23.1	77.52	199.968300	
		201	13.2	44.30	200.970277	
		202	29.8	100	201.970617	
		204	6.85	22.99	203.973467	
53	I	127	100	100	126.904476	126.9045
49	In	113	4.3	4.49	112.904061	114.82
		115	95.7	100	114.903880	
77	Ir	191	37.3	59.49	190.960584	192.22
		193	62.7	100	192.962917	
19	K	39	93.2581	100	38.963707	39.0983
		40	0.0117	0.0125	39.963999	
		41	6.7302	7.22	40.961825	
36	Kr	78	0.35	0.614	77.920401	83.80
		80	2.25	3.947	79.916380	
		82	11.6	20.35	81.913482	
		83	11.5	20.175	82.914135	
		84	57.0	100	83.911507	
		86	17.3	30.35	85.910610	
57	La	138	0.090	0.09	137.90711	138.91
		139	99.91	100	138.906347	
3	Li	6	7.5	8.0108	6.015121	6.941
		7	92.5	100	7.016003	
12	Mg	24	78.99	100	23.985042	24.3050
		25	10.00	12.66	24.985837	
		26	11.01	13.94	25.982593	
25	Mn	55	100	100	54.938046	54.9380
42	Mo	92	14.84	61.50	91.906808	95.94
		94	9.25	38.33	93.905085	
		95	15.92	65.98	94.905840	
		96	16.68	69.13	95.904678	
		97	9.55	39.58	96.906020	
		98	24.13	100	97.905406	
		100	9.63	39.91	99.907477	
7	N	14	99.63	100	14.003074	14.00674
		15	0.37	0.37	15.000108	
11	Na	23	100	100	22.989768	22.9898
41	Nb	93	100	100	92.906377	92.906
10	Ne	20	90.48	100	19.992435	20.1797
		21	0.27	0.298	20.993843	
		22	9.25	10.22	21.991264	
28	Ni	58	68.27	100	57.935346	58.6934
		60	26.10	38.23	59.930788	
		61	1.13	1.66	60.931058	
		62	3.59	5.26	61.928346	
		64	0.91	1.33	63.927968	

Z	Symbol	Nominal mass	%	Rel. %	Isotopic mass	Average mass
8	O	16	99.76	100	15.994915	15.9994
		17	0.04	0.04	16.999133	
		18	0.20	0.20	17.999160	
76	Os	184	0.02	0.05	183.952488	190.2
		186	1.58	3.85	185.953830	
		187	1.6	3.90	186.955741	
		188	13.3	32.44	187.955860	
		189	16.1	39.27	188.958137	
		190	26.4	64.39	189.958436	
		192	41.0	100	191.961467	
15	P	31	100	100	30.973762	30.9738
82	Pb	204	1.4	2.67	203.973020	207.2
		206	24.1	45.99	205.974440	
		207	22.1	42.18	206.975872	
		208	52.4	100	207.976627	
46	Pd	102	1.02	3.73	101.905634	106.42
		104	11.14	40.76	103.904029	
		105	22.33	81.71	104.905079	
		106	27.33	100	105.903478	
		108	26.46	96.82	107.903895	
		110	11.72	42.88	109.905167	
78	Pt	190	0.01	0.03	189.959917	195.08
		192	0.79	2.34	191.961019	
		194	32.9	97.34	193.962655	
		195	33.8	100	194.964766	
		196	25.3	74.85	195.964926	
		198	7.2	21.30	197.967869	
37	Rb	85	72.17	100	84.911794	85.4678
		87	27.83	38.562	86.909187	
75	Re	185	37.40	59.74	184.952951	186.207
		187	62.60	100	186.955744	
45	Rh	103	100	100	102.905500	102.905
44	Ru	96	5.54	17.53	95.907599	101.07
		98	1.86	5.89	97.905267	
		99	12.7	40.19	98.905939	
		100	12.6	38.87	99.904219	
		101	17.1	54.11	100.905582	
		102	31.6	100	101.904348	
		104	18.6	58.86	103.905424	
16	S	32	95.03	100	31.972070	32.066
		33	0.75	0.789	32.971456	
		34	4.22	4.44	33.967866	
		36	0.02	0.021	35.967080	
51	Sb	121	57.4	100	120.903821	121.752
		123	42.6	74.22	122.904216	
21	Sc	45	100	100	44.955911	44.956

(*continued overleaf*)

Z	Symbol	Nominal mass	%	Rel. %	Isotopic mass	Average mass
34	Se	74	0.9	1.80	73.922475	78.96
		76	9.1	18.24	75.919212	
		77	7.6	15.23	76.919912	
		78	23.6	47.29	77.917309	
		80	49.9	100	79.916520	
		82	8.9	17.84	81.916698	
14	Si	28	92.21	100	27.976927	28.0855
		29	4.67	5.065	28.976495	
		30	3.10	3.336	29.973770	
50	Sn	112	0.97	2.98	111.904826	118.710
		114	0.65	1.99	113.902784	
		115	0.36	1.10	114.903348	
		116	14.53	43.58	115.901747	
		117	7.68	23.57	116.902956	
		118	24.22	73.32	117.901609	
		119	8.58	26.33	118.903310	
		120	32.59	100	119.902200	
		122	4.63	14.21	121.903440	
		124	5.79	17.77	123.905274	
38	Sr	84	0.56	0.68	83.913431	87.62
		86	9.86	11.94	85.909267	
		87	7.00	8.5	86.908884	
		88	82.58	100	87.905619	
73	Ta	180	0.012	0.012	179.947462	180.948
		181	99.988	100	180.947992	
43	Tc			100		
52	Te	120	0.095	0.28	119.904048	127.60
		122	2.59	7.65	121.903054	
		123	0.905	2.67	122.904271	
		124	4.79	14.14	123.902823	
		125	7.12	21.02	124.904433	
		126	18.93	55.89	125.903314	
		128	31.70	93.59	127.904463	
		130	33.87	100	129.906229	
90	Th	232	100	100	232.038054	232.038
22	Ti	46	8.00	10.84	45.952629	47.88
		47	7.3	9.892	46.951764	
		48	73.8	100	47.947947	
		49	5.51	7.466	48.947871	
		50	5.4	7.317	49.944792	
81	Tl	203	29.524	41.89	202.972320	204.383
		205	70.476	100	204.974401	
92	U	234	0.0055	0.0055	234.040946	238.03
		235	0.720	0.725	235.043924	
		238	99.2745	100	238.050784	
23	V	50	0.25	0.251	49.947161	50.9415
		51	99.75	100	50.943962	

Z	Symbol	Nominal mass	%	Rel. %	Isotopic mass	Average mass
74	W	180	0.12	0.39	179.946701	183.85
		182	26.3	85.67	181.948202	
		183	14.28	46.51	182.950220	
		184	30.7	100	183.950928	
		186	28.6	93.16	185.954357	
54	Xe	124	0.10	0.37	123.905894	131.29
		126	0.09	0.33	125.904281	
		128	1.91	7.10	127.903531	
		129	26.4	98.14	128.904780	
		130	4.1	15.24	129.903509	
		131	21.2	78.81	130.905072	
		132	26.9	100	131.904144	
		134	10.4	38.866	133.905395	
		136	8.9	33.09	135.907214	
39	Y	89	100	100	88.905849	88.906
30	Zn	64	48.6	100	63.929145	65.39
		66	27.9	57.41	65.926034	
		67	4.1	8.44	66.927129	
		68	18.8	38.68	67.924846	
		70	0.6	1.23	69.925325	
40	Zr	90	51.45	100	89.904703	91.224
		91	11.22	21.73	90.905643	
		92	17.15	33.33	91.905039	
		94	17.38	33.78	93.906314	
		96	2.80	5.44	95.908275	

Source: *Handbook of Chemistry and Physics*, 71st edition, CRC Press, Boca Raton, FL, 1990.

5 Isotopic Abundances (in %) for Various Elemental Compositions CHON (M = 100%)

	M + 1	M + 2	Mass		M + 1	M + 2	Mass
12				*16*			
C	1.11	0.00	12.0000	O	0.04	0.20	15.9949
				H_2N	0.40	0.00	16.0187
13				CH_4	1.17	0.00	16.0313
CH	1.13	0.00	13.0078				
				17			
14				HO	0.06	0.20	17.0027
N	0.37	0.00	14.0031	H_3N	0.42	0.00	17.0266
CH_2	1.14	0.00	14.0157				
				18			
15				H_2O	0.07	0.20	18.0106
HN	0.39	0.00	15.0109				
CH_3	1.16	0.00	15.0235				

	M + 1	M + 2	Mass		M + 1	M + 2	Mass
24				*33 (cont.)*			
C_2	2.22	0.01	24.0000	H_3NO	0.46	0.20	33.0215
25				*34*			
C_2H	2.24	0.01	25.0078	H_2O_2	0.11	0.40	34.0054
26				*36*			
CN	1.48	0.00	26.0031	C_3	3.33	0.04	36.0000
C_2H_2	2.25	0.01	26.0157				
				37			
27				C_3H	3.35	0.04	37.0078
CHN	1.50	0.00	27.0109				
C_2H_3	2.27	0.01	27.0235				
				38			
28				C_2N	2.59	0.02	38.0031
N_2	0.74	0.00	28.0062	C_3H_2	3.36	0.04	38.0157
CO	1.15	0.20	27.9949				
CH_2N	1.51	0.00	28.0187	*39*			
C_2H_4	2.28	0.01	28.0313	C_2HN	2.61	0.02	39.0109
				C_3H_3	3.38	0.04	39.0235
29							
HN_2	0.76	0.00	29.0140	*40*			
CHO	1.17	0.20	29.0027	CN_2	1.85	0.01	40.0062
CH_3N	1.53	0.00	29.0266	C_2O	2.26	0.21	39.9949
C_2H_5	2.30	0.01	29.0391	C_2H_2N	2.62	0.02	40.0187
				C_3H_4	3.39	0.04	40.0313
30							
NO	0.41	0.20	29.9980	*41*			
H_2N_2	0.77	0.00	30.0218	CHN_2	1.87	0.01	41.0140
CH_2O	1.18	0.20	30.0106	C_2HO	2.28	0.21	41.0027
CH_4N	1.54	0.01	30.0344	C_2H_3N	2.64	0.02	41.0266
C_2H_6	2.31	0.01	30.0470	C_3H_5	3.41	0.04	41.0391
31				*42*			
HNO	0.43	0.20	31.0058	N_3	1.11	0.00	42.0093
H_3N_2	0.79	0.00	31.0297	CNO	1.52	0.21	41.9980
CH_3O	1.20	0.20	31.0184	CH_2N_2	1.88	0.01	42.0218
CH_5N	1.56	0.01	31.0422	C_2H_2O	2.29	0.21	42.0106
				C_2H_4N	2.65	0.02	42.0344
32				C_3H_6	3.42	0.04	42.0470
O_2	0.08	0.40	31.9898				
H_2NO	0.44	0.20	32.0136	*43*			
H_4N_2	0.80	0.00	32.0375	HN_3	1.13	0.00	43.0171
CH_4O	1.21	0.20	32.0262	$CHNO$	1.54	0.21	43.0058
				CH_3N_2	1.90	0.01	43.0297
33				C_2H_3O	2.31	0.21	43.0184
HO_2	0.10	0.40	32.9976	C_2H_5N	2.67	0.02	43.0422

	M + 1	M + 2	Mass		M + 1	M + 2	Mass
43 (cont.)				*49 (cont.)*			
C_3H_7	3.44	0.04	43.0548	H_3NO_2	0.50	0.40	49.0164
				C_4H	4.46	0.07	49.0078
44							
N_2O	0.78	0.20	44.0011	*50*			
H_2N_3	1.14	0.00	44.0249	H_2O_3	0.15	0.60	50.0003
CO_2	1.19	0.40	43.9898	C_3N	3.70	0.05	50.0031
CH_2NO	1.55	0.21	44.0136	C_4H_2	4.47	0.07	50.0157
CH_4N_2	1.91	0.01	44.0375				
C_2H_4O	2.32	0.21	44.0262	*51*			
C_2H_6N	2.68	0.02	44.0501	C_3HN	3.72	0.05	51.0109
C_3H_8	3.45	0.04	44.0626	C_4H_3	4.49	0.08	51.0235
45				*52*			
HN_2O	0.80	0.20	45.0089	C_2N_2	2.96	0.03	52.0062
H_3N_3	1.16	0.00	45.0328	C_3O	3.37	0.24	51.9949
CHO_2	1.21	0.40	44.9976	C_3H_2N	3.73	0.05	52.0187
CH_3NO	1.57	0.21	45.0215	C_4H_4	4.50	0.08	52.0313
CH_5N_2	1.93	0.01	45.0453				
C_2H_5O	2.34	0.21	45.0340	*53*			
C_2H_7N	2.70	0.02	45.0579	C_2HN_2	2.98	0.03	53.0140
				C_3HO	3.39	0.24	53.0027
46				C_3H_3N	3.75	0.05	53.0266
NO_2	0.45	0.40	45.9929	C_4H_5	4.52	0.08	53.0391
H_2N_2O	0.81	0.20	46.0167				
H_4N_3	1.17	0.01	46.0406	*54*			
CH_2O_2	1.22	0.40	46.0054	CN_3	2.22	0.02	54.0093
CH_4NO	1.58	0.21	46.0293	C_2NO	2.63	0.22	53.9980
CH_6N_2	1.94	0.01	46.0532	$C_2H_2N_2$	2.98	0.03	54.0218
C_2H_6O	2.35	0.22	46.0419	C_3H_2O	3.40	0.24	54.0106
				C_3H_4N	3.76	0.05	54.0344
47				C_4H_6	4.53	0.08	54.0470
HNO_2	0.47	0.40	47.0007				
H_3N_2O	0.83	0.20	47.0248	*55*			
H_5N_3	1.19	0.01	47.0484	CHN_3	2.24	0.02	55.0171
CH_3O_2	1.24	0.40	47.0133	C_2HNO	2.65	0.22	55.0058
CH_5NO	1.60	0.21	47.0371	$C_2H_3N_2$	3.01	0.03	55.0297
				C_3H_3O	3.42	0.24	55.0184
48				C_3H_5N	3.78	0.05	55.0422
O_3	0.12	0.60	47.9847	C_4H_7	4.55	0.08	55.0548
H_2NO_2	0.48	0.40	48.0085				
H_4N_2O	0.84	0.20	48.0324	*56*			
CH_4O_2	1.25	0.40	48.0211	N_4	1.48	0.01	56.0124
C_4	4.44	0.07	48.0000	CN_2O	1.89	0.21	56.0011
				CH_2N_3	2.25	0.02	56.0249
49				C_2O_2	2.30	0.41	55.9898
HO_3	0.14	0.60	48.9925	C_2H_2NO	2.66	0.22	56.0136

	M + 1	M + 2	Mass		M + 1	M + 2	Mass
56 (cont.)				*60 (cont.)*			
$C_2H_4N_2$	3.02	0.03	56.0375	CH_2NO_2	1.59	0.41	60.0085
C_3H_4O	3.43	0.24	56.0262	CH_4N_2O	1.95	0.21	60.0324
C_3H_6N	3.79	0.05	56.0501	CH_6N_3	2.31	0.02	60.0563
C_4H_8	4.56	0.08	56.0626	$C_2H_4O_2$	2.36	0.42	60.0211
				C_2H_6NO	2.72	0.22	60.0449
57				$C_2H_8N_2$	3.08	0.03	60.0688
HN_4	1.50	0.01	57.0202	C_3H_8O	3.49	0.24	60.0575
CHN_2O	1.91	0.21	57.0089	C_5	5.55	0.12	60.0000
CH_3N_3	2.27	0.02	57.0328				
C_2HO_2	2.32	0.41	56.9976	*61*			
C_2H_3NO	2.68	0.22	57.0215	HN_2O_2	0.84	0.40	61.0038
$C_2H_5N_2$	3.04	0.03	57.0453	H_3N_3O	1.20	0.21	61.0277
C_3H_5O	3.45	0.24	57.0340	H_5N_4	1.56	0.01	61.0515
C_3H_7N	3.81	0.05	57.0579	CHO_3	1.25	0.60	60.9925
C_4H_9	4.58	0.08	57.0705	CH_3NO_2	1.61	0.41	61.0164
				CH_5N_2O	1.97	0.21	61.0402
58				CH_7N_3	2.33	0.02	61.0641
N_3O	1.15	0.20	58.0042	$C_2H_5O_2$	2.38	0.42	61.0289
H_2N_4	1.51	0.01	58.0280	C_2H_7NO	2.74	0.22	61.0528
CNO_2	1.56	0.41	57.9929	C_5H	5.57	0.12	61.0078
CH_2N_2O	1.92	0.21	58.0167				
CH_4N_3	2.28	0.02	58.0406	*62*			
$C_2H_2O_2$	2.33	0.42	58.0054	NO_3	0.49	0.60	61.9878
C_2H_4NO	2.69	0.22	58.0293	$H_2N_2O_2$	0.85	0.40	62.0116
$C_2H_6N_2$	3.05	0.03	58.0532	H_4N_3O	1.21	0.42	62.0368
C_3H_6O	3.46	0.24	58.0419	H_6N_4	1.57	0.01	62.0594
C_3H_8N	3.82	0.05	58.0657	CH_2O_3	1.26	0.60	62.0003
C_4H_{10}	4.59	0.08	58.0783	CH_4NO_2	1.62	0.41	62.0242
				CH_6N_2O	1.98	0.21	62.0480
59				$C_2H_6O_2$	2.39	0.42	62.0368
HN_3O	1.17	0.20	59.0120	C_4N	4.81	0.09	62.0031
H_3N_4	1.53	0.01	59.0359	C_5H_2	5.58	0.12	62.0157
$CHNO_2$	1.58	0.41	59.0007				
CH_3N_2O	1.94	0.21	59.0246	*63*			
CH_5N_3	2.30	0.02	59.0484	HNO_3	0.51	0.60	62.9956
$C_2H_3O_2$	2.35	0.42	59.0133	$H_3N_2O_2$	0.87	0.40	63.0195
C_2H_5NO	2.71	0.22	59.0371	H_5N_3O	1.23	0.21	63.0433
$C_2H_7N_2$	3.07	0.03	59.0610	CH_3O_3	1.28	0.60	63.0082
C_3H_7O	3.48	0.24	59.0497	CH_5NO_2	1.64	0.41	63.0320
C_3H_9N	3.84	0.05	59.0736	C_4HN	4.83	0.09	63.0109
				C_5H_3	5.60	0.12	63.0235
60							
N_2O_2	0.82	0.40	59.9960	*64*			
H_2N_3O	1.18	0.20	60.0198	O_4	0.16	0.80	63.9796
H_4N_4	1.54	0.01	60.0437	H_2NO_3	0.52	0.60	64.0034
CO_3	1.23	0.60	59.9847	$H_4N_2O_2$	0.88	0.40	64.0273

	M + 1	M + 2	Mass		M + 1	M + 2	Mass
64 (cont.)				*69 (cont.)*			
CH4O3	1.29	0.60	64.0160	C3HO2	3.43	0.44	68.9976
C3N2	4.07	0.06	64.0062	C3H3NO	3.79	0.25	69.0215
C4O	4.48	0.27	63.9949	C3H5N2	4.15	0.06	69.0453
C4H2N	4.84	0.09	64.0187	C4H5O	4.56	0.28	69.0340
C5H4	5.61	0.12	64.0313	C4H7N	4.92	0.09	69.0579
				C5H9	5.69	0.13	69.0705
65							
HO4	0.18	0.80	64.9874	*70*			
H3NO3	0.54	0.60	65.0113	CN3O	2.26	0.22	70.0042
C3HN2	4.09	0.06	65.0140	CH2N4	2.62	0.03	70.0280
C4HO	4.50	0.27	65.0027	C2NO2	2.67	0.42	69.9929
C4H3N	4.86	0.09	65.0266	C2H2N2O	3.03	0.23	70.0167
C5H5	5.63	0.12	65.0391	C2H4N3	3.39	0.04	70.0406
				C3H2O2	3.44	0.44	70.0054
66				C3H4NO	3.80	0.25	70.0293
H2O4	0.19	0.80	65.9953	C3H6N2	4.16	0.07	70.0532
C2N3	3.33	0.04	66.0093	C4H6O	4.57	0.28	70.0419
C3NO	3.74	0.25	65.9980	C4H8N	4.93	0.09	70.0657
C3H2N2	4.10	0.06	66.0218	C5H10	5.70	0.13	70.0783
C4H2O	4.51	0.27	66.0106				
C4H4N	4.87	0.09	66.0344	*71*			
C5H6	5.64	0.12	66.0470	CHN3O	2.28	0.22	71.0120
				CH3N4	2.64	0.03	71.0359
67				C2HNO2	2.69	0.42	71.0007
C2HN3	3.35	0.04	67.0171	C2H3N2O	3.05	0.23	71.0246
C3HNO	3.76	0.25	67.0058	C2H5N3	3.41	0.04	71.0484
C3H3N2	4.12	0.06	67.0297	C3H3O2	3.46	0.44	71.0133
C4H3O	4.53	0.27	67.0184	C3H5NO	3.82	0.25	71.0371
C4H5N	4.89	0.09	67.0422	C3H7N2	4.18	0.07	71.0610
C5H7	5.66	0.12	67.0548	C4H7O	4.59	0.28	71.0497
				C4H9N	4.95	0.10	71.0736
68				C5H11	5.72	0.13	71.0861
CN4	2.59	0.02	68.0124				
C2N2O	3.00	0.23	68.0011	*72*			
C2H2N3	3.36	0.04	68.0249	N4O	1.52	0.21	72.0073
C3O2	3.41	0.44	67.9898	CN2O2	1.93	0.41	71.9960
C3H2NO	3.77	0.25	68.0136	CH2N3O	2.29	0.22	72.0198
C3H4N2	4.13	0.06	68.0375	CH4N4	2.65	0.03	72.0437
C4H4O	4.54	0.28	68.0262	C2O3	2.34	0.62	71.9847
C4H6N	4.90	0.09	68.0501	C2H2NO2	2.70	0.42	72.0085
C5H8	5.67	0.13	68.0626	C2H4N2O	3.06	0.23	72.0324
				C2H6N3	3.42	0.04	72.0563
69				C3H4O2	3.47	0.44	72.0211
CHN4	2.61	0.03	69.0202	C3H6NO	3.83	0.25	72.0449
C2HN2O	3.02	0.23	69.0089	C3H8N2	4.19	0.07	72.0688
C2H3N3	3.38	0.04	69.0328	C4H8O	4.60	0.28	72.0575

	M + 1	M + 2	Mass		M + 1	M + 2	Mass
72 (cont.)				*75 (cont.)*			
$C_4H_{10}N$	4.96	0.09	72.0814	$C_2H_5NO_2$	2.75	0.43	75.0320
C_5H_{12}	5.73	0.13	72.0939	$C_2H_7N_2O$	3.11	0.23	75.0559
C_6	6.66	0.18	72.0000	$C_2H_9N_3$	3.47	0.05	75.0798
				$C_3H_7O_2$	3.52	0.44	75.0446
73				C_3H_9NO	3.88	0.25	75.0684
HN_4O	1.54	0.21	73.0151	C_5HN	5.94	0.14	75.0109
CHN_2O_2	1.95	0.41	73.0038	C_6H_3	6.71	0.18	75.0235
CH_3N_3O	2.31	0.22	73.0277				
CH_5N_4	2.67	0.03	73.0515	*76*			
C_2HO_3	2.36	0.62	72.9925	N_2O_3	0.86	0.60	75.9909
$C_2H_3NO_2$	2.72	0.42	73.0164	$H_2N_3O_2$	1.22	0.41	76.0147
$C_2H_5N_2O$	3.08	0.23	73.0402	H_4N_4O	1.58	0.21	76.0386
$C_2H_7N_3$	3.44	0.04	73.0641	CO_4	1.27	0.80	75.9796
$C_3H_5O_2$	3.49	0.44	73.0289	CH_2NO_3	1.63	0.61	76.0034
C_3H_7NO	3.85	0.25	73.0528	$CH_4N_2O_2$	1.99	0.41	76.0273
$C_3H_9N_2$	4.21	0.07	73.0767	CH_6N_3O	2.35	0.22	76.0511
C_4H_9O	4.62	0.28	73.0653	CH_8N_4	2.71	0.03	76.0750
$C_4H_{11}N$	4.98	0.09	73.0892	$C_2H_4O_3$	2.40	0.62	76.0160
C_6H	6.68	0.18	73.0078	$C_2H_6NO_2$	2.76	0.43	76.0399
				$C_2H_8N_2O$	3.12	0.24	76.0637
74				$C_3H_8O_2$	3.53	0.44	76.0524
N_3O_2	1.19	0.41	73.9991	C_4N_2	5.18	0.10	76.0062
H_2N_4O	1.55	0.21	74.0229	C_5O	5.59	0.32	75.9949
CNO_3	1.60	0.61	73.9878	C_5H_2N	5.95	0.14	76.0187
$CH_2N_2O_2$	1.96	0.41	74.0116	C_6H_4	6.72	0.19	76.0313
CH_4N_3O	2.32	0.22	74.0355				
CH_6N_4	2.68	0.03	74.0594	*77*			
$C_2H_2O_3$	2.37	0.62	74.0003	HN_2O_3	0.88	0.60	76.9987
$C_2H_4NO_2$	2.73	0.42	74.0242	$H_3N_3O_2$	1.24	0.41	77.0226
$C_2H_6N_2O$	3.09	0.23	74.0480	H_5N_4O	1.60	0.21	77.0464
$C_2H_8N_3$	3.45	0.05	74.0719	CHO_4	1.29	0.80	76.9874
$C_3H_6O_2$	3.50	0.44	74.0368	CH_3NO_3	1.65	0.61	77.0113
C_3H_8NO	3.86	0.25	74.0606	$CH_5N_2O_2$	2.01	0.41	77.0351
$C_3H_{10}N_2$	4.22	0.07	74.0845	CH_7N_3O	2.37	0.22	77.0590
$C_4H_{10}O$	4.63	0.28	74.0732	$C_2H_5O_3$	2.42	0.62	77.0238
C_5N	5.92	0.14	74.0031	$C_2H_7NO_2$	2.78	0.43	77.0477
C_6H_2	6.69	0.18	74.0157	C_4HN_2	5.20	0.11	77.0140
				C_5HO	5.61	0.32	77.0027
75				C_5H_3N	5.97	0.15	77.0266
HN_3O_2	1.21	0.41	75.0069	C_6H_5	6.74	0.19	77.0391
H_3N_4O	1.57	0.21	75.0308				
$CHNO_3$	1.62	0.61	74.9956	*78*			
$CH_3N_2O_2$	1.98	0.41	75.0195	NO_4	0.53	0.80	77.9827
CH_5N_3O	2.34	0.22	75.0433	$H_2N_2O_3$	0.89	0.60	78.0065
CH_7N_4	2.70	0.03	75.0672	$H_4N_3O_2$	1.25	0.41	78.0304
$C_2H_3O_3$	2.39	0.62	75.0082	H_6N_4O	1.61	0.21	78.0542

	M + 1	M + 2	Mass		M + 1	M + 2	Mass
78 (cont.)				*81 (cont.)*			
CH$_2$O$_4$	1.30	0.80	77.9953	C$_4$H$_5$N$_2$	5.26	0.11	81.0453
CH$_4$NO$_3$	1.66	0.61	78.0191	C$_5$H$_5$O	5.67	0.32	81.0340
CH$_6$N$_2$O$_2$	2.02	0.41	78.0429	C$_5$H$_7$N	6.03	0.14	81.0579
C$_2$H$_6$O$_3$	2.43	0.62	78.0317	C$_6$H$_9$	6.80	0.19	81.0705
C$_3$N$_3$	4.44	0.08	78.0093				
C$_4$NO	4.85	0.29	77.9980	*82*			
C$_4$H$_2$N$_2$	5.21	0.11	78.0218	C$_2$N$_3$O	3.37	0.24	82.0042
C$_5$H$_2$O	5.62	0.32	78.0106	C$_2$H$_2$N$_4$	3.73	0.05	82.0280
C$_5$H$_4$N	5.98	0.14	78.0344	C$_3$NO$_2$	3.78	0.45	81.9929
C$_6$H$_6$	6.75	0.19	78.0470	C$_3$H$_2$N$_2$O	4.14	0.26	82.0167
				C$_3$H$_4$N$_3$	4.50	0.08	82.0406
79				C$_4$H$_2$O$_2$	4.55	0.48	82.0054
HNO$_4$	0.55	0.80	78.9905	C$_4$H$_4$NO	4.91	0.29	82.0293
H$_3$N$_2$O$_3$	0.91	0.60	79.0144	C$_4$H$_6$N$_2$	5.27	0.11	82.0532
H$_5$N$_3$O$_2$	1.27	0.41	79.0382	C$_5$H$_6$O	5.68	0.32	82.0419
CH$_3$O$_4$	1.32	0.80	79.0031	C$_5$H$_8$N	6.04	0.14	82.0657
CH$_5$NO$_3$	1.68	0.61	79.0269	C$_6$H$_{10}$	6.81	0.19	82.0783
C$_3$HN$_3$	4.46	0.08	79.0171				
C$_4$HNO	4.87	0.29	79.0058	*83*			
C$_4$H$_3$N$_2$	5.23	0.11	79.0297	C$_2$HN$_3$O	3.39	0.24	83.0120
C$_5$H$_3$O	5.64	0.32	79.0184	C$_2$H$_3$N$_4$	3.75	0.06	83.0359
C$_5$H$_5$N	6.00	0.14	79.0422	C$_3$HNO$_2$	3.80	0.45	83.0007
C$_6$H$_7$	6.77	0.19	79.0548	C$_3$H$_3$N$_2$O	4.16	0.27	83.0246
				C$_3$H$_5$N$_3$	4.52	0.08	83.0484
80				C$_4$H$_3$O$_2$	4.57	0.48	83.0133
H$_2$NO$_4$	0.56	0.80	79.9983	C$_4$H$_5$NO	4.93	0.29	83.0371
H$_4$N$_2$O$_3$	0.92	0.60	80.0222	C$_4$H$_7$N$_2$	5.29	0.11	83.0610
CH$_4$O$_4$	1.33	0.80	80.0109	C$_5$H$_7$O	5.70	0.33	83.0497
C$_2$N$_4$	3.70	0.05	80.0124	C$_5$H$_9$N	6.06	0.15	83.0736
C$_3$N$_2$O	4.11	0.26	80.0011	C$_6$H$_{11}$	6.83	0.19	83.0861
C$_3$H$_2$N$_3$	4.47	0.08	80.0249				
C$_4$O$_2$	4.52	0.47	79.9898	*84*			
C$_4$O$_2$	4.52	0.47	79.9898	CN$_4$O	2.63	0.23	84.0073
C$_4$H$_2$NO	4.88	0.29	80.0136	C$_2$H$_2$N$_3$O	3.40	0.24	84.0198
C$_4$H$_4$N$_2$	5.24	0.11	80.0375	C$_2$N$_2$O$_2$	3.04	0.43	83.9960
C$_5$H$_4$O	5.65	0.32	80.0262	C$_2$H$_4$N$_4$	3.76	0.06	84.0437
C$_5$H$_6$N	6.01	0.14	80.0501	C$_3$O$_3$	3.45	0.64	83.9847
C$_6$H$_8$	6.78	0.19	80.0626	C$_3$H$_2$NO$_2$	3.81	0.45	84.0085
				C$_3$H$_4$N$_2$O	4.17	0.27	84.0324
81				C$_3$H$_6$N$_3$	4.53	0.08	84.0563
H$_3$NO$_4$	0.58	0.80	81.0062	C$_4$H$_4$O$_2$	4.58	0.48	84.0211
C$_2$HN$_4$	3.72	0.05	81.0202	C$_4$H$_6$NO	4.94	0.29	84.0449
C$_3$HN$_2$O	4.13	0.26	81.0089	C$_4$H$_8$N$_2$	5.30	0.11	84.0688
C$_3$H$_3$N$_3$	4.49	0.08	81.0328	C$_5$H$_8$O	5.71	0.33	84.0575
C$_4$HO$_2$	4.54	0.48	80.9976	C$_5$H$_{10}$N	6.07	0.15	84.0814
C$_4$H$_3$NO	4.90	0.29	81.0215	C$_6$H$_{12}$	6.84	0.19	84.0939

	M + 1	M + 2	Mass		M + 1	M + 2	Mass
84 (cont.)				*87 (cont.)*			
C_7	7.77	0.26	84.0000	$C_3H_3O_3$	3.50	0.64	87.0082
				$C_3H_5NO_2$	3.86	0.45	87.0320
85				$C_3H_7N_2O$	4.22	0.27	87.0559
CHN_4O	2.65	0.23	85.0151	$C_3H_9N_3$	4.58	0.08	87.0798
$C_2HN_2O_2$	3.06	0.43	85.0038	$C_4H_7O_2$	4.63	0.48	87.0446
$C_2H_3N_3O$	3.42	0.24	85.0277	C_4H_9NO	4.99	0.30	87.0684
$C_2H_5N_4$	3.78	0.06	85.0515	$C_4H_{11}N_2$	5.35	0.11	87.0923
C_3HO_3	3.47	0.64	84.9925	$C_5H_{11}O$	5.76	0.33	87.0810
$C_3H_3NO_2$	3.83	0.45	85.0164	$C_5H_{13}N$	6.12	0.15	87.1049
$C_3H_5N_2O$	4.19	0.27	85.0402	C_6HN	7.05	0.21	87.0109
$C_3H_7N_3$	4.55	0.08	85.0641	C_7H_3	7.82	0.26	87.0235
$C_4H_5O_2$	4.60	0.48	85.0289				
C_4H_7NO	4.96	0.29	85.0528	*88*			
$C_4H_9N_2$	5.32	0.11	85.0767	N_4O_2	1.56	0.41	88.0022
C_5H_9O	5.73	0.33	85.0653	CN_2O_3	1.97	0.61	87.9909
$C_5H_{11}N$	6.09	0.16	85.0892	$CH_2N_3O_2$	2.33	0.42	88.0147
C_6H_{13}	6.86	0.20	85.1018	CH_4N_4O	2.69	0.23	88.0386
C_7H	7.79	0.26	85.0078	C_2O_4	2.38	0.82	87.9796
				$C_2H_2NO_3$	2.74	0.63	88.0034
86				$C_2H_4N_2O_2$	3.10	0.43	88.0273
CN_3O_2	2.30	0.41	85.9991	$C_2H_6N_3O$	3.46	0.25	88.0511
CH_2N_4O	2.66	0.21	86.0229	$C_2H_8N_4$	3.82	0.06	88.0750
C_2NO_3	2.71	0.62	85.9878	$C_3H_4O_3$	3.51	0.64	88.0160
$C_2H_2N_2O_2$	3.07	0.43	86.0116	$C_3H_6NO_2$	3.87	0.45	88.0399
$C_2H_4N_3O$	3.43	0.24	86.0355	$C_3H_8N_2O$	4.23	0.27	88.0637
$C_2H_6N_4$	3.79	0.06	86.0594	$C_3H_{10}N_3$	4.59	0.08	88.0876
$C_3H_2O_3$	3.48	0.64	86.0003	$C_4H_8O_2$	4.64	0.48	88.0524
$C_3H_4NO_2$	3.84	0.45	86.0242	$C_4H_{10}NO$	5.00	0.30	88.0763
$C_3H_6N_2O$	4.20	0.27	86.0480	$C_4H_{12}N_2$	5.36	0.11	88.1001
$C_3H_8N_3$	4.56	0.08	86.0719	$C_5H_{12}O$	5.77	0.33	88.0888
$C_4H_6O_2$	4.61	0.48	86.0368	C_5N_2	6.29	0.16	88.0062
C_4H_8NO	4.97	0.30	86.0606	C_6O	6.70	0.38	87.9949
$C_4H_{10}N_2$	5.33	0.11	86.0845	C_6H_2N	7.06	0.21	88.0187
$C_5H_{10}O$	5.74	0.33	86.0732	C_7H_4	7.83	0.26	88.0313
$C_5H_{12}N$	6.10	0.16	86.0970				
C_6H_{14}	6.87	0.21	86.1096	*89*			
C_6N	7.03	0.21	86.0031	HN_4O_2	1.58	0.41	89.0100
C_7H_2	7.80	0.26	86.0157	CHN_2O_3	1.99	0.61	88.9987
				$CH_3N_3O_2$	2.35	0.42	89.0226
87				CH_5N_4O	2.71	0.23	89.0464
CHN_3O_2	2.32	0.42	87.0069	C_2HO_4	2.40	0.82	88.9874
CH_3N_4O	2.68	0.23	87.0308	$C_2H_3NO_3$	2.76	0.63	89.0113
C_2HNO_3	2.73	0.62	86.9956	$C_2H_5N_2O_2$	3.12	0.44	89.0351
$C_2H_3N_2O_2$	3.09	0.43	87.0195	$C_2H_7N_3O$	3.48	0.25	89.0590
$C_2H_5N_3O$	3.45	0.25	87.0433	$C_2H_9N_4$	3.84	0.06	89.0829
$C_2H_7N_4$	3.81	0.06	87.0672	$C_3H_5O_3$	3.53	0.64	89.0238

	M + 1	M + 2	Mass		M + 1	M + 2	Mass
89 (cont.)				*91 (cont.)*			
$C_3H_7NO_2$	3.89	0.46	89.0477	C_4HN_3	5.57	0.13	91.0171
$C_3H_9N_2O$	4.25	0.27	89.0715	C_5HNO	5.98	0.34	91.0058
$C_3H_{11}N_3$	4.61	0.08	89.0954	$C_5H_3N_2$	6.34	0.17	91.0297
$C_4H_9O_2$	4.66	0.48	89.0603	C_6H_3O	6.75	0.38	91.0184
$C_4H_{11}NO$	5.02	0.30	89.0841	C_6H_5N	7.11	0.21	91.0422
C_5HN_2	6.31	0.16	89.0140	C_7H_7	7.88	0.26	91.0548
C_6HO	6.72	0.38	89.0027				
C_6H_3N	7.08	0.21	89.0266	*92*			
C_7H_5	7.85	0.26	89.0391	N_2O_4	0.90	0.80	91.9858
				$H_2N_3O_3$	1.26	0.60	92.0096
90				$H_4N_4O_2$	1.62	0.41	92.0335
N_3O_3	1.23	0.60	89.9940	CH_2NO_4	1.67	0.81	91.9983
$H_2N_4O_2$	1.59	0.40	90.0178	$CH_4N_2O_3$	2.03	0.61	92.0222
CNO_4	1.64	0.80	89.9827	$CH_6N_3O_2$	2.39	0.42	92.0460
$CH_2N_2O_3$	2.00	0.61	90.0065	CH_8N_4O	2.75	0.23	92.0699
$CH_4N_3O_2$	2.36	0.42	90.0304	$C_2H_4O_4$	2.44	0.82	92.0109
CH_6N_4O	2.72	0.23	90.0542	$C_2H_6NO_3$	2.80	0.63	92.0348
$C_2H_2O_4$	2.41	0.82	89.9953	$C_2H_8N_2O_2$	3.16	0.44	92.0586
$C_2H_4NO_3$	2.77	0.63	90.0191	$C_3H_8O_3$	3.57	0.64	92.0473
$C_2H_6N_2O_2$	3.13	0.44	90.0429	C_3N_4	4.81	0.09	92.0124
$C_2H_8N_3O$	3.49	0.25	90.0668	C_6H_4N	7.09	0.21	90.0344
$C_2H_{10}N_4$	3.85	0.06	90.0907	C_7H_6	7.86	0.26	90.0470
$C_3H_6O_3$	3.54	0.64	90.0317	C_4N_2O	5.22	0.31	92.0011
$C_3H_8NO_2$	3.90	0.46	90.0555	$C_4H_2N_3$	5.58	0.13	92.0249
$C_3H_{10}N_2O$	4.26	0.27	90.0794	C_5O_2	5.63	0.52	91.9898
$C_4H_{10}O_2$	4.67	0.48	90.0681	C_5H_2NO	5.99	0.34	92.0136
C_4N_3	5.55	0.13	90.0093	$C_5H_4N_2$	6.35	0.17	92.0375
C_5NO	5.96	0.34	89.9980	C_6H_4O	5.76	0.38	92.0262
$C_5H_2N_2$	6.32	0.17	90.0218	C_6H_6N	7.12	0.21	92.0501
C_6H_2O	6.73	0.38	90.0106	C_7H_8	7.89	0.27	92.0626
C_6H_4N	7.09	0.21	90.0344				
C_7H_6	7.86	0.26	90.0470	*93*			
				HN_2O_4	0.92	0.80	92.9936
91				$H_3N_3O_3$	1.28	0.60	93.0175
HN_3O_3	1.25	0.60	91.0018	$H_5N_4O_2$	1.64	0.41	93.0413
$H_3N_4O_2$	1.61	0.41	91.0257	CH_3NO_4	1.69	0.81	93.0062
$CHNO_4$	1.66	0.81	90.9905	$CH_5N_2O_3$	2.05	0.61	93.0300
$CH_3N_2O_3$	2.02	0.61	91.0144	$CH_7N_3O_2$	2.41	0.42	93.0539
$CH_5N_3O_2$	2.38	0.42	91.0382	$C_2H_5O_4$	2.46	0.82	93.0187
CH_7N_4O	2.74	0.23	91.0621	$C_2H_7NO_3$	2.82	0.63	93.0426
$C_2H_3O_4$	2.43	0.82	91.0031	C_3HN_4	4.83	0.09	93.0202
$C_2H_5NO_3$	2.79	0.63	91.0269	C_4HN_2O	5.24	0.31	93.0089
$C_2H_7N_2O_2$	3.15	0.44	91.0508	$C_4H_3N_3$	5.60	0.13	93.0328
$C_2H_9N_3O$	3.51	0.25	91.0746	C_5HO_2	5.65	0.52	92.9976
$C_3H_7O_3$	3.56	0.64	91.0395	C_5H_3NO	6.01	0.35	93.0215
$C_3H_9NO_2$	3.92	0.46	91.0634	$C_5H_5N_2$	6.37	0.17	93.0453

	M + 1	M + 2	Mass		M + 1	M + 2	Mass
93 (cont.)				96 (cont.)			
C_6H_5O	6.78	0.38	93.0340	$C_4H_2NO_2$	4.92	0.49	96.0085
C_6H_7N	7.14	0.22	93.0579	$C_4H_4N_2O$	5.28	0.31	96.0324
C_7H_9	7.91	0.27	93.0705	$C_4H_6N_3$	5.64	0.13	96.0563
				$C_5H_4O_2$	5.69	0.53	96.0211
94				C_5H_6NO	6.05	0.35	96.0449
$H_2N_2O_4$	0.93	0.80	94.0014	$C_5H_8N_2$	6.41	0.17	96.0688
$H_4N_3O_3$	1.29	0.61	94.0253	C_6H_8O	6.82	0.39	96.0575
$H_6N_4O_2$	1.65	0.41	94.0491	$C_6H_{10}N$	7.18	0.22	96.0814
CH_4NO_4	1.70	0.81	94.0140	C_7H_{12}	7.95	0.27	96.0939
$CH_6N_2O_3$	2.06	0.62	94.0379	C_8	8.88	0.34	96.0000
$C_2H_6O_4$	2.47	0.82	94.0266				
C_3N_3O	4.48	0.28	94.0042	97			
$C_3H_2N_4$	4.84	0.09	94.0280	C_2HN_4O	3.76	0.26	97.0151
C_4NO_2	4.89	0.49	93.9929	$C_3HN_2O_2$	4.17	0.47	97.0038
$C_4H_2N_2O$	5.25	0.31	94.0167	$C_3H_3N_3O$	4.53	0.28	97.0277
$C_4H_4N_3$	5.61	0.13	94.0406	$C_3H_5N_4$	4.89	0.10	97.0515
$C_5H_2O_2$	5.66	0.52	94.0054	C_4HO_3	4.58	0.68	96.9925
C_5H_4NO	6.02	0.35	94.0293	$C_4H_3NO_2$	4.94	0.49	97.0164
$C_5H_6N_2$	6.38	0.17	94.0532	$C_4H_5N_2O$	5.30	0.31	97.0402
C_6H_6O	6.79	0.38	94.0419	$C_4H_7N_3$	5.66	0.13	97.0641
C_6H_8N	7.15	0.22	94.0657	$C_5H_5O_2$	5.71	0.53	97.0289
C_7H_{10}	7.92	0.27	94.0783	C_5H_7NO	6.07	0.35	97.0528
				$C_5H_9N_2$	6.43	0.17	97.0767
95				C_6H_9O	6.84	0.39	97.0653
$H_3N_2O_4$	0.95	0.80	95.0093	$C_6H_{11}N$	7.20	0.22	97.0892
$H_5N_3O_3$	1.31	0.60	95.0331	C_7H_{13}	7.97	0.27	97.1018
CH_5NO_4	1.72	0.81	95.0218	C_8H	8.90	0.34	97.0078
C_3HN_3O	4.50	0.28	95.0120				
$C_3H_3N_4$	4.86	0.10	95.0359	98			
C_4HNO_2	4.91	0.49	95.0007	$C_2N_3O_2$	3.41	0.44	97.9991
$C_4H_3N_2O$	5.27	0.31	95.0246	$C_2H_2N_4O$	3.77	0.26	98.0229
$C_4H_5N_3$	5.63	0.13	95.0484	C_3NO_3	3.82	0.65	97.9878
$C_5H_3O_2$	5.68	0.52	95.0133	$C_3H_2N_2O_2$	4.18	0.47	98.0116
C_5H_5NO	6.04	0.35	95.0371	$C_3H_4N_3O$	4.54	0.28	98.0355
$C_5H_7N_2$	6.40	0.17	95.0610	$C_3H_6N_4$	4.90	0.10	98.0594
C_6H_7O	6.81	0.39	95.0497	$C_4H_2O_3$	4.59	0.68	98.0003
C_6H_9N	7.17	0.22	95.0736	$C_4H_4NO_2$	4.95	0.49	98.0242
C_7H_{11}	7.94	0.27	95.0861	$C_4H_6N_2O$	5.31	0.31	98.0480
				$C_4H_8N_3$	5.67	0.13	98.0719
96				$C_5H_6O_2$	5.72	0.53	98.0368
$H_4N_2O_4$	0.96	0.80	96.0171	C_5H_8NO	6.08	0.35	98.0606
C_2N_4O	3.74	0.26	96.0073	$C_5H_{10}N_2$	6.44	0.17	98.0645
$C_3N_2O_2$	4.15	0.47	95.9960	$C_6H_{10}O$	6.85	0.39	98.0732
$C_3H_2N_3O$	4.51	0.28	96.0198	$C_6H_{12}N$	7.21	0.21	98.0970
$C_3H_4N_4$	4.87	0.10	96.0437	C_7H_{14}	7.98	0.26	98.1096
C_4O_3	4.56	0.67	95.9847	C_7N	8.14	0.27	98.0031

	M + 1	M + 2	Mass		M + 1	M + 2	Mass
98 (cont.)				*100 (cont.)*			
C_8H_2	8.91	0.33	98.0157	$C_2N_2O_3$	3.08	0.63	99.9909
				$C_2H_2N_3O_2$	3.44	0.45	100.0147
99				$C_2H_4N_4O$	3.80	0.26	100.0386
$C_2HN_3O_2$	3.43	0.44	99.0069	C_3O_4	3.45	0.84	99.9796
$C_2H_3N_4O$	3.79	0.25	99.0308	$C_3H_2NO_3$	3.85	0.65	100.0034
C_3HNO_3	3.84	0.65	98.9956	$C_3H_4N_2O_2$	4.21	0.47	100.0273
$C_3H_3N_2O_2$	4.20	0.47	99.0195	$C_3H_6N_3O$	4.57	0.28	100.0511
$C_3H_5N_3O$	4.56	0.28	99.0433	$C_3H_8N_4$	4.94	0.10	100.0750
$C_3H_7N_4$	4.92	0.10	99.0672	$C_4H_4O_3$	4.62	0.68	100.0160
$C_4H_3O_3$	4.61	0.68	99.0082	$C_4H_6NO_2$	4.98	0.49	100.0399
$C_4H_5NO_2$	4.97	0.49	99.0320	$C_4H_8N_2O$	5.34	0.31	100.0637
$C_4H_7N_2O$	5.33	0.31	99.0559	$C_4H_{10}N_3$	5.70	0.13	100.0876
$C_4H_9N_3$	5.69	0.13	99.0798	$C_5H_8O_2$	5.76	0.53	100.0524
$C_5H_7O_2$	5.74	0.53	99.0446	$C_5H_{10}NO$	6.11	0.35	100.0763
C_5H_9NO	6.11	0.35	99.0684	$C_5H_{12}N_2$	6.47	0.18	100.1001
$C_5H_{11}N_2$	6.46	0.17	99.0923	$C_6H_{12}O$	6.88	0.39	100.0888
$C_6H_{11}O$	6.86	0.39	99.0810	$C_6H_{14}N$	7.24	0.22	100.1127
$C_6H_{13}N$	7.23	0.22	99.1049	C_6N_2	7.40	0.23	100.0062
C_7H_{15}	8.00	0.27	99.1174	C_7H_{16}	8.01	0.28	100.1253
C_7HN	8.16	0.29	99.0109	C_7O	7.81	0.46	99.9949
C_8H_3	8.93	0.35	99.0235	C_7H_2N	8.17	0.29	100.0187
				C_8H_4	8.94	0.35	100.0313
100							
CN_4O_2	2.67	0.43	100.0022				

6 Gas-phase Ion Thermochemical Data of Molecules

Molecules	IE (eV)[a]	EA (eV)[b]	PA (kJ mol^{-1})[c]	GB (kJ mole^{-1})[c]	ΔH°_{acid} (kJ mole^{-1})[d]	ΔG°_{acid} (kJ mole^{-1})[d]
Ar	15.6	—	369	346	—	—
Br_2	10.5	2.4	—	—	—	—
CO	14	1.4	426	402	—	—
CO_2	13.8	−0.6	540	515	—	—
CS_2	10	0.5	682	658	—	—
Cl_2	11.5	2.3	—	—	—	—
I_2	9.3	2.3	—	—	—	—
F_2	15.7	3.1	332	305	—	—
HBr	11.7	—	584	558	1353	1332
HCl	12.7	—	557	530	—	—
HF	16	—	484	457	1554	1530
HI	10.4	—	627	601	1315	1293

(continued overleaf)

Molecules	IE $(eV)^a$	EA $(eV)^b$	PA $(kJ \ mol^{-1})^c$	GB $(kJ \ mole^{-1})^c$	ΔH°_{acid} $(kJ \ mole^{-1})^d$	ΔG°_{acid} $(kJ \ mole^{-1})^d$
H_2	15.4	—	422	394	—	—
H_2O	12.6	—	691	660	1634	1607
H_2S	10.4	—	705	674	1469	1446
He	24.6	—	178	148	—	—
NH_3	10	—	854	819	1689	1660
N_2	15.6	—	494	464	—	—
O_2	12	0.4	421	396	—	—
PH_3	9.8	—	785	751	1551	1520
CH_4	12.6	—	543	521	1749	1715
C_2H_6	11.5	—	590	570	1758	1721
C_3H_8	10.9	—	625	607	1755	1721
$n\text{-}C_4H_{10}$	10.5	—	—	—	1739	1703
$i\text{-}C_4H_{10}$	10.7	—	677	671	1728	1697
$n\text{-}C_6H_{14}$	10.1	—	—	—	—	—
Cyclohexane	9.9	—	687	667	—	—
$CH_2{=}CH_2$	10.5	—	680	651	1713	1678
$CH_3CH{=}CH_2$	9.7	—	751	722	1635	1607
$CH_3CH{=}CHCH_3$	9.1	—	747	720	—	—
$CH_2{=}CHCH{=}CH_2CH_3$	8.6	—	834	804	1545	1525
$CH_2{=}CHCH{=}CH_2$	9	—	783	757	1672	1637
$HC{\equiv}CH$	12.4	—	641	616	1582	1547
$CH_3C{\equiv}CH$	10.4	—	748	723	1595	1562
$CH_3C{\equiv}CCH_3$	9.6	—	776	745	—	—
Benzene	9.2	—	750	725	1681	1644
$C_6H_5CH_3$	8.8	—	784	756	1593	1564
$C_6H_5CH{=}CH_2$	8.2	—	840	809	1636	1604
Biphenyl	8.2	—	814	783	—	—
Naphthalene	8.1	−0.2	803	779	—	—
CH_3F	12.5	—	599	571	1711	1676
CH_3Cl	11.3	—	647	621	1657	1628
CH_3Br	10.5	—	664	638	1660	1631
$CH_2{=}CHF$	10.4	—	729	700	1618	1586
$CH_2{=}CHCl$	10	—	—	—	—	—
$CH_2{=}CHBr$	9.8	—	—	—	—	—
CH_3OH	10.8	1.6	754	724	1592	1565
C_2H_5OH	10.5	1.7	776	746	1583	1555
$n\text{-}C_3H_7OH$	10.2	—	786	756	1572	1544
$i\text{-}C_3H_7OH$	10.2	—	793	763	1573	1545
C_6H_5OH	8.5	—	817	786	1456	1437
$C_6H_5CH_2OH$	8.3	2.1	778	748	1548	1520
CH_3SH	9.4	—	773	742	1654	1624
$(CH_3)_2O$	10.1	—	792	764	1703	1666
$(C_2H_5)_2O$	9.5	—	828	801	—	—
$C_6H_5OCH_3$	8.2	—	840	807	1679	1648
Furan	8.9	—	803	771	1624	1590
Tetrahydrofuran	9.4	—	822	795	—	—

Molecules	IE (eV)[a]	EA (eV)[b]	PA (kJ mol^{-1})[c]	GB (kJ mole^{-1})[c]	ΔH°_{acid} (kJ mole^{-1})[d]	ΔG°_{acid} (kJ mole^{-1})[d]
$(CH_3)_2S$	8.7	—	831	801	1645	1615
$C_2H_5NH_2$	8.9	—	899	864	1687	1656
n-$C_4H_9NH_2$	8.9	0.7	912	878	1671	1639
$C_6H_5NH_2$	8.7	—	921	886	—	—
$(C_2H_5)_2NH$	7.7	—	882	851	1533	1502
$(C_2H_5)_3N$	7.8	—	952	919	—	—
Pyrrole	7.5	—	981	951	—	—
Pyridine	8.2	2.3	875	843	1500	1468
H_2CHO	10.9	—	713	683	1646	1613
CH_3CHO	10.2	—	768	736	1531	1502
n-C_3H_7CHO	9.8	—	793	761	—	—
C_6H_5CHO	9.5	0.4	834	802	—	—
CH_2CO	9.6	—	825	794	1526	1497
CH_3COCH_3	9.7	—	812	782	1544	1514
$C_2H_5COCH_3$	9.5	—	827	795	1536	1508
$C_6H_5COCH_3$	9.3	0.3	861	829	1512	1483
HCOOH	11.3	—	742	710	1445	1415
CH_3COOH	10.6	3.1	784	753	1456	1427
n-C_3H_7COOH	10.2	3.2	—	—	1450	1420
C_6H_5COOH	9.3	3.5	821	790	1423	1393
CH_3COOCH_3	10.3	1.5	821	790	1556	1528
CH_3CONH_2	9.7	—	864	833	1515	1485
$CH_3CON(CH_3)_2$	9.2	—	908	877	1568	1540

Sources: Thermodynamic data from Lias S.G., Bartmess J.E., Liebman J.F., *et al.*, *J. Phys. Chem. Ref. Data*, **17**, Suppl. 1 (1988); also available on CDROM from NIST (National Institute of Standards and Technology, Washington, DC, USA); Mallard W.G. and Linstrom P.J., *NIST Chemistry WebBook*, NIST Standard Reference Database Number 69, National Institute of Standards and Technology, Gaithersburg, MD, 1998 (http://webbook.nist.gov).

[a]Ionization energy (IE) is defined as the 0 K enthalpy change required to remove an electron from a molecule. It is possible to have either adiabatic or vertical ionization energy, with the value of the vertical ionization energy being greater than or equal to the adiabatic ionization energy:

$$M \longrightarrow M^+ + e^- \quad \Delta H = IE$$

[b]Electron affinity (EA) is defined as the negative of the 0 K enthalpy change for the electron attachment reaction from a molecule. As with the ionization energy, adiabatic or vertical electron affinity can be possible:

$$M + e^- \longrightarrow M^- \quad \Delta H = -EA$$

[c]Proton affinity (PA) and gas-phase basicity (GB) are, respectively, the negative of the enthalpy change or the Gibbs energy change defined at 298 K for the protonation reaction:

$$M + H^+ \rightleftharpoons MH^+ \quad \Delta H = -PA \text{ and } \Delta G = -GB$$

[d]Gas-phase Acidity: ΔG°_{acid} and ΔH°_{acid} are the Gibbs energy change and the enthalpy change defined at 298 K to remove a proton from a molecule, ΔH°_{acid} and ΔG°_{acid} are, respectively, the proton affinity and the gas-phase basicity of the anion:

$$AH \rightleftharpoons A^- + H^+$$

7 Gas-phase Ion Thermochemical Data of Radicals

Radicals	IE (eV)	EA (eV)	PA (kJ mol^{-1})	GB (kJ mol^{-1})
F$^{\bullet}$	17.4	3.4	340	315
Cl$^{\bullet}$	13	3.6	513	490
Br$^{\bullet}$	11.8	3.4	554	531
I$^{\bullet}$	10.4	3.1	608	583
H$^{\bullet}$	13.6	0.75	–	–
HO$^{\bullet}$	13	1.8	593	564
NC$^{\bullet}$	14.2	3.9	–	–
H$_2$N$^{\bullet}$	10.8	0.8	–	–
HS$^{\bullet}$	10.4	2.3	–	–
CH$_3^{\bullet}$	9.8	0.1	671	639
C$_2$H$_5^{\bullet}$	8.1	−0.3	616	583
i-C$_3$H$_7^{\bullet}$	7.4	−0.3	–	–
n-C$_3$H$_7^{\bullet}$	8.1	−0.1	–	–
n-, i-C$_4$H$_9^{\bullet}$	8	0.1	–	–
s-C$_4$H$_9^{\bullet}$	7.2	−0.1	–	–
t-C$_4$H$_9^{\bullet}$	6.7	−0.2	–	–
CH≡C$^{\bullet}$	11.6	2.9	753	721
CH≡CCH$_2^{\bullet}$	8.7	–	741	708
CH$_2$=CH$^{\bullet}$	8.3	0.7	755	720
CH$_2$=CHCH$_2^{\bullet}$	8.2	0.4	736	707
C$_6$H$_5^{\bullet}$	8.3	1.1	884	831
C$_6$H$_5$CH$_2^{\bullet}$	7.2	0.9	831	800
Tropyl$^{\bullet}$	6.3	0.5	832	800
BrCH$_2^{\bullet}$	8.6	0.8	–	–
ClCH$_2^{\bullet}$	8.7	0.8	–	–
FCH$_2^{\bullet}$	9	0.2	–	–
ICH$_2^{\bullet}$	8.4	–	–	–
F$_3$C$^{\bullet}$	8.7	–	–	–
Cl$_3$C$^{\bullet}$	8.1	–	–	–
Br$_3$C$^{\bullet}$	7.5	–	–	–
CH$_3$O$^{\bullet}$	10.7	–	–	–
HOCH$_2^{\bullet}$	7.6	–	695	662
CH$_3$S$^{\bullet}$	9.3	1.9	–	–
HSCH$_2^{\bullet}$	7.5	0.8	734	701
C$_6$H$_5$O$^{\bullet}$	8.6	2.2	858	827
CH$_3$OCH$_2^{\bullet}$	6.9	−0.1	756	724
CH$_3$SCH$_2^{\bullet}$	6.8	0.9	–	–
H$_2$NCH$_2^{\bullet}$	6.3	–	832	802
HCO$^{\bullet}$	8.1	0.3	636	601
CH$_3$CO$^{\bullet}$	7	0.4	653	620
HCOO$^{\bullet}$	8.2	3.2	623	590

Sources: Thermodynamic data from Lias S.G., Bartmess J.E., Liebman J.F., *et al.*, *Phys. Chem. Ref. Data*, **17**, Suppl. 1 (1988); also available on CDROM from NIST (National Institute of Standards and Technology, Washington, DC, USA); Mallard W.G. and Linstrom P.J., *NIST Chemistry WebBook*, NIST Standard Reference Database Number 69, National Institute of Standards and Technology, Gaithersburg, MD, 1998 (http://webbook.nist.gov).

8 Literature on Mass Spectrometry

The literature on mass spectrometry may be divided into three broad categories: journals, periodicals and books. Another important category is the compilations of mass spectra, which have been mentioned already in Chapter 5.

Journals devoted only to mass spectrometry

1. *International Journal of Mass Spectrometry* (Elsevier), formerly *International Journal of Mass Spectrometry and Ion Processes* before 1998.

2. *Journal of Mass Spectrometry* (Wiley), formerly *Organic Mass Spectrometry* incorporating *Biological Mass Spectrometry* before 1995. Each issue of this journal contains a Special Feature including 'Perspective' and 'Tutorial' articles that present authoritative materials on a featured topic in a succinct format.

3. *Journal of the American Society for Mass Spectrometry* (Elsevier).

4. *Rapid Communications in Mass Spectrometry* (Wiley).

5. *European Mass Spectrometry* (IM Publications).

Journals containing substantial papers on mass spectrometry (not exhaustive)

1. *Analytical Chemistry* (American Chemical Society).

2. *Analytical Biochemistry* (Academic Press).

3. *Journal of Chromatography* (Elsevier).

Abstracting journals devoted only to mass spectrometry

1. *Mass Spectrometry Bulletin* (Royal Society of Chemistry) contains a comprehensive list of references with abstracts that are relevant to mass spectrometry. It also exists on diskettes with a search program.

2. *GC/MS Update, LC/MS Update and MALDI Update* (HD Science) are more specialized abstracting services. They contain abstracts that emphasize the mass spectrometric aspects of the original publications. They also exist on diskettes with a search program.

The most important periodicals on mass spectrometry

1. *Mass Spectrometry Reviews* (Wiley).

2. *Specialist Periodical Reports on Mass Spectrometry* (Royal Society of Chemistry) reports on progress in mass spectrometry. Published every 2 years.

3. *Fundamental Reviews of Analytical Chemistry* (American Chemical Society) covers all mass spectrometry aspects in a condensed fashion. Published biennially in each even year.

4. *Proceedings of the nth ASMS Conference on Mass Spectrometry and Allied Topics* (American Society for Mass Spectrometry). These reports are published every year.

5. *Advances in Mass Spectrometry* (Heyden, Wiley and Elsevier). This series publishes reports of the *x*th International Mass Spectrometry Conferences that occur every 3 years (15th in Barcelona in 2000).

Some useful mass spectrometry books

Adams F., Gijbels R. and Van Grieken R., *Inorganic Mass Spectrometry*, Wiley, New York, 1988.

Adams N.G. and Babcock L.M., *Advances in Gas Phase Ion Chemistry*, Vol. 1, JAI Press, Greenwich, CT, 1992.

Adrey B., *Liquid Chromatography/Mass Spectrometry*, VCH, New York, 1993.

Asamoto B. and Dunbar R.C., *Analytical Applications of Fourier Transform Ion Cyclotron Resonance Mass Spectrometry*, VCH, New York, 1991.

Ashcroft A.E., *Ionization Methods in Organic Mass Spectrometry*, RSC Analytical Spectroscopy Monographs, Royal Society of Chemistry, Cambridge, 1997.

Ausloos P., *Kinetics of Ion–Molecule Reactions*, Plenum Press, New York, 1979.

Beckey H.D., *Principles of Field Ionization and Field Desorption Mass Spectrometry*, Pergamon Press, Oxford, 1977.

Benninghoven A., *Ion Formation from Organic Solids*, Wiley, New York, 1989.

Beynon J.H. and Gilbert J.R., *Applications of Transition State Theory to Unimolecular Reactions*, Wiley, New York, 1984.

Bowers M.T., *Gas Phase Ion Chemistry* (3 vols), Academic Press, New York, 1979.

Burlingame A.L., *Mass Spectrometry in the Life and Health Sciences*, Elsevier, Amsterdam, 1985.

Burlingame A.L. and Carr S.A., *Mass Spectrometry in the Biological Sciences,* Humana Press, Totowa, NJ, 1996.

Burlingame A.L. and McCloskey J.A., *Biological Mass Spectrometry*, Elsevier, Amsterdam, 1990.

Bush K.L., Glish G.L. and McLuckey S.A., *Mass Spectrometry/Mass Spectrometry*, VCH, New York, 1988.

Bush K.L. and Lehman T.A., *Guide to Mass Spectrometry*, VCH Publishers, New York, 1995.

Caprioli R.M., *Continuous-flow Fast Atom Bombardment Mass Spectrometry*, Wiley, New York, 1990.

Caprioli R.M., Malorni A. and Sindona G., *Mass Spectrometry in the Biomolecular Sciences*, Kluwer Academic, Dordrecht, 1996.

Chapman J.R., *Practical Organic Mass Spectrometry*, 2nd edition, Wiley, New York, 1993.

Cole R.B., *Electrospray Ionization Mass Spectrometry: Fundamentals, Instrumentation, and Applications*, Wiley, New York, 1997.

Cotter R.J., *Time-of-flight Mass Spectrometry*, American Chemical Society, Washington, DC, 1994.

Cotter R.J., *Time-of-flight Mass Spectrometry: Instrumentation and Applications in Biological Research*, American Chemical Society, Washington, DC, 1997.

Dawson P.H., *Quadrupole Mass Spectrometry and its Applications*, American Institute of Physics, Woodbury, New York, 1994.

Desiderio D.M., *Mass Spectrometry: Clinical and Biomedical Applications (Modern Analytical Chemistry)*, Vol. 1, Plenum Press, New York, 1993.

Desiderio D.M., *Mass Spectrometry: Clinical and Biomedical Applications (Modern Analytical Chemistry)*, Vol. 2, Plenum Press, New York, 1994.

Farrar J.M. and Saunders W.H., *Techniques for the Study of Ion–Molecule Reactions*, Wiley, New York, 1988.

Fenselau C., *Mass Spectrometry for the Characterization of Microoganisms*, American Chemical Society, Washington, DC, 1993.

Forst W., *Theory of Unimolecular Reactions*, Academic Press, New York, 1973.

Gaskell S.J., *Mass Spectrometry in Biomedical Research*, Wiley, Chichester, 1986.

Gilbert J., *Applications of Mass Spectrometry in Food Science*, Elsevier, London, 1987.

Goodman S.I. and Markey S.P., *Diagnosis of Organic Acidemias by Gas Chromatography–Mass Spectrometry*, Alan R. Liss, New York, 1981.

Gross L.M., *Mass Spectrometry in the Biological Sciences*, Kluwer, Dordrecht, 1992.

Harrison A.G., *Chemical Ionization Mass Spectrometry*, 2nd edition, CRC Press, Boca Raton, FL, 1992.

Hites R.A., *Handbook of Mass Spectra of Environmental Contaminants*, 2nd edition, Lewis Publishers, CRC Press, Boca Raton, FL, 1992.

Holland G., *Applications of Plasma Source Mass Spectrometry*, Royal Society of Chemistry, Cambridge, 1993.

Jarvis K.E., *Plasma Source Mass Spectrometry*, Royal Society of Chemistry, Cambridge, 1990.

Jarvis K.E., Grays A.L. and Houk R.S., *Handbook of Inductively Coupled Plasma Mass Spectrometry*, Chapman Hall, New York, 1991.

Johnstone R.A. and Rose M.E., *Mass Spectrometry for Chemists and Biochemists*, 2nd edition, Cambridge University Press, New York, 1996.

Kienitz H., *Massenspectrometrie*, Verlag Chemie, Weinheim, 1968.

Kitson F.G., Larsen B.S. and McEwen C.N., *Gas Chromatography and Mass Spectrometry: a Practical Guide*, Academic Press, New York, 1996.

Larsen B.S. and McEwen C.N., *Mass Spectrometry of Biological Materials*, 2nd edition, Marcel Dekker, New York, 1990.

Lee T.A., *A Beginner's Guide to Mass Spectral Interpretation*, Wiley, New York, 1998.

Lias S.G., Bartmess J.E., Liebman J.F., *et al.*, *Gas-phase Ion and Neutral Thermochemistry*, American Chemical Society, American Institute of Physics and NBS Publishers, *J. Phys. Chem. Ref. Data*, Vol. 17, Supplement, 1988.

Lubman D.H., *Lasers in Mass Spectrometry*, Oxford University Press, Oxford, 1990.

Maier J.P., *Ion and Cluster Ion Spectrometry and Structure*, Elsevier, Amsterdam, 1989.

March R.E. and Hughes R.J., *Quadrupole Storage Mass Spectrometry*, Wiley, New York, 1989.

March R.E. and Todd J.F.J., *Practical Aspects of Ion Trap Mass Spectrometry: Fundamentals of Ion Trap Mass Spectrometry (Modern Mass Spectrometry)*, Vol. 1, CRC Press, Boca Raton, FL, 1995.

March R.E. and Todd J.F.J., *Practical Aspects of Ion Trap Mass Spectrometry (Modern Mass Spectrometry)*, Vol. 2, CRC Press, Boca Raton, FL, 1995.

March R.E. and Todd J.F.J., *Practical Aspects of Ion Trap Mass Spectrometry: Chemical, Environmental, and Biomedical Applications (Practical Aspects of Ion Trap Mass spectrometry)*, Vol. 3, CRC Press, Boca Raton, FL, 1995.

Mark T.D. and Dunn G.H., *Electron Impact Ionization*, Springer, Berlin, 1985.

Marshall A.G. and Verdun F.T., *Fourier Transform in NMR, Optical and Mass Spectrometry*, Elsevier, Amsterdam, 1990.

Massey H.S.W., *Negative Ions*, Cambridge University Press, New York, 1976.

Matsuo T., Seyama Y., Caprioli R.M., *et al.*, *Biological Mass Spectrometry. Present and Future*, Wiley, Chichester, 1994.

McCloskey J.A., *Methods in Enzymology, Vol. 193, Mass Spectrometry*, Academic Press, New York, 1990.

McEwen C.N. and Larsen B.S., eds., *Mass Spectrometry of Biological Materials*, Marcel Dekker, New York, 1990.

McLafferty F.W., ed., *Tandem Mass Spectrometry*, Wiley, New York, 1983.

McLafferty F.W., *The Wiley/NBS Registry of Mass Spectral Data*, Wiley–Interscience, New York, 1991.

McLafferty F.W. and Stauffer D.B., *Important Peak Index of the Registry of Mass Spectral Data*, Wiley, New York, 1991.

McLafferty F.W. and Turecek F., *Interpretation of Mass Spectra*, 4th edition, University Science Books, Mill Valley, CA, 1993.

McLafferty F.W. and Venkataraghavan R., *Mass Spectral Correlations*, 2nd edition, Advances in Chemistry Series No. 40, American Chemical Society, Washington, DC, 1982.

McMaster M.C. and McMaster C., *GC/MS: a Practical User's Guide*, Wiley, New York, 1998.

Middletich B.S., Missler S.R. and Hines H.B., *Mass Spectrometry of Priority Pollutants*, Plenum Press, New York, 1981.

Millard B.J., *Quantitative Mass Spectrometry*, Heyden, London, 1978.

Montaser A., *Inductively Coupled Plasma Mass Spectrometry*, VCH, New York, 1998.

Morris H.R., *Soft Ionization Biological Mass Spectrometry*, Heyden, London, 1981.

Murphy, R.C., *Mass Spectrometry of Lipids*, Plenum Press, New York, 1993.

Niessen W.M.A. and Voyksner R.D., *Current Practice of Liquid Chromatography–Mass Spectrometry*, Elsevier Science, New York, 1998.

Pfleger K., Maurer H.H., and Weber A., *Mass Spectral and GC Data of Drugs, Poisons, Pesticides, Pollutants and Their Metabolites*, VCH, New York, 1992.

Porter Q.N. and Baldas J., *Mass Spectrometry of Heterocyclic Compounds*, 2nd edition, Wiley, New York, 1985.

Russel D.H., *Experimental Mass Spectrometry*, Plenum Press, New York, 1994.

Siuzdak G., *Mass Spectrometry for Biotechnology*, Academic Press, San Diego, CA, 1996.

Schlag E.W., *Time-of-flight Mass Spectrometry and its Applications*, Elsevier, Amsterdam, 1994.

Silverstein R.M., Bassler G.C. and Morril T.C., *Spectrometric Identification of Organic Compounds*, 6th edition, Wiley, New York, 1997.

Snyder A.P., *Biomedical and Biotechnological Applications of Electrospray Ionization Mass Spectrometry*, American Chemical Society, Washington, DC, 1995.

Splitter J.G. and Turecek F., *Applications of Mass Spectrometry to Organic Stereochemistry*, VCH, New York, 1994.

Standing K.G. and Ens W., *Methods and Mechanisms for Producing Ions from Large Molecules*, Plenum Press, New York, 1991.

Suelter C.H. and Watson J.T., *Biomedical Applications of Mass Spectrometry*, Wiley, New York, 1990.

Tuniz C., Tuniz J.R. and Bird D.F., *Accelerator Mass Spectrometry: Ultrasensitive Analysis for Global Science*, CRC Press, Boca Raton, FL, 1998.

Vertes A., Gijbels R. and Adams F., *Laser Ionization Mass Analysis*, Wiley, New York, 1993.

Waller G.R., *Biochemical Applications of Mass Spectrometry*, Wiley, New York, 1972.

Waller G.R., *Biochemical Applications of Mass Spectrometry*, First Supplementary Volume, Wiley, New York, 1980.

Watson J.T., *Introduction to Mass Spectrometry*, 3rd edition, Lippincott-Raven, Philadelphia, 1997.

Williams R., *Spectroscopy and the Fourier Transform: an Interactive Tutorial*, VCH, New York, 1995.

Yergey A.L., Edmonds C.G., Lewis I.A.S., *et al.*, *Liquid Chromatography/Mass Spectrometry*, Plenum Press, New York, 1990.

Yinon J., *Forensic Applications of Mass Spectrometry (Modern Mass Spectrometry)*, CRC Press, Boca Raton, FL, 1995.

9 Mass Spectrometry on the Internet

The Base Peak site (http://base-peak.wiley.com) is the most comprehensive web resource for mass spectrometrists. Developed by Kermit Murray and Wiley, it contains a collection of links to mass spectrometry Internet sites. A paper giving an introduction to the Internet resources for mass spectrometry was published: Murray, K.K., *J. Mass Spectrom.*, **34**, 1 (1999).

WWW sites for national and international mass spectrometry societies

Typical information that can be found on the society sites includes contact information, membership applications, information on meetings, etc. Furthermore, member directories and employment information can be available at some of these sites:

American Society for Mass Spectrometry (ASMS):

  http://www.asms.org/

Australian and New Zealand Society for Mass Spectrometry (ANZSMS):

  http://www.latrobe.edu.au/www/anzsms/

British Mass Spectrometry Society (BMSS):

   http://www.bmss.org.uk/

Belgian Society for Mass Spectrometry (BSMS):

   http://wwwmasse.cico.ucl.ac.be/

Canadian Society for Mass Spectrometry (CSMS):

   http://www.csms.inter.ab.ca/

European Society for Mass Spectrometry (EMS):

   http://masseroute.cico.ucl.ac.be/esms/esms.html

International Mass Spectrometry Society (IMSS):

   http://www.chem.purdue.edu/imss

Dutch Society for Mass Spectrometry (NVMS):

   http://www.xs4all.nl/~pjacobs/nvms.html

Mass Spectrometry Society of Japan (MSSJ):

   http://wwwsoc.nacsis.ac.jp/mass/index.html

South African Association for Mass Spectrometry (SAAMS):

   http://www.up.ac.za/academic/acadorgs/saams/

Swiss Group for Mass Spectrometry (SGMS):

   http://www.sgms.ch/

WWW sites for mass spectrometry journals

Most of the mass spectrometry journal sites contain tables of contents and abstracts. Many of these sites offer on-line access to the full articles although it is typically restricted to subscribers only. General information such as aims and scope, editorial board, instructions to authors, contact and subscription information are also found on these sites:

Journal of the American Society for Mass Spectrometry (Elsevier):

   http://www.elsevier.com/homepage/saa/webjam/

International Journal of Mass Spectrometry (Elsevier):

   http://www.elsevier.com/homepage/saa/ijmsip/

European Mass Spectrometry (IM Publications):

   http://www.impub.co.uk/ems.html

Journal of Mass Spectrometry (Wiley):

   http://www.interscience.wiley.com/jpages/1076-5174/

Mass Spectrometry Reviews (Wiley):

   http://www.interscience.wiley.com/jpages/0277-7037/

Rapid Communications in Mass Spectrometry (Wiley):

   http://www.interscience.wiley.com/jpages/0951-4198/

WWW sites for useful mass spectrometry databases

The National Institute of Standards and Technology (NIST) provides thermochemical, thermophysical and ion energetics data for chemical species through the Internet. Indeed, the NIST Chemistry WebBook (http://webbook.nist.gov/chemistry/) contains, among other data, mass spectra for over 10 000 compounds and ion energetics data for over 14 000 compounds.

   The WebElements site (http://www.webelements.com/) contains chemical and physical data of the elements. This site also has on-line isotope pattern calculators.

   The PubMed site (http://www.ncbi.nlm.nih.gov/Entrez) is developed by the National Center for Biotechnology Information (NCBI) at the National Library of Medicine (NLM), located at the National Institutes of Health (NIH). It has been developed in conjunction with publishers of biomedical literature as a search tool for accessing literature citations and linking to full-text journals at Web sites of participating publishers. This site allows on-line access to the Medline database containing bibliographic citations and abstracts from journals dating back to 1966.

Index